Philip Korman

Lectures on Linear Algebra and its Applications

Also of Interest

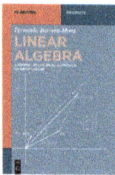

Linear Algebra
A Minimal Polynomial Approach to Eigen Theory
Fernando Barrera-Mora, 2023
ISBN 978-3-11-113589-2, e-ISBN (PDF) 978-3-11-066699-1

Abstract Algebra
An Introduction with Applications
Derek J. S. Robinson, 2022
ISBN 978-3-11-068610-4, e-ISBN (PDF) 978-3-11-069116-0

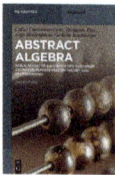

Abstract Algebra
Applications to Galois Theory, Algebraic Geometry, Representation Theory and Cryptography
Celine Carstensen-Opitz, Benjamin Fine, Anja Moldenhauer,
Gerhard Rosenberger, 2019
ISBN 978-3-11-060393-4, e-ISBN (PDF) 978-3-11-060399-6

The Structure of Compact Groups
A Primer for the Student – A Handbook for the Expert
Karl H. Hofmann, Sidney A. Morris, 2020
ISBN 978-3-11-069595-3, e-ISBN (PDF) 978-3-11-069599-1

Commutative Algebra
Aron Simis, 2020
ISBN 978-3-11-061697-2, e-ISBN (PDF) 978-3-11-061698-9

Philip Korman

Lectures on Linear Algebra and its Applications

—

DE GRUYTER

Mathematics Subject Classification 2020
15-01, 15A, 47-01, 47A10, 47A30, 47A35, 47A56, 47A60, 47B07, 47B25, 46-01, 46A30, 46A35, 46A45, 46E15

Author
Prof. Philip Korman
University of Cincinnati
Department of Mathematics
2600 Clifton Avenue
Cincinnati OH 45221-0025
USA
kormanp@ucmail.uc.edu

ISBN 978-3-11-108540-1
e-ISBN (PDF) 978-3-11-108650-7
e-ISBN (EPUB) 978-3-11-108662-0

Library of Congress Control Number: 2023942983

Bibliographic information published by the Deutsche Nationalbibliothek
The Deutsche Nationalbibliothek lists this publication in the Deutsche Nationalbibliografie;
detailed bibliographic data are available on the Internet at http://dnb.dnb.de.

Cover image: background: fototrav / E+ / Getty Images; equation: author
Typesetting: VTeX UAB, Lithuania
Printing and binding: CPI books GmbH, Leck

www.degruyter.com

Introduction

How do you cover a semester long course of "Linear Algebra" in half the time? That is what happened in the Fall of 2020 when classroom capacities were reduced due to Covid. I was teaching an 80-minute lecture to half of the class on Tuesdays, and repeating the same lecture to the other half on Thursdays. I had to concentrate on the basics, trying to explain concepts on simple examples, and to cover several concepts with each example. Detailed notes were produced (with lines numbered), which I projected on a screen, and made them available to students. Questions were encouraged, but not of a review nature (students were very cooperative). Pictures were drawn on a white board, and the most crucial concepts were also discussed there. On "free days" students were directed to specific resources on the web, particularly to lectures of G. Strang at MIT [16], and 3blue1brown.com [19] that contains nice visualizations. I managed to cover the basics, Sections 1.1–5.5 (although many sections were thinner then).

Chapters 1–5 represent mostly the transcripts of my lectures in a situation when every minute counted. Toward the end of the sections, and in exercises, nontrivial and useful applications are covered, like Fredholm alternative, Hadamard's inequality, Gram's determinant, Hilbert's matrices, etc. I tried to make use of any theory developed in this book, and thus avoid "blind alleys." For example, the QR factorization was used in the proofs of the law of inertia, and of Hadamard's inequality. Diagonalization had many uses, including the Raleigh quotient, which in turn led us to principal curvatures. Quadratic forms were developed in some detail, and then applied to Calculus and Differential Geometry. Gram–Schmidt process led us to Legendre's polynomials.

I tried to keep the presentation focused. For example, only the Euclidean norm of matrices is covered. It gives a natural generalization of length for vectors, and it is sufficient for elementary applications, like convergence of Jacoby's iterations. Other norms, seminorms, definition of a norm, etc., are left out.

Chapters 6 and 7 contain applications to Differential Equations, Calculus, and Differential Geometry. They are also based on classroom presentations, although in different courses. In Differential Equations, after intuitive presentation of the basics, we cover the case of repeated eigenvalues of deficiency greater than one, which is hard to find in the literature. The presentation is based on the matrix exponentials developed in the preceding section, and it leads to the theory of the Jordan normal form. Detailed discussion of systems with periodic coefficients allowed us to showcase the Fredholm alternative.

Applications to Differential Geometry is a unique feature of this book. Some readers may be surprised to find discussion of Gaussian curvature in a Linear Algebra book. However, the connection is very strong as is explained next. Principal curvatures are the eigenvalues of the generalized eigenvalue problem $Ax = \lambda Bx$, where A and B are matrices of the second and first fundamental quadratic forms, respectively. The corresponding generalized eigenvectors give coordinates of the principal directions in the tangent plane with respect to the basis consisting of tangent vectors to the coordinate curves. This involves several key concepts of Linear Algebra.

https://doi.org/10.1515/9783111086507-201

One of the central results of Linear Algebra says that every symmetric matrix is diagonalizable. We include a very nice proof, due to I. M. Gelfand [9]. In addition to its simplicity and clarity, Gelfand's proof shows the power of abstract reasoning, when it is advantageous to work with the transformation that the matrix represents, rather than the matrix itself. Generally though, we tried to keep the presentation concrete.

A detailed solution manual, written by the author, is meant to enhance the text. In addition to routine problems, it covers more challenging and theoretical ones. In particular, it contains discussion of Perron–Frobenius theorem, and of Gram determinants. A number of excellent books and online resources are listed in the references [2–5, 7, 11–15, 17, 18]. Our book has different applications, and tends to be more concise.

A word on notation. It is customary to use boldface letters to denote vectors $\mathbf{a}, \mathbf{b}$, etc. Instructors use bars $\bar{a}, \bar{b}$, when writing on a board. Roman letters are also used, if there is no danger of confusing vectors with scalars. We begin by using boldface letters, then gradually shift to the Roman ones, but still occasionally use boldface letters, particularly for the zero vector $\mathbf{0}$. When discussing Differential Geometry, we use boldface letters for vectors in the tangent plane, Roman letters for their coordinate vectors, while $\bar{N}$ is reserved for the unit normal to the tangent plane.

It is a pleasure to thank my colleagues Robbie Buckingham, Ken Meyer, and Dieter Schmidt for a number of useful comments.

Contents

1 Systems of linear equations

In this chapter we develop *Gaussian elimination*, a systematic and practical way for solving systems of linear equations. This technique turns out to be an important theoretical cornerstone of the entire subject.

1.1 Gaussian elimination

The following equation with two variables x and y:

$$2x - y = 3$$

is an example of a *linear equation*. Geometrically, this equation describes a straight line of slope 2 (write it as $y = 2x - 3$). The point $(2, 1)$ with $x = 2$ and $y = 1$ is a solution of our equation so that it lies on this line, while the point $(3, 1)$ does not satisfy the equation, and it lies off our line. The equation has infinitely many solutions representing geometrically a line. Similarly, the equation

$$4x + y = 9$$

has infinitely many solutions. Now let us put these equations together, and solve the following *system of two equations with two unknowns*:

$$2x - y = 3,$$
$$4x + y = 9.$$

We need to find the point (or points) that lie on both lines, or the point of intersection. The lines are not parallel, so that there is a unique point of intersection. To find its coordinates, we solve this system by adding the equations:

$$6x = 12,$$

so that $x = 2$. To find y, use the value of $x = 2$ in the first equation:

$$2 \cdot 2 - y = 3,$$

so that $y = 1$.

We used an opportunity to eliminate y when solving this system. A more systematic approach will be needed to solve larger systems, say a system of four equations with five unknowns. We indicate such an approach for the same system next. Observe that multiplying one of the equations by a number will not change the solution set. Similarly, the solution set is preserved when adding or subtracting the equations. For example, if the first equation is multiplied by 2 (to get $4x - 2y = 6$), the solution set is not changed.

https://doi.org/10.1515/9783111086507-001

From the second equation we subtract the first one, multiplied by 2 (subtract $4x - 2y$ from the left side of the second equation, and subtract 6 from the right side of the second equation). The new system

$$2x - y = 3,$$
$$3y = 3$$

has the same solution set (we obtained an *equivalent system*). The x variable is now eliminated in the second equation. From the second equation, we obtain $y = 1$, and substituting this value of y back into the first equation gives $2x - 1 = 3$, or $x = 2$. The answer is $x = 2$ and $y = 1$. (The lines intersect at the point $(2, 1)$.)

Proceeding similarly, the system

$$2x + y = 3,$$
$$-8x - 4y = -12$$

is solved by adding to the second equation the first multiplied by 4:

$$2x + y = 3,$$
$$0 = 0.$$

The second equation carries no information, and it is discarded, leaving only the first equation:

$$2x + y = 3.$$

Answer: this system has infinitely many solutions, consisting of all pairs (x, y) (points (x, y)) lying on the line $2x + y = 3$. One can present the answer in several ways: $y = -2x + 3$ with x arbitrary, $x = -\frac{1}{2}y + \frac{3}{2}$ with y arbitrary, or $y = t$ and $x = -\frac{1}{2}t + \frac{3}{2}$, with t arbitrary. Geometrically, both equations of this system define the same line. That line intersects itself at all of its points.

For the system

$$2x - 3y = -1,$$
$$2x - 3y = 0,$$

subtracting from the second equation the first gives

$$2x - 3y = -1,$$
$$0 = 1.$$

The second equation will never be true, no matter what x and y are. Answer: this system has no solutions. One says that *this system is inconsistent.* Geometrically, the lines $2x - 3y = -1$ and $2x - 3y = 0$ are parallel, and have no points of intersection.

The system

$$2x - y = 3,$$
$$4x + y = 9,$$
$$x - y = -\frac{1}{2}$$

has three equations, but only two unknowns. If one considers only the first two equations, one recognizes the system of two equations with two unknowns that was solved above. The solution was $x = 2$ and $y = 1$. The point $(2, 1)$ is the only one with a chance to be a solution of the entire system. For that, it must lie on the third line $x - y = -\frac{1}{2}$. It does not. Answer: this system has no solutions, it is inconsistent. Geometrically, the third line misses the point of intersection of the first two lines.

The system of two equations

$$2x - y + 5z = 1,$$
$$x + y + z = -2$$

affords us a "luxury" of three variables x, y, and z to satisfy it. To eliminate x in the second equation, we need to subtract from it the first equation multiplied by $\frac{1}{2}$. (From the second equation subtract $x - \frac{1}{2}y + \frac{5}{2}z = \frac{1}{2}$.) To avoid working with fractions, let us switch the order of equations

$$x + y + z = -2,$$
$$2x - y + 5z = 1,$$

which clearly results in an equivalent system. Now to eliminate x in the second equation, we subtract from it the first equation multiplied by 2. We obtain

$$x + y + z = -2,$$
$$-3y + 3z = 5.$$

Set $z = t$, an arbitrary number. Then from the second equation we shall obtain y as a function of t. Finally, from the first equation x is expressed as a function of t. Details: from the second equation

$$-3y + 3t = 5,$$

giving $y = t - \frac{5}{3}$. Substituting this expression for y, and $z = t$, into the first equation yields

$$x + t - \frac{5}{3} + t = -2,$$

so that $x = -2t - \frac{1}{3}$. Answer: this system has infinitely many solutions of the form $x = -2t - \frac{1}{3}, y = t - \frac{5}{3}, z = t$, and t is an arbitrary number. One can present this answer in *vector form*

$$\begin{bmatrix} x \\ y \\ z \end{bmatrix} = \begin{bmatrix} -2t - \frac{1}{3} \\ t - \frac{5}{3} \\ t \end{bmatrix} = t \begin{bmatrix} -2 \\ 1 \\ 1 \end{bmatrix} - \frac{1}{3} \begin{bmatrix} 1 \\ 5 \\ 0 \end{bmatrix}.$$

The next example involves a three-by-three system

$$x - y + z = 4,$$
$$-2x + y - z = -5,$$
$$3x + 4z = 11$$

of three equations with three unknowns. Our plan is to eliminate x in the second and third equations. These two operations are independent of each other and can be performed simultaneously (in parallel). To the second equation we add the first multiplied by 2, and from the third equation subtract the first multiplied by 3. We thus obtain

$$x - y + z = 4,$$
$$-y + z = 3,$$
$$3y + z = -1.$$

Our next goal is to eliminate y in the third equation. To the third equation we add the second multiplied by 3. (If we used the first equation to do this task, then x would reappear in the third equation, negating our work to eliminate it.) We obtain

$$x - y + z = 4,$$
$$-y + z = 3,$$
$$4z = 8.$$

We are finished with the elimination process, also called *forward elimination*. Now the system can be quickly solved by *back-substitution*: from the third equation calculate $z = 2$. Using this value of z in the second equation, one finds y. Using these values of y and z in the first equation, one finds x. Details: from the second equation $-y + 2 = 3$, giving $y = -1$. From the first equation $x + 1 + 2 = 4$, so that $x = 1$. Answer: $x = 1, y = -1$, and $z = 2$. Geometrically, the three planes defined by the three equations intersect at the point $(1, -1, 2)$.

Our examples suggest the following *rule of thumb*: if there are more variables than equations, the system is likely to have infinitely many solutions. If there are more equations than variables, the system is likely to have no solutions. And if the numbers of variables and equations are the same, the system is likely to have a unique solution. This rule *does not always apply*. For example, the system

$$x - y = 2,$$
$$-2x + 2y = -4,$$
$$3x - 3y = 6$$

has more equations than unknowns, but the number of solutions is infinite, because all three equations define the same line. On the other hand, the system

$$x - 2y + 3z = 2,$$
$$2x - 4y + 6z = -4$$

has more variables than equations, but there are no solutions, because the equations of this system define two parallel planes.

The method for solving linear systems described in this section is known as *Gaussian elimination*, named in honor of C. F. Gauss, a famous German mathematician.

Exercises

1. Solve the following systems by back-substitution:
   (a)

$$x + 3y = -1,$$
$$-2y = 1.$$

   Answer: $x = \frac{1}{2}, y = -\frac{1}{2}$.
   (b)

$$x + y + 3z = 1,$$
$$y - z = 2,$$
$$2z = -2.$$

   Answer: $x = 3, y = 1, z = -1$.
   (c)

$$x + 4z = 2,$$
$$2y - z = 5,$$
$$-3z = -3.$$

Answer: $x = -2, y = 3, z = 1$.

(d)

$$x - y + 2z = 0,$$
$$y - z = 3.$$

Answer: $x = -t + 3, y = t + 3, z = t$, where t is arbitrary.

(e)

$$x + y - z - u = 2,$$
$$3y - 3z + 5u = 3,$$
$$2u = 0.$$

Answer: $x = 1, y = t + 1, z = t, u = 0$, where t is arbitrary.

2. Solve the following systems by Gaussian elimination (or otherwise), and if possible interpret your answer geometrically:

(a)

$$x + 3y = -1,$$
$$-x - 2y = 3.$$

Answer: $x = -7, y = 2$. Two lines intersecting at the point $(-7, 2)$.

(b)

$$2x - y = 3,$$
$$x + 2y = 4,$$
$$-x + 5y = 3.$$

Answer: $x = 2, y = 1$. Three lines intersecting at the point $(2, 1)$.

(c)

$$x + 2y = -1,$$
$$-2x - 4y = 3.$$

Answer: There is no solution, the system is inconsistent. The lines are parallel.

(d)

$$x + 2y = -1,$$
$$-2x - 4y = 2.$$

Answer: There are infinitely many solutions, consisting of all points on the line $x + 2y = -1$.

(e)

$$x + y + z = -2,$$
$$x + 2y = -3,$$
$$x - y - z = 4.$$

Answer: $x = 1, y = -2, z = -1$. Three planes intersect at the unique point $(1, -2, -1)$.

(f)

$$x - y + 2z = 0,$$
$$x + z = 3,$$
$$2x - y + 3z = 3.$$

Answer: $x = -t + 3, y = t + 3, z = t$, where t is arbitrary.

(g)

$$x - 2y + z = 1,$$
$$2x - 4y + 2z = 3,$$
$$4x - y + 3z = 5.$$

Answer: There are no solutions (the system is inconsistent). The first two planes are parallel.

3. Three points, not lying on the same line, uniquely determine the plane passing through them. Find an equation of the plane passing through the points $(1, 0, 2)$, $(0, 1, 5)$, $(2, 1, 1)$.
   Answer: $2x - y + z = 4$. Hint. Starting with $ax + by + cz = d$, obtain three equations for $a, b, c,$ and d. There are infinitely many solutions, depending on a parameter t. Select the value of t giving the simplest looking answer.

4. Find the number a, so that the system

$$2x - 3y = -1,$$
$$ax - 6y = 5$$

has no solution. Can one find a number a, so that this system has infinitely many solutions?

5. Find all solutions of the equation

$$5x - 3y = 1,$$

where x and y are integers (*Diophantine equation*).
Hint. Solve for y: $y = \frac{5x-1}{3} = 2x - \frac{x+1}{3}$. Set $\frac{x+1}{3} = n$. Then $x = 3n - 1$, leading to $y = 5n - 2$, where n is an arbitrary integer.

1.2 Using matrices

We shall deal with linear systems possibly involving a large number of unknowns. Instead of denoting the variables by $x, y, z, \ldots$, we shall write $x_1, x_2, x_3, \ldots, x_n$, where n is the number of variables. Our next example is

$$x_1 - x_2 + x_3 = -1,$$
$$2x_1 - x_2 + 2x_3 = 0,$$
$$-3x_1 + 4x_3 = -10.$$

The first step of Gaussian elimination is to subtract from the second equation the first multiplied by 2. This will involve working with the coefficients of x_1, x_2, x_3. So let us put these coefficients into a *matrix* (or a table)

$$\begin{bmatrix} 1 & -1 & 1 \\ 2 & -1 & 2 \\ -3 & 0 & 4 \end{bmatrix}$$

called the *matrix of the system*. It has 3 rows and 3 columns. When this matrix is augmented with the right-hand sides of the equations,

$$\left[\begin{array}{ccc|c} 1 & -1 & 1 & -1 \\ 2 & -1 & 2 & 0 \\ -3 & 0 & 4 & -10 \end{array}\right],$$

one obtains the *augmented matrix*. Subtracting from the second equation the first multiplied by 2 is the same as subtracting from the second row of the augmented matrix the first multiplied by 2. Then, to the third row we add the first multiplied by 3. We thus obtain

$$\left[\begin{array}{ccc|c} ① & -1 & 1 & -1 \\ 0 & 1 & 0 & 2 \\ 0 & -3 & 7 & -13 \end{array}\right].$$

We circled the element, called *pivot*, used to produce two zeroes in the first column of the augmented matrix. Next, to the third row add 3 times the second:

$$\left[\begin{array}{ccc|c} ① & -1 & 1 & -1 \\ 0 & ① & 0 & 2 \\ 0 & 0 & ⑦ & -7 \end{array}\right].$$

Two more pivots are circled. All elements under the diagonal are now zero. The Gaussian elimination is complete. Restoring the system corresponding to the last augmented matrix (a step that will be skipped later) gives

$$x_1 - x_2 + x_3 = -1,$$
$$x_2 = 2,$$
$$7x_3 = -7.$$

This system is equivalent to the original one. Back-substitution produces $x_3 = -1$, $x_2 = 2$, and from the first equation

$$x_1 - 2 - 1 = -1,$$

or $x_1 = 2$.

The next example is

$$3x_1 + 2x_2 - 4x_3 = 1,$$
$$x_1 - x_2 + x_3 = 2,$$
$$5x_2 - 7x_3 = -1,$$

with the augmented matrix

$$\left[\begin{array}{ccc|c} 3 & 2 & -4 & 1 \\ 1 & -1 & 1 & 2 \\ 0 & 5 & -7 & -1 \end{array}\right].$$

(Observe that we could have started this example with the augmented matrix, as well.) The first step is to subtract from the second row the first multiplied by $-\frac{1}{3}$. To avoid working with fractions, we interchange the first and the second rows (this changes the order of equations, giving an equivalent system):

$$\left[\begin{array}{ccc|c} 1 & -1 & 1 & 2 \\ 3 & 2 & -4 & 1 \\ 0 & 5 & -7 & -1 \end{array}\right].$$

Subtract from the second row the first multiplied by 3. We shall denote this operation by $R_2 - 3R_1$, for short. (R_2 and R_1 refer to row 2 and row 1, respectively.) We obtain

$$\left[\begin{array}{ccc|c} ① & -1 & 1 & 2 \\ 0 & 5 & -7 & -5 \\ 0 & 5 & -7 & -1 \end{array}\right].$$

There is a "free" zero at the beginning of third row R_3, so we move on to the second column and perform $R_3 - R_2$:

$$\left[\begin{array}{ccc|c} ① & -1 & 1 & 2 \\ 0 & ⑤ & -7 & -5 \\ 0 & 0 & 0 & 4 \end{array}\right].$$

The third equation says $0x_1 + 0x_2 + 0x_3 = 4$, or

$$0 = 4.$$

The system *is inconsistent*, there is no solution.

The next example we begin with the augmented matrix

$$\left[\begin{array}{ccc|c} 3 & 2 & -4 & 1 \\ 1 & -1 & 1 & 2 \\ 0 & 5 & -7 & -5 \end{array}\right].$$

This system is a small modification of the preceding one, with only the right-hand side of the third equation being different. The same steps of forward elimination lead to

$$\left[\begin{array}{ccc|c} ① & -1 & 1 & 2 \\ 0 & ⑤ & -7 & -5 \\ 0 & 0 & 0 & 0 \end{array}\right].$$

The third equation now says $0 = 0$, and it is discarded. There are pivots in columns 1 and 2 corresponding to the variables x_1 and x_2, respectively. We call x_1 and x_2 the *pivot variables*. In column 3 there is no pivot (pivot is a nonzero element, used to produce zeroes). The corresponding variable x_3 is called a *free variable*. We now restore the system, move the terms involving the free variable x_3 to the right, let x_3 be arbitrary, and then solve for the pivot variables x_1 and x_2 in terms of x_3. Details:

$$x_1 - x_2 + x_3 = 2,$$
$$5x_2 - 7x_3 = -5,$$

or

$$x_1 - x_2 = -x_3 + 2,$$
$$5x_2 = 7x_3 - 5. \tag{2.1}$$

From the second equation

$$x_2 = \frac{7}{5}x_3 - 1.$$

From the first equation of the system (2.1) express x_1 to get

$$x_1 = x_2 - x_3 + 2 = \frac{7}{5}x_3 - 1 - x_3 + 2 = \frac{2}{5}x_3 + 1.$$

Answer: $x_1 = \frac{2}{5}x_3 + 1$, $x_2 = \frac{7}{5}x_3 - 1$, and x_3 is arbitrary ("free"). We can set $x_3 = t$, an arbitrary number, and present the answer in the form $x_1 = \frac{2}{5}t + 1$, $x_2 = \frac{7}{5}t - 1$, $x_3 = t$.

Moving on to larger systems, consider a four-by-four system

$$x_2 - x_3 + x_4 = 2,$$
$$2x_1 + 6x_2 - 2x_4 = 4,$$
$$x_1 + 2x_2 + x_3 - 2x_4 = 0,$$
$$x_1 + 3x_2 - x_4 = 2,$$

with the augmented matrix

$$\left[\begin{array}{cccc|c} 0 & 1 & -1 & 1 & 2 \\ 2 & 6 & 0 & -2 & 4 \\ 1 & 2 & 1 & -2 & 0 \\ 1 & 3 & 0 & -1 & 2 \end{array}\right].$$

We need a nonzero element (or pivot) at the beginning of row 1. For that, we may switch row 1 with any other row, but to avoid fractions we do not switch with row 2. Let us switch row 1 with row 3. We shall denote this operation by $R_1 \leftrightarrow R_3$, for short. We obtain

$$\left[\begin{array}{cccc|c} 1 & 2 & 1 & -2 & 0 \\ 2 & 6 & 0 & -2 & 4 \\ 0 & 1 & -1 & 1 & 2 \\ 1 & 3 & 0 & -1 & 2 \end{array}\right].$$

Now perform $R_2 - 2R_1$ and $R_4 - R_1$ to obtain

$$\left[\begin{array}{cccc|c} ① & 2 & 1 & -2 & 0 \\ 0 & ② & -2 & 2 & 4 \\ 0 & 1 & -1 & 1 & 2 \\ 0 & 1 & -1 & 1 & 2 \end{array}\right].$$

To produce zeroes in the second column under the diagonal, perform $R_3 - \frac{1}{2}R_2$ and $R_4 - \frac{1}{2}R_2$. We thus obtain

$$\left[\begin{array}{cccc|c} ① & 2 & 1 & -2 & 0 \\ 0 & ② & -2 & 2 & 4 \\ 0 & 0 & 0 & 0 & 0 \\ 0 & 0 & 0 & 0 & 0 \end{array}\right].$$

The next step is optional: multiply the second row by $\frac{1}{2}$. We shall denote this operation by $\frac{1}{2}R_2$. This produces a little simpler matrix

$$\begin{bmatrix} ① & 2 & 1 & -2 & \vline & 0 \\ 0 & ① & -1 & 1 & \vline & 2 \\ 0 & 0 & 0 & 0 & \vline & 0 \\ 0 & 0 & 0 & 0 & \vline & 0 \end{bmatrix}.$$

The pivot variables are x_1 and x_2, while x_3 and x_4 are free. Restore the system (the third and fourth equations are discarded), move the free variables to the right, and solve for pivot variables to get

$$x_1 + 2x_2 + x_3 - 2x_4 = 0,$$
$$x_2 - x_3 + x_4 = 2,$$

or

$$x_1 + 2x_2 = -x_3 + 2x_4,$$
$$x_2 = x_3 - x_4 + 2.$$

The second equation gives us x_2. Then from the first equation we get

$$x_1 = -2x_2 - x_3 + 2x_4 = -2(x_3 - x_4 + 2) - x_3 + 2x_4 = -3x_3 + 4x_4 - 4.$$

Answer: $x_1 = -3x_3 + 4x_4 - 4$, $x_2 = x_3 - x_4 + 2$, x_3 and x_4 are two arbitrary numbers. We can set $x_3 = t$ and $x_4 = s$, two arbitrary numbers, and present the answer in the form $x_1 = -3t + 4s - 4$, $x_2 = t - s + 2$, $x_3 = t$, $x_4 = s$.

The next system of three equations with four unknowns is given by its augmented matrix

$$\begin{bmatrix} 1 & -1 & 0 & 2 & \vline & 3 \\ -1 & 1 & 2 & 1 & \vline & -1 \\ 2 & -2 & 4 & 0 & \vline & 10 \end{bmatrix}.$$

Performing $R_2 + R_1$ and $R_3 - 2R_1$ produces zeroes in the first column under the diagonal term (the pivot), namely

$$\begin{bmatrix} ① & -1 & 0 & 2 & \vline & 3 \\ 0 & 0 & 2 & 3 & \vline & 2 \\ 0 & 0 & 4 & -4 & \vline & 4 \end{bmatrix}.$$

Moving on to the second column, there is zero in the diagonal position. We look under this zero for a nonzero element, in order to change rows and obtain a pivot. There is no such nonzero element, so we move on to the third column (the second column is left without a pivot), and perform $R_3 - 2R_2$:

$$\begin{bmatrix} ① & -1 & 0 & 2 & | & 3 \\ 0 & 0 & ② & 3 & | & 2 \\ 0 & 0 & 0 & ⑩ & | & 0 \end{bmatrix}.$$

The augmented matrix is reduced to its *row echelon form*. Looking at this matrix from the left, one sees in each row zeroes followed by a pivot. Observe that no two pivots occupy the same row or the same column (*each pivot occupies its own row, and its own column*). Here the pivot variables are x_1, x_3, and x_4, while x_2 is free variable. The last equation $-10x_4 = 0$ implies that $x_4 = 0$. Restore the system, keeping in mind that $x_4 = 0$, then take the free variable x_2 to the right to get

$$x_1 = 3 + x_2,$$
$$2x_3 = 2.$$

Answer: $x_1 = 3 + x_2$, $x_3 = 1$, $x_4 = 0$, and x_2 is arbitrary.

We summarize *the strategy for solving linear systems*. If a diagonal element is nonzero, use it as a pivot to produce zeroes underneath it, then work on the next column. If a diagonal element is zero, look underneath it for a nonzero element to perform a switch of rows. If a diagonal element is zero, and all elements underneath it are also zeroes, this column has no pivot; move on to the next column. After the matrix is reduced to the row echelon form, move the free variables to the right-hand side, and let them be arbitrary numbers. Then solve for the pivot variables.

1.2.1 Complete forward elimination

Let us revisit the system

$$\begin{bmatrix} 1 & -1 & 1 & | & -1 \\ 2 & -1 & 2 & | & 0 \\ -3 & 0 & 4 & | & -10 \end{bmatrix}.$$

Forward elimination ($R_2 - 2R_1$, $R_3 + 3R_1$, followed by $R_3 + 3R_2$) gave us

$$\begin{bmatrix} ① & -1 & 1 & | & -1 \\ 0 & ① & 0 & | & 2 \\ 0 & 0 & ⑦ & | & -7 \end{bmatrix}.$$

Then we restored the system, and quickly solved it by back-substitution. However, one can continue to simplify the matrix of the system. First, we shall make all pivots equal to 1. To that end, the third row is multiplied by $\frac{1}{7}$, an elementary operation denoted by $\frac{1}{7}R_3$, to obtain

$$\begin{bmatrix} ① & -1 & 1 & | & -1 \\ 0 & ① & 0 & | & 2 \\ 0 & 0 & ① & | & -1 \end{bmatrix}.$$

Now we shall use the third pivot to produce zeroes in the third column above it, and then use the second pivot to produce a zero above it. (In this order!) Performing $R_1 - R_3$ gives

$$\begin{bmatrix} ① & -1 & 0 & | & 0 \\ 0 & ① & 0 & | & 2 \\ 0 & 0 & ① & | & -1 \end{bmatrix}.$$

(The other zero in the third column we got for free.) Now perform $R_1 + R_2$ to have

$$\begin{bmatrix} ① & 0 & 0 & | & 2 \\ 0 & ① & 0 & | & 2 \\ 0 & 0 & ① & | & -1 \end{bmatrix}.$$

The point of the extra elimination steps is that restoring the system, immediately produces the answer $x_1 = 2$, $x_2 = 2$, $x_3 = -1$.

Complete forward elimination produces a matrix that has ones on the diagonal, and all off-diagonal elements are zeros.

Exercises

1. The following augmented matrices are in row echelon form. Circle the pivots, then restore the corresponding systems and solve them by back-substitution:

(a)

$$\begin{bmatrix} 2 & -1 & | & 0 \\ 0 & 3 & | & 6 \end{bmatrix};$$

(b)

$$\begin{bmatrix} 2 & -2 & | & 4 \\ 0 & 0 & | & 0 \end{bmatrix};$$

(c)

$$\begin{bmatrix} 4 & -1 & | & 5 \\ 0 & 0 & | & 3 \end{bmatrix}.$$

Answer: No solution.

(d)

$$\begin{bmatrix} 2 & -1 & 0 & | & 3 \\ 0 & 3 & 1 & | & 1 \\ 0 & 0 & 2 & | & -4 \end{bmatrix}.$$

Answer: $x_1 = 2$, $x_2 = 1$, $x_3 = -2$.

(e)

$$\begin{bmatrix} 1 & -1 & 1 & | & 3 \\ 0 & 1 & 2 & | & -1 \end{bmatrix}.$$

Answer: $x_1 = -3t + 2$, $x_2 = -2t - 1$, $x_3 = t$, t is arbitrary.

(f)

$$\begin{bmatrix} 2 & -1 & 0 & | & 2 \\ 0 & 0 & 1 & | & -4 \end{bmatrix}.$$

Answer: $x_1 = \frac{1}{2}t + 1$, $x_2 = t$, $x_3 = -4$.

(g)

$$\begin{bmatrix} 5 & -1 & 2 & | & 3 \\ 0 & 3 & 1 & | & -1 \\ 0 & 0 & 0 & | & -4 \end{bmatrix}.$$

Answer: The system is inconsistent (no solution).

(h)

$$\begin{bmatrix} 1 & -1 & 1 & 1 & | & 0 \\ 0 & 0 & 1 & 2 & | & 5 \\ 0 & 0 & 0 & 0 & | & 0 \end{bmatrix}.$$

Answer: $x_1 = x_2 + x_4 - 5$, $x_3 = -2x_4 + 5$, x_2 and x_4 are arbitrary.

2. For the following systems write down the augmented matrix, reduce it to the row echelon form, then solve the system by back-substitution. Which variables are pivot variables, and which are free? Circle the pivots.

(a)

$$\frac{1}{3}x_1 - \frac{1}{3}x_2 = 1,$$
$$2x_1 + 6x_2 = -2.$$

(b)

$$x_2 - x_3 = 1,$$
$$x_1 + 2x_2 + x_3 = 0,$$
$$3x_1 + x_2 + 2x_3 = 1.$$

Answer: $x = 1$, $y = 0$, $z = -1$.

(c)

$$3x_1 - 2x_2 - x_3 = 0,$$

$$x_1 + 2x_2 + x_3 = -1,$$
$$x_1 - 6x_2 - 3x_3 = 5.$$

Answer. No solution.

(d)

$$3x_1 - 2x_2 - x_3 = 0,$$
$$x_1 + 2x_2 + x_3 = -1,$$
$$x_1 - 6x_2 - 3x_3 = 2.$$

Answer: $x_1 = -\frac{1}{4}$, $x_2 = -\frac{1}{2}t - \frac{3}{8}$, $x_3 = t$.

(e)

$$x_1 - x_2 + x_4 = 1,$$
$$2x_1 - x_2 + x_3 + x_4 = -3,$$
$$x_2 + x_3 - x_4 = -5.$$

Answer: $x_1 = -t - 4$, $x_2 = -t + s - 5$, $x_3 = t$, $x_4 = s$.

3. Solve the following systems given by their augmented matrices:

(a)

$$\begin{bmatrix} 1 & -2 & 0 & | & 2 \\ 2 & 3 & 1 & | & -4 \\ 1 & 5 & 1 & | & -5 \end{bmatrix}.$$

Answer: No solution.

(b)

$$\begin{bmatrix} 1 & -2 & -3 & | & 1 \\ 2 & -3 & -1 & | & 4 \\ 3 & -5 & -4 & | & 5 \end{bmatrix}.$$

Answer: $x = -7t + 5$, $y = -5t + 2$, $z = t$.

(c)

$$\begin{bmatrix} 1 & -2 & -1 & 3 & | & 1 \\ 2 & -4 & 1 & 0 & | & 5 \\ 1 & -2 & 2 & -3 & | & 4 \end{bmatrix}.$$

Answer: $x_1 = -t + 2s + 2$, $x_2 = s$, $x_3 = 2t + 1$, $x_4 = t$.

(d)

$$\begin{bmatrix} 1 & -1 & 0 & 1 & | & 0 \\ 2 & -2 & 1 & -1 & | & 1 \\ 3 & -3 & 2 & 0 & | & 2 \end{bmatrix}.$$

Answer: $x_1 = t, x_2 = t, x_3 = 1, x_4 = 0$.

(e)

$$\begin{bmatrix} 0 & 0 & 3 & | & 6 \\ 0 & 1 & -2 & | & 0 \\ 1 & 0 & 1 & | & 1 \end{bmatrix}.$$

Answer: $x_1 = -1, x_2 = 4, x_3 = 2$.

4. Solve again the systems in Problems 2(a) and 2(b) by performing complete Gaussian elimination.

5. Find the number a for which the following system has infinitely many solutions, then find these solutions:

$$\begin{bmatrix} 1 & -1 & 2 & | & 3 \\ 0 & 1 & -1 & | & -2 \\ 1 & 0 & a & | & 1 \end{bmatrix}.$$

Answer: $a = 1, x_1 = -x_3 + 1, x_2 = x_3 - 2, x_2$ is arbitrary.

6. What is the maximal possible number of pivots for the matrices of the following sizes: (a) 5×6; (b) 11×3; (c) 7×1; (d) 1×8; (e) $n \times n$.

1.3 Vector interpretation of linear systems

In this section we discuss geometrical interpretation of systems of linear equations in terms of vectors.

Given two three-dimensional vectors $C_1 = \begin{bmatrix} 1 \\ -1 \\ 3 \end{bmatrix}$ and $C_2 = \begin{bmatrix} 5 \\ -4 \\ 2 \end{bmatrix}$, we may add them by adding the corresponding components $C_1 + C_2 = \begin{bmatrix} 6 \\ -5 \\ 5 \end{bmatrix}$, or multiply C_1 by a number x_1 (componentwise), $x_1 C_1 = \begin{bmatrix} x_1 \\ -x_1 \\ 3x_1 \end{bmatrix}$, or calculate their *linear combination*

$$x_1 C_1 + x_2 C_2 = \begin{bmatrix} x_1 + 5x_2 \\ -x_1 - 4x_2 \\ 3x_1 + 2x_2 \end{bmatrix},$$

where x_2 is another scalar (number). Recall that the vector C_1 joins the origin $(0, 0, 0)$ to the point with coordinates $(1, -1, 3)$. The vector $x_1 C_1$ points in the same direction as C_1 if $x_1 > 0$, and in the opposite direction in case $x_1 < 0$. The sum $C_1 + C_2$ corresponds to the parallelogram rule of addition of vectors.

Given a vector $b = \begin{bmatrix} -3 \\ 2 \\ 4 \end{bmatrix}$, let us try to find the numbers x_1 and x_2, so that

$$x_1 C_1 + x_2 C_2 = b.$$

In components, we need

$$x_1 + 5x_2 = -3,$$
$$-x_1 - 4x_2 = 2,$$
$$3x_1 + 2x_2 = 4.$$

But that is just a three-by-two system of equations! It has a unique solution $x_1 = 2$ and $x_2 = -1$, found by Gaussian elimination, so that

$$b = 2C_1 - C_2.$$

The vector b is a *linear combination of the vectors* C_1 *and* C_2. Geometrically, the vector b lies in the plane determined by the vectors C_1 and C_2 (this plane passes through the origin). One also says that b belongs to the *span of the vectors* C_1 *and* C_2, denoted by Span$\{C_1, C_2\}$, and defined to be *the set of all possible linear combinations* $x_1 C_1 + x_2 C_2$. The columns of the augmented matrix of this system

$$\begin{bmatrix} 1 & 5 & -3 \\ -1 & -4 & 2 \\ 3 & 2 & 4 \end{bmatrix}$$

are precisely the vectors C_1, C_2, and b. We can write the augmented matrix as $[C_1 \ C_2 \vdots b]$ by listing its columns.

In place of b, let us consider another vector $B = \begin{bmatrix} -3 \\ 2 \\ 1 \end{bmatrix}$, and again try to find the numbers x_1 and x_2, so that

$$x_1 C_1 + x_2 C_2 = B.$$

In components, this time we need

$$x_1 + 5x_2 = -3,$$
$$-x_1 - 4x_2 = 2,$$
$$3x_1 + 2x_2 = 1.$$

This three-by-two system of equations has no solutions, since the third equation does not hold at the solution $x_1 = 2$, $x_2 = -1$ of the first two equations. The vector B does not lie in the plane determined by the vectors C_1 and C_2 (equivalently, B is not a linear combination of the vectors C_1 and C_2, so that B does not belong to Span$\{C_1, C_2\}$). The columns of the augmented matrix for the last system

$$\begin{bmatrix} 1 & 5 & \vdots & -3 \\ -1 & -4 & \vdots & 2 \\ 3 & 2 & \vdots & 1 \end{bmatrix} = [C_1 \; C_2 \vdots B]$$

are the vectors C_1, C_2, and B.

The above examples illustrate that a system with the augmented matrix $[C_1 \; C_2 \vdots b]$ has a solution exactly when (if and only if) the vector of the right-hand sides b belongs to Span$\{C_1, C_2\}$. Observe that C_1 and C_2 are the columns of the matrix of the system.

Similarly, a system of three equations with three unknowns and the augmented matrix $[C_1 \; C_2 \; C_3 \vdots b]$ has a solution if and only if the vector of the right-hand sides b belongs to Span$\{C_1, C_2, C_3\}$. In other words, b is a linear combination of C_1, C_2, and C_3 if and only if the system with the augmented matrix $[C_1 \; C_2 \; C_3 \vdots b]$ is consistent (has solutions). The same is true for systems of arbitrary size, say a system of seven equations with eleven unknowns (the columns of its matrix will be seven-dimensional vectors). We discuss vectors of arbitrary dimension next.

In Calculus and Physics, one deals with either two- or three-dimensional vectors. The set of all possible two-dimensional vectors is denoted by R^2, while R^3 denotes all vectors in the three-dimensional space we live in. By analogy, R^n is the set of all possible n-dimensional vectors of the form $\begin{bmatrix} a_1 \\ a_2 \\ \vdots \\ a_n \end{bmatrix}$, which can be added or multiplied by a scalar the same way as in R^2 or in R^3. For example, one adds two vectors in R^4 as follows:

$$\begin{bmatrix} a_1 \\ a_2 \\ a_3 \\ a_4 \end{bmatrix} + \begin{bmatrix} b_1 \\ b_2 \\ b_3 \\ b_4 \end{bmatrix} = \begin{bmatrix} a_1 + b_1 \\ a_2 + b_2 \\ a_3 + b_3 \\ a_4 + b_4 \end{bmatrix},$$

by adding the corresponding components. If c is a scalar, then

$$c \begin{bmatrix} a_1 \\ a_2 \\ a_3 \\ a_4 \end{bmatrix} = \begin{bmatrix} c\,a_1 \\ c\,a_2 \\ c\,a_3 \\ c\,a_4 \end{bmatrix}.$$

It is customary to use boldface (or capital) letters when denoting vectors, for example, $\mathbf{a} = \begin{bmatrix} a_1 \\ a_2 \\ a_3 \\ a_4 \end{bmatrix}$, $\mathbf{b} = \begin{bmatrix} b_1 \\ b_2 \\ b_3 \\ b_4 \end{bmatrix}$. (We shall also write $a = \begin{bmatrix} a_1 \\ a_2 \\ a_3 \\ a_4 \end{bmatrix}$, when it is clear from context that $a \in R^4$ is a vector.) Usual algebra rules apply to vectors, for example,

$$\mathbf{b} + \mathbf{a} = \mathbf{a} + \mathbf{b},$$
$$c(\mathbf{a} + \mathbf{b}) = c\,\mathbf{a} + c\,\mathbf{b},$$

for any scalar c.

Recall that *matrix* is a rectangular array (a table) of numbers. We say that a matrix A is *of size* (or of *type*) $m \times n$ if it has m rows and n columns. For example, the matrix

$$A = \begin{bmatrix} 1 & -1 & 2 \\ -1 & 0 & 4 \end{bmatrix}$$

is of size 2×3. It has three columns $\mathbf{a_1} = \begin{bmatrix} 1 \\ -1 \end{bmatrix}$, $\mathbf{a_2} = \begin{bmatrix} -1 \\ 0 \end{bmatrix}$, and $\mathbf{a_3} = \begin{bmatrix} 2 \\ 4 \end{bmatrix}$, which are vectors in R^2. One can write the matrix A through its columns

$$A = [\mathbf{a_1} \ \mathbf{a_2} \ \mathbf{a_3}].$$

A matrix A of size $m \times n$,

$$A = [\mathbf{a_1} \ \mathbf{a_2} \ \dots \ \mathbf{a_n}],$$

has n columns, and each of them is a vector in R^m.

The augmented matrix for a system of m equations with n unknowns has the form $[\mathbf{a_1} \ \mathbf{a_2} \ \dots \ \mathbf{a_n} \vdots \ b]$, and each column is a vector in R^m. The system is *consistent* (it has a solution) if and only if the vector of the right-hand sides b belongs to $\mathrm{Span}\{\mathbf{a_1}, \mathbf{a_2}, \dots, \mathbf{a_n}\}$, which is defined as *the set of all possible linear combinations* $x_1\mathbf{a_1} + x_2\mathbf{a_2} + \dots + x_n\mathbf{a_n}$.

One defines *the product Ax* of an $m \times n$ matrix $A = [\mathbf{a_1} \ \mathbf{a_2} \ \dots \ \mathbf{a_n}]$ and of vector $x = \begin{bmatrix} x_1 \\ x_2 \\ \vdots \\ x_n \end{bmatrix}$ in R^n as the following linear combination of columns of A:

$$A x = [\mathbf{a_1} \ \mathbf{a_2} \ \dots \ \mathbf{a_n}] \begin{bmatrix} x_1 \\ x_2 \\ \vdots \\ x_n \end{bmatrix} = x_1\mathbf{a_1} + x_2\mathbf{a_2} + \dots + x_n\mathbf{a_n}.$$

The vector Ax belongs to R^m. For example,

$$\begin{bmatrix} 1 & -1 & 2 \\ -1 & 0 & 4 \end{bmatrix} \begin{bmatrix} 3 \\ -2 \\ 1 \end{bmatrix} = 3 \begin{bmatrix} 1 \\ -1 \end{bmatrix} + (-2) \begin{bmatrix} -1 \\ 0 \end{bmatrix} + 1 \begin{bmatrix} 2 \\ 4 \end{bmatrix} = \begin{bmatrix} 7 \\ 1 \end{bmatrix}.$$

If $y = \begin{bmatrix} y_1 \\ y_2 \\ \vdots \\ y_n \end{bmatrix}$ is another vector in R^n, it is straightforward to verify that

$$A(x + y) = Ax + Ay.$$

Indeed,

$$A(x+y) = (x_1+y_1)\mathbf{a_1} + \cdots + (x_n+y_n)\mathbf{a_n}$$
$$= x_1\mathbf{a_1} + \cdots + x_n\mathbf{a_n} + y_1\mathbf{a_1} + \cdots + y_n\mathbf{a_n} = Ax + Ay.$$

One also checks that

$$A(cx) = cAx,$$

for any scalar c.

We now connect the product Ax to linear systems. The matrix of the system

$$x_1 - x_2 + 3x_3 = 2,$$
$$2x_1 + 6x_2 - 2x_3 = 4,$$
$$5x_1 + 2x_2 + x_3 = 0 \tag{3.1}$$

is $A = \begin{bmatrix} 1 & -1 & 3 \\ 2 & 6 & -2 \\ 5 & 2 & 1 \end{bmatrix}$, and the vector of right-hand sides is $b = \begin{bmatrix} 2 \\ 4 \\ 0 \end{bmatrix}$. Define $x = \begin{bmatrix} x_1 \\ x_2 \\ x_3 \end{bmatrix}$, the *vector of unknowns*. (Here we do not use boldface letters to denote the vectors b and x.) Calculate

$$Ax = \begin{bmatrix} 1 & -1 & 3 \\ 2 & 6 & -2 \\ 5 & 2 & 1 \end{bmatrix} \begin{bmatrix} x_1 \\ x_2 \\ x_3 \end{bmatrix} = \begin{bmatrix} x_1 - x_2 + 3x_3 \\ 2x_1 + 6x_2 - 2x_3 \\ 5x_1 + 2x_2 + x_3 \end{bmatrix}.$$

It follows that the system (3.1) can be written in the matrix form

$$Ax = b. \tag{3.2}$$

Any $m \times n$ linear system can be written in the form (3.2), where A is the $m \times n$ matrix of the system, $b \in R^m$ is the vector of right-hand sides, and $x \in R^n$ is the vector of unknowns.

Analogy is a key concept when dealing with objects in dimensions greater than three. Suppose a four-dimensional spaceship of the form of four-dimensional ball ($x_1^2 + x_2^2 + x_3^2 + x_4^2 \le R^2$) passes by us. What will we see? By analogy, imagine people living in a plane (or flatland) and a three-dimensional ball passes by. At first they see nothing (the ball is out of their plane), then they see a point, then an expanding disc, then a contracting disc, followed by a point, and then they see nothing again. Can you now answer the original question? (One will see: nothing, one point, expanding balls, contracting balls, one point, nothing.)

Exercises

1. Express the vector $b = \begin{bmatrix} 1 \\ 0 \\ 4 \end{bmatrix}$ as a linear combination of the vectors $C_1 = \begin{bmatrix} 1 \\ 0 \\ 1 \end{bmatrix}$, $C_2 = \begin{bmatrix} 0 \\ -1 \\ 1 \end{bmatrix}$, and $C_3 = \begin{bmatrix} 1 \\ 2 \\ 3 \end{bmatrix}$. In other words, find the numbers x_1, x_2, x_3 so that $b = x_1C_1 + x_2C_2 + x_3C_3$. Write down the augmented matrix for the corresponding system of equations.

Answer: $x_1 = \frac{1}{4}, x_2 = \frac{3}{2}, x_3 = \frac{3}{4}$.

2. Is it possible to express the vector $b = \begin{bmatrix} 5 \\ 3 \\ -3 \end{bmatrix}$ as a linear combination of the vectors

$C_1 = \begin{bmatrix} 1 \\ 1 \\ -1 \end{bmatrix}, C_2 = \begin{bmatrix} 2 \\ 1 \\ -1 \end{bmatrix}$, and $C_3 = \begin{bmatrix} 3 \\ 2 \\ -2 \end{bmatrix}$?

Answer: Yes.

3. Is it possible to express the vector $b = \begin{bmatrix} 5 \\ 4 \\ -3 \end{bmatrix}$ as a linear combination of the vectors

$C_1 = \begin{bmatrix} 0 \\ 1 \\ -1 \end{bmatrix}, C_2 = \begin{bmatrix} 0 \\ -2 \\ 1 \\ -1 \end{bmatrix}$, and $C_3 = \begin{bmatrix} 0 \\ 1 \\ 2 \\ -2 \end{bmatrix}$?

Answer: No.

4. Calculate the following products involving a matrix and a vector:

(a) $\begin{bmatrix} 1 & 2 \\ -1 & 1 \end{bmatrix} \begin{bmatrix} 3 \\ -2 \end{bmatrix}$.

Answer: $\begin{bmatrix} -1 \\ -5 \end{bmatrix}$.

(b) $\begin{bmatrix} 1 & 2 & 0 \\ 0 & -1 & 1 \\ 1 & -2 & 1 \end{bmatrix} \begin{bmatrix} x_1 \\ x_2 \\ x_3 \end{bmatrix}$.

Answer: $\begin{bmatrix} x_1 + 2x_2 \\ -x_2 + x_3 \\ x_1 - 2x_2 + x_3 \end{bmatrix}$.

(c) $\begin{bmatrix} 1 & -2 & 0 \\ 3 & -1 & 1 \end{bmatrix} \begin{bmatrix} 1 \\ 2 \\ 3 \end{bmatrix}$.

(d) $\begin{bmatrix} -1 & 2 \\ 0 & -1 \\ 1 & 4 \\ 3 & 0 \end{bmatrix} \begin{bmatrix} -1 \\ 2 \end{bmatrix}$.

Answer: $\begin{bmatrix} 5 \\ -2 \\ 7 \\ -3 \end{bmatrix}$.

(e) $\begin{bmatrix} 3 & -1 & 1 \end{bmatrix} \begin{bmatrix} 1 \\ 0 \\ 3 \end{bmatrix}$.

Answer: 6.

(f) $\begin{bmatrix} 1 & -2 & 0 \\ 3 & -1 & 1 \end{bmatrix} \begin{bmatrix} 0 \\ 0 \\ 0 \end{bmatrix}$.

Answer: $\begin{bmatrix} 0 \\ 0 \end{bmatrix}$.

5. Does the vector b lie in the plane determined by the vectors C_1 and C_2?

(a) $b = \begin{bmatrix} 0 \\ 1 \\ -4 \end{bmatrix}, C_1 = \begin{bmatrix} 2 \\ 1 \\ -2 \end{bmatrix}, C_2 = \begin{bmatrix} 1 \\ 0 \\ 1 \end{bmatrix}$.

Answer: Yes.

(b) $b = \begin{bmatrix} 5 \\ 1 \\ -4 \end{bmatrix}, C_1 = \begin{bmatrix} 2 \\ -1 \\ 0 \end{bmatrix}, C_2 = \begin{bmatrix} 1 \\ -3 \\ 0 \end{bmatrix}$.

Answer: No.

(c) $b = \begin{bmatrix} 2 \\ 1 \\ -2 \end{bmatrix}, C_1 = \begin{bmatrix} 2 \\ -1 \\ 3 \end{bmatrix}, C_2 = \begin{bmatrix} -4 \\ -2 \\ 4 \end{bmatrix}$.

Answer: Yes.

(d) $b = \begin{bmatrix} 2 \\ -4 \\ 5 \end{bmatrix}, C_1 = \begin{bmatrix} 3 \\ -1 \\ 1 \end{bmatrix}, C_2 = \begin{bmatrix} -1 \\ -3 \\ 2 \end{bmatrix}$.

Answer: No.

6. Does the vector b belong to $\mathrm{Span}\{C_1, C_2, C_3\}$?

(a) $b = \begin{bmatrix} 1 \\ 1 \\ 1 \end{bmatrix}$, $C_1 = \begin{bmatrix} 1 \\ 0 \\ 0 \end{bmatrix}$, $C_2 = \begin{bmatrix} 0 \\ 1 \\ 0 \end{bmatrix}$, $C_3 = \begin{bmatrix} 1 \\ 1 \\ 0 \end{bmatrix}$.

Answer: No.

(b) $b = \begin{bmatrix} 1 \\ 1 \\ 1 \end{bmatrix}$, $C_1 = \begin{bmatrix} 1 \\ 0 \\ 0 \end{bmatrix}$, $C_2 = \begin{bmatrix} 0 \\ 1 \\ 0 \end{bmatrix}$, $C_3 = \begin{bmatrix} 0 \\ 0 \\ 1 \end{bmatrix}$.

Answer: Yes.

7. Let A be of size 4×5, and x be in R^4. Is the product Ax defined?
8. Let A be of size 7×8, and $x \in R^8$. Is the product Ax defined?
9. Let A be of size $m \times n$, and $\mathbf{0}$ the zero vector in R^n (all components of $\mathbf{0}$ are zero). Calculate the product $A\,\mathbf{0}$, and show that it is the zero vector in R^m.

1.4 Solution set of a linear system $Ax = b$

When all right-hand sides are zero, the system is called *homogeneous*,

$$Ax = 0. \tag{4.1}$$

On the right-hand side in (4.1) is *the zero vector*, or a vector with all components equal to zero (often denoted by $\mathbf{0}$). Here the matrix A is of size $m \times n$. The vector of unknowns x is in R^n. The system (4.1) always has a solution $x = 0$, or $x_1 = x_2 = \cdots = x_n = 0$, called *the trivial solution*. We wish to find all solutions.

Our first example is the homogeneous system

$$
\begin{aligned}
x_1 - x_2 + x_3 &= 0, \\
-2x_1 + x_2 - x_3 &= 0, \\
3x_1 - 2x_2 + 4x_3 &= 0,
\end{aligned}
$$

with the augmented matrix

$$
\left[\begin{array}{ccc|c}
1 & -1 & 1 & 0 \\
-2 & 1 & -1 & 0 \\
3 & -2 & 4 & 0
\end{array}\right].
$$

Forward elimination ($R_2 + 2R_1$, $R_3 - 3R_1$, followed by $R_3 + R_2$) leads to

$$
\left[\begin{array}{ccc|c}
① & -1 & 1 & 0 \\
0 & ① & 1 & 0 \\
0 & 0 & ② & 0
\end{array}\right],
$$

or

$$
\begin{aligned}
x_1 - x_2 + x_3 &= 0, \\
-x_2 + x_3 &= 0, \\
2x_3 &= 0.
\end{aligned}
$$

Back-substitution gives $x_1 = x_2 = x_3 = 0$, the trivial solution. There are three pivot variables, and no free variables. The trivial solution is the only solution of this system. *Homogeneous system must have free variables, in order to have nontrivial solutions.*

Our next example has the augmented matrix

$$\left[\begin{array}{ccc|c} 1 & -1 & 1 & 0 \\ -2 & 1 & -1 & 0 \\ 3 & -2 & 2 & 0 \end{array}\right],$$

which is a small modification of the preceding system, with only one entry of the third row changed. The same steps of forward elimination ($R_2 + 2R_1$, $R_3 - 3R_1$, followed by $R_3 + R_2$) lead to

$$\left[\begin{array}{ccc|c} ① & -1 & 1 & 0 \\ 0 & ① & 1 & 0 \\ 0 & 0 & 0 & 0 \end{array}\right],$$

or

$$x_1 - x_2 + x_3 = 0,$$
$$-x_2 + x_3 = 0,$$

after discarding a row of zeroes. Solving for the pivot variables x_1, x_2 in terms of the free variable x_3, we obtain infinitely many solutions: $x_1 = 0$, $x_2 = x_3$, and x_3 is an arbitrary number. Write this solution in vector form

$$\begin{bmatrix} x_1 \\ x_2 \\ x_3 \end{bmatrix} = \begin{bmatrix} 0 \\ x_3 \\ x_3 \end{bmatrix} = x_3 \begin{bmatrix} 0 \\ 1 \\ 1 \end{bmatrix} = x_3\, u,$$

where $u = \begin{bmatrix} 0 \\ 1 \\ 1 \end{bmatrix}$. It is customary to set $x_3 = t$, then the solution set of this system is given by $t\,u$, all possible multiples of the vector u. Geometrically, the solution set consists of all vectors lying on the line through the origin parallel to u, or Span$\{u\}$.

The next example is a homogeneous system of four equations with four unknowns given by its augmented matrix

$$\left[\begin{array}{cccc|c} 1 & 0 & -1 & 1 & 0 \\ -2 & 1 & 3 & 4 & 0 \\ -1 & 1 & 2 & 5 & 0 \\ 5 & -2 & -7 & -7 & 0 \end{array}\right].$$

Forward elimination steps $R_2 + 2R_1$, $R_3 + R_1$, $R_4 - 5R_1$ give

$$\begin{bmatrix} 1 & 0 & -1 & 1 & \vdots & 0 \\ 0 & 1 & 1 & 6 & \vdots & 0 \\ 0 & 1 & 1 & 6 & \vdots & 0 \\ 0 & -2 & -2 & -12 & \vdots & 0 \end{bmatrix}.$$

Then perform $R_3 - R_2$ and $R_4 + 2R_2$ to obtain

$$\begin{bmatrix} ① & 0 & -1 & 1 & \vdots & 0 \\ 0 & ① & 1 & 6 & \vdots & 0 \\ 0 & 0 & 0 & 0 & \vdots & 0 \\ 0 & 0 & 0 & 0 & \vdots & 0 \end{bmatrix}.$$

Restoring the system to get

$$x_1 - x_3 + x_4 = 0,$$
$$x_2 + x_3 + 6x_4 = 0,$$

expressing the pivot variables x_1, x_2 in terms of the free x_3, x_4, then setting $x_3 = t$ and $x_4 = s$, two arbitrary numbers, we obtain infinitely many solutions: $x_1 = t - s$, $x_2 = -t - 6s$, $x_3 = t$, and $x_4 = s$. Writing this solution in vector form

$$\begin{bmatrix} t - s \\ -t - 6s \\ t \\ s \end{bmatrix} = t \begin{bmatrix} 1 \\ -1 \\ 1 \\ 0 \end{bmatrix} + s \begin{bmatrix} -1 \\ -6 \\ 0 \\ 1 \end{bmatrix} = t\,u + s\,v,$$

we see that the solution set is a linear combination of the vectors $u = \begin{bmatrix} 1 \\ -1 \\ 1 \\ 0 \end{bmatrix}$ and $v = \begin{bmatrix} -1 \\ -6 \\ 0 \\ 1 \end{bmatrix}$,

or Span$\{u, v\}$.

In general, if the number of free variables is k, then the solution set of an $m \times n$ homogeneous system $Ax = 0$ has the form Span$\{u_1, u_2, \ldots, u_k\}$ for some vectors $u_1, u_2, \ldots, u_k$ that are solutions of this system.

An $m \times n$ homogeneous system $Ax = 0$ has at most m pivots, so that there are at most m pivot variables. That is because each pivot occupies its own row, and the number of rows is m. If $n > m$, there are more variables in total than the number of pivot variables. Hence some variables are free, and the system $Ax = 0$ has infinitely many solutions. For future reference, this fact is stated as a theorem.

Theorem 1.4.1. *An $m \times n$ homogeneous system $Ax = 0$, with $n > m$, has infinitely many solutions.*

Turning to *nonhomogeneous systems* $Ax = b$, with vector $b \neq 0$, let us revisit the system

$$2x_1 - x_2 + 5x_3 = 1,$$
$$x_1 + x_2 + x_3 = -2,$$

for which we calculated in Section 1.1 the solution set to be

$$\begin{bmatrix} x_1 \\ x_2 \\ x_3 \end{bmatrix} = t \begin{bmatrix} -2 \\ 1 \\ 1 \end{bmatrix} - \frac{1}{3} \begin{bmatrix} 1 \\ 5 \\ 0 \end{bmatrix} = t u + p,$$

denoting $u = \begin{bmatrix} -2 \\ 1 \\ 1 \end{bmatrix}$ and $p = -\frac{1}{3} \begin{bmatrix} 1 \\ 5 \\ 0 \end{bmatrix}$. Recall that $t u$ represents vectors on a line through the origin parallel to the vector u (with t arbitrary). The vector p translates this line to a parallel one, off the origin. Let us consider the *corresponding homogeneous system*:

$$2x_1 - x_2 + 5x_3 = 0,$$
$$x_1 + x_2 + x_3 = 0,$$

with the right-hand sides changed to zero. One calculates its solution set to be $t u$, with the same u. In general, the solution set of the system $Ax = b$ is a translation by some vector p of the solution set of the corresponding homogeneous system $Ay = 0$. Indeed, if p is any particular solution of the nonhomogeneous system, so that $Ap = b$, then $A(p + y) = Ap + Ay = Ap = b$. It follows that $p + y$ gives the solution set of the nonhomogeneous system.

We conclude this section with a "book-keeping" remark. Suppose one needs to solve three systems $Ax = b_1$, $Ax = b_2$, and $Ax = b_3$, all with the same matrix A. Calculations can be done in parallel by considering a "long" augmented matrix $[\ A\ |\ b_1\ |\ b_2\ |\ b_3\ ]$. If the first step in the row reduction of A is, say $R_2 - 2R_1$, this step is performed on the entire "long" second row. Once A is reduced to the row echelon form, we restore each of the systems separately, and perform back-substitution.

Exercises

1. Let $A = \begin{bmatrix} 1 & 2 & -1 \\ 1 & 2 & 0 \\ 1 & 2 & -1 \end{bmatrix}$, $b_1 = \begin{bmatrix} 2 \\ 3 \\ 2 \end{bmatrix}$, $b_2 = \begin{bmatrix} -1 \\ 0 \\ 2 \end{bmatrix}$. Determine the solution set of the following systems. (Calculations for all three cases can be done in parallel.)

    (a) $Ax = 0$.

      Answer: $x = t \begin{bmatrix} -2 \\ 1 \\ 0 \end{bmatrix}$.

    (b) $Ax = b_1$.

      Answer: $x = t \begin{bmatrix} -2 \\ 1 \\ 0 \end{bmatrix} + \begin{bmatrix} 3 \\ 0 \\ 1 \end{bmatrix}$.

    (c) $Ax = b_2$.

      Answer: The system is inconsistent (no solutions).

2. Let A be a 4×5 matrix. Does the homogeneous system $Ax = 0$ have nontrivial solutions?

3. Let A be an $n \times n$ matrix, with n pivots. Are there any solutions of the system $Ax = 0$, in addition to the trivial one?

4. Let $x_1 = 2$, $x_2 = 1$ be a solution of some system $Ax = b$, with a 2×2 matrix A. Assume that the solution set of the corresponding homogeneous system $Ax = 0$ is $t \begin{bmatrix} 1 \\ -3 \end{bmatrix}$, with arbitrary t. Describe geometrically the solution set of $Ax = b$.
   Answer: The line of slope -3 passing through the point $(2, 1)$, or $x_2 = -3x_1 + 7$.

5. Show that the system $Ax = b$ has at most one solution if the corresponding homogeneous system $Ax = 0$ has only the trivial solution.
   Hint. Show that the difference of any two solutions of $Ax = b$ satisfies the corresponding homogeneous system.

6. Let x and y be two solutions of the homogeneous system $Ax = 0$.
   (a) Show that $x + y$ is also a solution of this system.
   (b) Show that $c_1 x + c_2 y$ is a solution of this system, for any scalars c_1, c_2.

7. Let x and y be two solutions of a nonhomogeneous system $Ax = b$, with nonzero vector b. Show that $x + y$ is not a solution of this system.

8. True or false?
   (a) If a linear system of equations has a trivial solution, this system is homogeneous.
   (b) If A of size 5×5 has 4 pivots, then the system $Ax = 0$ has nontrivial solutions.
   (c) If A is a 4×5 matrix with 3 pivots, then the solution set of $Ax = 0$ involves one arbitrary constant.
   Answer: False.
   (d) If A is a 5×6 matrix, then for any b the system $Ax = b$ is consistent (has solutions).
   Answer: False.

1.5 Linear dependence and independence

Given a set of vectors $u_1, u_2, \ldots, u_n$ in R^m, we look for the scalars (coefficients) $x_1, x_2, \ldots, x_n$ which will make their linear combination to be equal to the zero vector,

$$x_1 u_1 + x_2 u_2 + \cdots + x_n u_n = 0. \tag{5.1}$$

The trivial combination $x_1 = x_2 = \cdots = x_n = 0$ clearly works. If the trivial combination is the only way to produce the zero vector, we say that the vectors $u_1, u_2, \ldots, u_n$ are *linearly independent*. If any nontrivial combination is equal to the zero vector, we say that the vectors $u_1, u_2, \ldots, u_n$ are *linearly dependent*.

Suppose that the vectors $u_1, u_2, \ldots, u_n$ are linearly dependent. Then (5.1) holds, with at least one of the coefficients not zero. Let us say, $x_1 \neq 0$. Writing $x_1 u_1 = -x_2 u_2 - \cdots - x_n u_n$, express

$$u_1 = -\frac{x_2}{x_1} u_2 - \cdots - \frac{x_n}{x_1} u_n,$$

so that u_1 is a linear combination of the other vectors. Conversely, suppose that u_1 is a linear combination of the other vectors $u_1 = y_2 u_2 + \cdots + y_n u_n$, with some coefficients $y_2, \ldots, y_n$. Then

$$(-1)u_1 + y_2 u_2 + \cdots + y_n u_n = 0.$$

We have a nontrivial linear combination, with at least one of the coefficients nonzero (namely, $(-1) \neq 0$), producing the zero vector. The vectors $u_1, u_2, \ldots, u_n$ are linearly dependent. Conclusion: *a set of vectors is linearly dependent if and only if (exactly when) one of the vectors is a linear combination of the others.*

For two vectors u_1, u_2, linear dependence means that $u_1 = y_2 u_2$, for some scalar y_2, so that the vectors are proportional, and they go along the same line (in case of R^2 or R^3). For three vectors u_1, u_2, u_3, linear dependence implies that $u_1 = y_2 u_2 + y_3 u_3$ (geometrically, if these vectors are in R^3 they lie in the same plane).

For example, $a_1 = \begin{bmatrix} 1 \\ -1 \\ 2 \end{bmatrix}$, $a_2 = \begin{bmatrix} 1 \\ -3 \\ 3 \end{bmatrix}$, and $a_3 = \begin{bmatrix} 1 \\ 1 \\ 1 \end{bmatrix}$ are linearly dependent, because

$$a_2 = 2a_1 - a_3,$$

while the vectors $b_1 = \begin{bmatrix} 1 \\ -1 \\ 2 \end{bmatrix}$, $b_2 = \begin{bmatrix} -2 \\ 2 \\ -4 \end{bmatrix}$, and $b_3 = \begin{bmatrix} 1 \\ 4 \\ -5 \end{bmatrix}$ are linearly dependent, because

$$b_1 = \left(-\frac{1}{2}\right) b_2 + 0\, b_3.$$

The vectors $u_1 = \begin{bmatrix} 2 \\ 0 \\ 0 \end{bmatrix}$, $u_2 = \begin{bmatrix} 1 \\ -3 \\ 0 \end{bmatrix}$, and $u_3 = \begin{bmatrix} -1 \\ 1 \\ 3 \end{bmatrix}$, are linearly independent, because none of these vectors is a linear combination of the other two. Let us see why u_2 is not a linear combination of u_1 and u_3. Indeed, if we had $u_2 = x_1 u_1 + x_2 u_3$, or

$$\begin{bmatrix} 1 \\ -3 \\ 0 \end{bmatrix} = x_1 \begin{bmatrix} 2 \\ 0 \\ 0 \end{bmatrix} + x_2 \begin{bmatrix} -1 \\ 1 \\ 3 \end{bmatrix},$$

then comparing the third components would give $x_2 = 0$, so that

$$\begin{bmatrix} 1 \\ -3 \\ 0 \end{bmatrix} = x_1 \begin{bmatrix} 2 \\ 0 \\ 0 \end{bmatrix},$$

which is not possible. One shows similarly that u_1 and u_3 are not linear combinations of the other two vectors. A more systematic approach to decide on linear dependence or independence is developed next.

Vectors $u_1, u_2, \ldots, u_n$ in R^m are linearly dependent if the vector equation (5.1) has a nontrivial solution. In components, the vector equation (5.1) is an $m \times n$ homogeneous system with the augmented matrix $[u_1 \ u_2 \ \ldots \ u_n \vdots 0]$. Apply forward elimination. Nontrivial solutions will exist if and only if there are free (nonpivot) variables. If there are no free variables (all columns have pivots), then the trivial solution is the only one. Since we are only interested in pivots, there is no need to carry a column of zeroes in the augmented matrix when performing row reduction.

Algorithm: perform row reduction on the matrix $[u_1 \ u_2 \ \ldots \ u_n]$. If the number of pivots is less than n, the vectors $u_1, u_2, \ldots, u_n$ are linearly dependent. If the number of pivots is equal to n, the vectors $u_1, u_2, \ldots, u_n$ are linearly independent. (The number of pivots cannot exceed the number of columns n, because each pivot occupies its own column.)

Example 1. Determine whether the vectors $u_1 = \begin{bmatrix} 1 \\ 2 \\ 3 \end{bmatrix}$, $u_2 = \begin{bmatrix} 4 \\ 5 \\ 6 \end{bmatrix}$, and $u_3 = \begin{bmatrix} 0 \\ 1 \\ 1 \end{bmatrix}$ are linearly dependent or independent.

Using these vectors as columns, form the matrix

$$\begin{bmatrix} 1 & 4 & 0 \\ 2 & 5 & 1 \\ 3 & 6 & 1 \end{bmatrix}.$$

Performing row reduction ($R_2 - 2R_1, R_3 - 3R_1$, followed by $R_3 - 2R_2$) gives

$$\begin{bmatrix} ① & 4 & 0 \\ 0 & ③ & 1 \\ 0 & 0 & ① \end{bmatrix}.$$

All three columns have pivots. The vectors u_1, u_2, u_3 are linearly independent.

Example 2. Let us revisit the vectors $u_1 = \begin{bmatrix} 2 \\ 0 \\ 0 \end{bmatrix}$, $u_2 = \begin{bmatrix} 1 \\ -3 \\ 0 \end{bmatrix}$, and $u_3 = \begin{bmatrix} -1 \\ 1 \\ 3 \end{bmatrix}$ from a previous example. Using these vectors as columns, form the matrix

$$\begin{bmatrix} ② & 1 & -1 \\ 0 & ③ & 1 \\ 0 & 0 & ③ \end{bmatrix},$$

which is already in row echelon form, with three pivots. The vectors are linearly independent.

Example 3. Determine whether the vectors $v_1 = \begin{bmatrix} 1 \\ 0 \\ -1 \\ 2 \end{bmatrix}$, $v_2 = \begin{bmatrix} 0 \\ -1 \\ 1 \\ 3 \end{bmatrix}$, and $v_3 = \begin{bmatrix} 1 \\ -1 \\ 0 \\ 5 \end{bmatrix}$ are linearly dependent or independent.

Using these vectors as columns, form the matrix

$$\begin{bmatrix} 1 & 0 & 1 \\ 0 & -1 & -1 \\ -1 & 1 & 0 \\ 2 & 3 & 5 \end{bmatrix}.$$

Performing row reduction ($R_3 + R_1$, $R_4 - 2R_1$, followed by $R_3 + R_2$, $R_4 + 3R_2$) gives

$$\begin{bmatrix} ① & 0 & 1 \\ 0 & ① & -1 \\ 0 & 0 & 0 \\ 0 & 0 & 0 \end{bmatrix}.$$

There is no pivot in the third column. The vectors v_1, v_2, and v_3 are linearly dependent. In fact, $v_3 = v_1 + v_2$.

If $n > m$, any vectors $u_1, u_2, \ldots, u_n$ in R^m are linearly dependent. Indeed, row reduction on the matrix $[u_1\ u_2\ \ldots\ u_n]$ will produce no more than m pivots (each pivot occupies its own row), and hence there will be columns without pivots. For example, *any* three (or more) vectors in R^2 are linearly dependent. In R^3 *any* four (or more) vectors are linearly dependent.

There are other instances when linear dependence can be recognized at a glance. For example, if a set of vectors $\mathbf{0}, u_1, u_2, \ldots, u_n$ contains the zero vector $\mathbf{0}$, then this set is linearly dependent. Indeed,

$$1 \cdot \mathbf{0} + 0 \cdot u_1 + 0 \cdot u_2 + \cdots + 0 \cdot u_n = \mathbf{0}$$

is a nontrivial combination producing the zero vector. Another example: the set $u_1, 2u_1, u_3, \ldots, u_n$ is linearly dependent. Indeed,

$$(-2) \cdot u_1 + 1 \cdot 2u_1 + 0 \cdot u_3 + \cdots + 0 \cdot u_n = \mathbf{0}$$

is a nontrivial combination producing the zero vector. More generally, if a subset is linearly dependent, the entire set is linearly dependent.

We shall need the following theorem.

Theorem 1.5.1. *Assume that the vectors $u_1, u_2, \ldots, u_n$ in R^m are linearly independent, and a vector w in R^m is not in their span. Then the vectors $u_1, u_2, \ldots, u_n, w$ are also linearly independent.*

Proof. Assume, *on the contrary*, that the vectors $u_1, u_2, \ldots, u_n, w$ are linearly dependent. Then one can arrange for

$$x_1 u_1 + x_2 u_2 + \cdots + x_n u_n + x_{n+1} w = 0, \tag{5.2}$$

with at least one of the x_i's not zero. If $x_{n+1} \neq 0$, we may solve this relation for w in terms of $u_1, u_2, \dots, u_n$:

$$ w = -\frac{x_1}{x_{n+1}} u_1 - \frac{x_2}{x_{n+1}} u_2 - \cdots - \frac{x_n}{x_{n+1}} u_n, $$

contradicting the assumption that w is not in the span of $u_1, u_2, \dots, u_n$. In the other case when $x_{n+1} = 0$, it follows from (5.2) that

$$ x_1 u_1 + x_2 u_2 + \cdots + x_n u_n = 0, $$

with at least one of the x_i's not zero, contradicting the linear independence of $u_1, u_2, \dots, u_n$.

So that assuming that the theorem is not true leads to a contradiction (an impossible situation). Hence, the theorem is true. □

The method of proof we just used is known as *proof by contradiction*.

Exercises

1. Determine if the following vectors are linearly dependent or independent:

    (a) $\begin{bmatrix} 2 \\ -1 \\ 0 \\ 3 \end{bmatrix}, \begin{bmatrix} -4 \\ 2 \\ 0 \\ -6 \end{bmatrix}$.

    Answer: Dependent.

    (b) $\begin{bmatrix} -1 \\ 1 \\ 3 \end{bmatrix}, \begin{bmatrix} -2 \\ 2 \\ 7 \end{bmatrix}$.

    Answer: Independent.

    (c) $\begin{bmatrix} 1 \\ -1 \\ 2 \\ -3 \\ 4 \end{bmatrix}, \begin{bmatrix} 0 \\ 0 \\ 2 \\ -4 \\ 5 \end{bmatrix}, \begin{bmatrix} 0 \\ 0 \\ 0 \\ 0 \\ 0 \end{bmatrix}$.

    Answer: Dependent.

    (d) $\begin{bmatrix} -1 \\ 2 \\ -3 \end{bmatrix}, \begin{bmatrix} 0 \\ 2 \\ -4 \end{bmatrix}, \begin{bmatrix} -2 \\ 2 \\ -2 \end{bmatrix}$.

    Answer: Dependent.

    (e) $\begin{bmatrix} 1 \\ 1 \\ 1 \\ 1 \end{bmatrix}, \begin{bmatrix} 1 \\ 1 \\ 1 \\ 2 \end{bmatrix}, \begin{bmatrix} 1 \\ 1 \\ 2 \\ 2 \end{bmatrix}, \begin{bmatrix} 1 \\ 2 \\ 2 \\ 2 \end{bmatrix}$.

    Answer: Independent.

    (f) $\begin{bmatrix} 2 \\ -3 \end{bmatrix}, \begin{bmatrix} 0 \\ -4 \end{bmatrix}, \begin{bmatrix} -2 \\ 2 \end{bmatrix}$.

    Answer: Dependent.

    (g) $\begin{bmatrix} -1 \\ 0 \end{bmatrix}, \begin{bmatrix} 2 \\ -3 \end{bmatrix}$.

    Answer: Independent.

(h) $\begin{bmatrix} -2 \\ 0 \\ 0 \end{bmatrix}, \begin{bmatrix} 1 \\ 2 \\ 0 \end{bmatrix}, \begin{bmatrix} 0 \\ 2 \\ 1 \end{bmatrix}.$

Answer: Independent.

(i) $\begin{bmatrix} -2 \\ 0 \\ 0 \\ 0 \end{bmatrix}, \begin{bmatrix} -1 \\ 2 \\ 0 \\ 0 \end{bmatrix}, \begin{bmatrix} 3 \\ 2 \\ -7 \\ 0 \end{bmatrix}, \begin{bmatrix} -3 \\ 2 \\ 5 \\ 4 \end{bmatrix}.$

Answer: Independent.

(j) $\begin{bmatrix} -2 \\ 0 \\ 0 \\ 0 \end{bmatrix}, \begin{bmatrix} -1 \\ 3 \\ 0 \\ 0 \end{bmatrix}, \begin{bmatrix} 4 \\ -2 \\ -7 \\ 0 \end{bmatrix}, \begin{bmatrix} -3 \\ 2 \\ 1 \\ 0 \end{bmatrix}.$

Answer: Dependent.

(k) $\begin{bmatrix} 1 \\ 1 \\ 1 \\ 1 \end{bmatrix}, \begin{bmatrix} -1 \\ -1 \\ -2 \\ 3 \end{bmatrix}, \begin{bmatrix} 2 \\ 2 \\ 0 \\ 1 \end{bmatrix}.$

Answer: Independent.

(l) $\begin{bmatrix} -2 \\ 1 \\ 0 \end{bmatrix}, \begin{bmatrix} 0 \\ 0 \\ 0 \end{bmatrix}, \begin{bmatrix} -1 \\ 2 \end{bmatrix}.$

Answer: The concept of linear dependence or independence is defined only for vectors of the same dimension.

2. Suppose that u_1 and u_2 are linearly independent vectors in R^3.
   (a) Show that the vectors $u_1 + u_2$ and $u_1 - u_2$ are also linearly independent.
   (b) Explain geometrically why this is true.
3. Suppose that the vectors $u_1 + u_2$ and $u_1 - u_2$ are linearly dependent. Show that the vectors u_1 and u_2 are also linearly dependent.
4. Assume that the vectors u_1, u_2, u_3, u_4 in R^n ($n \geq 4$) are linearly independent. Show that the same is true for the vectors $u_1, u_1 + u_2, u_1 + u_2 + u_3$, and $u_1 + u_2 + u_3 + u_4$.
5. Given vectors u_1, u_2, u_3 in R^3, suppose that the following three pairs $(u_1, u_2), (u_1, u_3)$, and (u_2, u_3) are linearly independent. Does it follow that the vectors u_1, u_2, u_3 are linearly independent? Explain.
6. Show that any vectors $u_1, u_2, u_1 + u_2, u_4$ in R^8 are linearly dependent.
7. Suppose that some vectors u_1, u_2, u_3 in R^n are linearly dependent. Show that the same is true for u_1, u_2, u_3, u_4, no matter what the vector $u_4 \in R^n$ is.
8. Suppose that some vectors u_1, u_2, u_3, u_4 in R^n are linearly independent. Show that the same is true for u_1, u_2, u_3.
9. Assume that u_1, u_2, u_3, u_4 are vectors in R^5 and $u_2 = 0$. Justify that these vectors are linearly dependent. (Starting from the definition of linear dependence.)
10*. The following example serves to illustrate possible pitfalls when doing proofs.

For any positive integer n,

$$n^2 = n + n + \cdots + n,$$

where the sum on the right has n terms. Differentiate both sides with respect to the variable n,

$$2n = 1 + 1 + \cdots + 1,$$

which gives

$$2n = n.$$

Dividing by $n > 0$, obtain

$$2 = 1.$$

Is there anything wrong with this argument? Explain.

2 Matrix algebra

In this chapter we develop the central *concept of matrices*, and study their basic properties, including the notions of inverse matrices, elementary matrices, null spaces, and column spaces.

2.1 Matrix operations

A general matrix of size 2×3 can be written as

$$A = \begin{bmatrix} a_{11} & a_{12} & a_{13} \\ a_{21} & a_{22} & a_{23} \end{bmatrix}.$$

Each element has two indices. The first index identifies the row, and the second index refers to the column number. All of the elements of the first row have the first index 1, while all elements of the third column have the second index 3. For example, the matrix $\begin{bmatrix} 1 & -2 & 0 \\ 3 & \frac{1}{2} & \pi \end{bmatrix}$ has $a_{11} = 1$, $a_{12} = -2$, $a_{13} = 0$, $a_{21} = 3$, $a_{22} = \frac{1}{2}$, $a_{23} = \pi$. A 1×1 matrix is just the scalar a_{11}.

Any matrix can be multiplied by a scalar, and any two matrices of the same size can be added. Both operations are performed componentwise, similarly to vectors. For example,

$$\begin{bmatrix} a_{11} & a_{12} \\ a_{21} & a_{22} \\ a_{31} & a_{32} \end{bmatrix} + \begin{bmatrix} b_{11} & b_{12} \\ b_{21} & b_{22} \\ b_{31} & b_{32} \end{bmatrix} = \begin{bmatrix} a_{11} + b_{11} & a_{12} + b_{12} \\ a_{21} + b_{21} & a_{22} + b_{22} \\ a_{31} + b_{31} & a_{32} + b_{32} \end{bmatrix},$$

$$5A = 5 \begin{bmatrix} a_{11} & a_{12} & a_{13} \\ a_{21} & a_{22} & a_{23} \end{bmatrix} = \begin{bmatrix} 5a_{11} & 5a_{12} & 5a_{13} \\ 5a_{21} & 5a_{22} & 5a_{23} \end{bmatrix}.$$

If A is an $m \times n$ matrix, given by its columns $A = [a_1 \, a_2 \, \ldots \, a_n]$, and $x = \begin{bmatrix} x_1 \\ x_2 \\ \vdots \\ x_n \end{bmatrix}$ is a vector in R^n, recall that their product

$$Ax = x_1 a_1 + x_2 a_2 + \cdots + x_n a_n \tag{1.1}$$

is a vector in R^m. Let B be an $n \times p$ matrix, given by its columns $B = [b_1 \, b_2 \, \ldots \, b_p]$. Each of these columns is a vector in R^n. Define the *product of two matrices* as the following matrix, given by its columns

$$AB = [Ab_1 \, Ab_2 \, \ldots \, Ab_p],$$

so that the first column of AB is the vector Ab_1 in R^m (calculated using (1.1)), and so on. Not every two matrices can be multiplied. If the size of A is $m \times n$, then the size of B must

https://doi.org/10.1515/9783111086507-002

be $n \times p$, with the *same n* (m and p are arbitrary). The size of AB is $m \times p$ (one sees from the definition that AB has m rows and p columns).

For example,

$$\begin{bmatrix} 1 & -1 & 1 \\ 0 & -3 & 2 \\ -4 & 2 & 0 \end{bmatrix} \begin{bmatrix} 2 & -1 & 2 \\ 1 & -1 & 2 \\ -3 & 2 & -2 \end{bmatrix} = \begin{bmatrix} -2 & 2 & -2 \\ -9 & 7 & -10 \\ -6 & 2 & -4 \end{bmatrix},$$

because the first column of the product is

$$\begin{bmatrix} 1 & -1 & 1 \\ 0 & -3 & 2 \\ -4 & 2 & 0 \end{bmatrix} \begin{bmatrix} 2 \\ 1 \\ -3 \end{bmatrix} = 2 \begin{bmatrix} 1 \\ 0 \\ -4 \end{bmatrix} + 1 \begin{bmatrix} -1 \\ -3 \\ 2 \end{bmatrix} + (-3) \begin{bmatrix} 1 \\ 2 \\ 0 \end{bmatrix} = \begin{bmatrix} -2 \\ -9 \\ -6 \end{bmatrix},$$

and the second and third columns of the product matrix are calculated similarly.

If a matrix A has size 2×3 and B is of size 3×4, their product AB of size 2×4 is defined, while the product BA is not defined (because the second index of the first matrix B does not match the first index of A). For a matrix C of size 3×4 and a matrix D of size 4×3, both products CD and DC are defined, but CD has size 3×3, while DC is of size 4×4. Again, the *order of the matrices matters*.

Matrices of size $n \times n$ are called *square matrices of size n*. For two square matrices of size n, both products AB and BA are defined, both are square matrices of size n, but even then

$$BA \neq AB,$$

in most cases. In a rare case when $BA = AB$, one says that *the matrices A and B commute*.

Aside from $BA \neq AB$, the usual rules of algebra apply, which is straightforward to verify. For example (assuming that all products are defined),

$$A(BC) = (AB)C,$$
$$((AB)C)D = A(BC)D = (AB)(CD).$$

It does not matter in which order you multiply (or pair the matrices), so long as the order in which the matrices appear is preserved. Also,

$$A(B + C) = AB + AC,$$
$$(A + B)C = AC + BC,$$
$$2A(-3B) = -6AB.$$

A square matrix $I = \begin{bmatrix} 1 & 0 & 0 \\ 0 & 1 & 0 \\ 0 & 0 & 1 \end{bmatrix}$ is called the *identity matrix of size 3* (identity matrices come in all sizes). If A is any square matrix of size 3, then one calculates

$$IA = AI = A,$$

and the same is true for the unit matrix of any size.

A square matrix $D = \begin{bmatrix} 2&0&0 \\ 0&3&0 \\ 0&0&4 \end{bmatrix}$ is an example of a *diagonal matrix*, which is a square matrix with all off-diagonal entries equal to zero. Let A be any 3×3 matrix, given by its columns $A = [a_1\ a_2\ a_3]$. One calculates

$$AD = [2a_1\quad 3a_2\quad 4a_3],$$

so that to produce AD, the columns of A are multiplied by the corresponding diagonal entries of D. Indeed, the first column of AD is

$$A \begin{bmatrix} 2 \\ 0 \\ 0 \end{bmatrix} = [a_1\ a_2\ a_3] \begin{bmatrix} 2 \\ 0 \\ 0 \end{bmatrix} = 2a_1 + 0a_2 + 0a_3 = 2a_1,$$

and the other columns of AD are calculated similarly. In particular, if $A = \begin{bmatrix} p&0&0 \\ 0&q&0 \\ 0&0&r \end{bmatrix}$ is another diagonal matrix, then

$$AD = \begin{bmatrix} p&0&0 \\ 0&q&0 \\ 0&0&r \end{bmatrix} \begin{bmatrix} 2&0&0 \\ 0&3&0 \\ 0&0&4 \end{bmatrix} = \begin{bmatrix} 2p&0&0 \\ 0&3q&0 \\ 0&0&4r \end{bmatrix}.$$

In general, the product of two diagonal matrices of the same size is the diagonal matrix obtained by multiplying the corresponding diagonal entries.

A *row vector* $R = [2\ 3\ 4]$ can be viewed as a 1×3 matrix. Similarly, the column vector $C = \begin{bmatrix} 1 \\ -2 \\ 5 \end{bmatrix}$ is a matrix of size 3×1. Their product RC is defined, it has size 1×1, which is a scalar,

$$RC = [2\ 3\ 4] \begin{bmatrix} 1 \\ -2 \\ 5 \end{bmatrix} = 2 \cdot 1 + 3 \cdot (-2) + 4 \cdot 5 = 16.$$

We now describe an equivalent alternative way to multiply an $m \times n$ matrix A and an $n \times p$ matrix B. The row i of A is

$$R_i = [a_{i1}\ a_{i2}\ \dots\ a_{in}],$$

while the column j of B is

$$C_j = \begin{bmatrix} b_{1j} \\ b_{2j} \\ \vdots \\ b_{nj} \end{bmatrix}.$$

To calculate the *ij* element of the product AB, denoted by $(AB)_{ij}$, just multiply R_i and C_j,

$$(AB)_{ij} = R_i C_j = a_{i1} b_{1j} + a_{i2} b_{2j} + \cdots + a_{in} b_{nj}.$$

For example,

$$\begin{bmatrix} 1 & 2 \\ 3 & 1 \end{bmatrix} \begin{bmatrix} 0 & -3 \\ -2 & 2 \end{bmatrix} = \begin{bmatrix} -4 & 1 \\ -2 & -7 \end{bmatrix},$$

because

$$\begin{bmatrix} 1 & 2 \end{bmatrix} \begin{bmatrix} 0 \\ -2 \end{bmatrix} = 1 \cdot 0 + 2(-2) = -4,$$

$$\begin{bmatrix} 1 & 2 \end{bmatrix} \begin{bmatrix} -3 \\ 2 \end{bmatrix} = 1(-3) + 2 \cdot 2 = 1,$$

$$\begin{bmatrix} 3 & 1 \end{bmatrix} \begin{bmatrix} 0 \\ -2 \end{bmatrix} = 3 \cdot 0 + 1(-2) = -2,$$

$$\begin{bmatrix} 3 & 1 \end{bmatrix} \begin{bmatrix} -3 \\ 2 \end{bmatrix} = 3(-3) + 1 \cdot 2 = -7.$$

If $A = \begin{bmatrix} a_{11} & a_{12} & a_{13} \\ a_{21} & a_{22} & a_{23} \end{bmatrix}$, the *transpose of A* is defined to be

$$A^T = \begin{bmatrix} a_{11} & a_{21} \\ a_{12} & a_{22} \\ a_{13} & a_{23} \end{bmatrix}.$$

To calculate A^T, one turns the first row of A into the first column of A^T, the second row of A into the second column of A^T, and so on. (*Observe that in the process the columns of A become the rows of A^T.*) If A is of size $m \times n$, then the size of A^T is $n \times m$. It is straightforward to verify that

$$(A^T)^T = A$$

and

$$(AB)^T = B^T A^T,$$

provided that the matrix product AB is defined.

A matrix with all entries equal to zero is called *the zero matrix*, and is denoted by O. For example, $O = \begin{bmatrix} 0 & 0 \\ 0 & 0 \\ 0 & 0 \end{bmatrix}$ is the 3×2 zero matrix. If the matrices A and O are of the same size, then $A + O = A$. If the product AO is defined, it is equal to the zero matrix.

Powers of a square matrix A are defined as follows: $A^2 = AA$, $A^3 = A^2A$, and so on; A^n is a square matrix of the same size as A.

Exercises

1. Determine the 3×2 matrix X from the relation

$$2X + \begin{bmatrix} 1 & -1 \\ 0 & 2 \\ 3 & 0 \end{bmatrix} = -3 \begin{bmatrix} 0 & 1 \\ -1 & 0 \\ 0 & 2 \end{bmatrix}.$$

2. Determine the 3×3 matrix X from the relation

$$3X + I = O.$$

Answer: $X = \begin{bmatrix} -\frac{1}{3} & 0 & 0 \\ 0 & -\frac{1}{3} & 0 \\ 0 & 0 & -\frac{1}{3} \end{bmatrix}$.

3. Calculate the products AB and BA, and compare.

   (a) $A = \begin{bmatrix} 1 & -1 \\ 0 & 2 \\ 3 & 0 \end{bmatrix}$, $B = \begin{bmatrix} 1 & -1 & 2 \\ 0 & 2 & 1 \end{bmatrix}$.

   Answer: $AB = \begin{bmatrix} 1 & -3 & 1 \\ 0 & 4 & 2 \\ 3 & -3 & 6 \end{bmatrix}$, $BA = \begin{bmatrix} 7 & -3 \\ 3 & 4 \end{bmatrix}$.

   (b) $A = [1 \; -1 \; 4]$, $B = \begin{bmatrix} 1 \\ -1 \\ 2 \end{bmatrix}$.

   Answer: $AB = 10$, $BA = \begin{bmatrix} 1 & -1 & 4 \\ -1 & 1 & -4 \\ 2 & -2 & 8 \end{bmatrix}$.

   (c) $A = \begin{bmatrix} 1 & -1 \\ 3 & 0 \end{bmatrix}$, $B = \begin{bmatrix} -1 & 2 \\ 2 & 1 \end{bmatrix}$.

   (d) $A = \begin{bmatrix} 2 & -1 \\ 3 & 1 \end{bmatrix}$, $B = \begin{bmatrix} 1 & -1 & 2 \\ 3 & 2 & 1 \end{bmatrix}$.

   Hint. The product BA is not defined.

   (e) $A = \begin{bmatrix} 1 & 1 & 1 \\ 1 & 1 & 1 \\ 1 & 1 & 1 \end{bmatrix}$, $B = \begin{bmatrix} 2 & 0 & 0 \\ 0 & 3 & 0 \\ 0 & 0 & 4 \end{bmatrix}$.

   (f) $A = \begin{bmatrix} a & 0 & 0 & 0 \\ 0 & b & 0 & 0 \\ 0 & 0 & c & 0 \\ 0 & 0 & 0 & d \end{bmatrix}$, $B = \begin{bmatrix} 2 & 0 & 0 & 0 \\ 0 & 3 & 0 & 0 \\ 0 & 0 & 4 & 0 \\ 0 & 0 & 0 & 5 \end{bmatrix}$.

   Answer: $AB = BA = \begin{bmatrix} 2a & 0 & 0 & 0 \\ 0 & 3b & 0 & 0 \\ 0 & 0 & 4c & 0 \\ 0 & 0 & 0 & 5d \end{bmatrix}$.

   (g) $A = \begin{bmatrix} 1 & 1 & 1 \\ 1 & 1 & 1 \\ 1 & 1 & 1 \end{bmatrix}$, $B = \begin{bmatrix} 2 & 0 & 0 \\ 0 & -1 & 0 \\ 0 & 0 & 0 \end{bmatrix}$.

   Hint. B is diagonal matrix.

Answer: $BA = \begin{bmatrix} 2 & 2 & 2 \\ -1 & -1 & -1 \\ 0 & 0 & 0 \end{bmatrix}$. Observe the general fact: multiplying A by a diagonal matrix B from the left results in rows of A being multiplied by the corresponding diagonal entries of B.

4. Let A and B be square matrices of the same size. Can one assert the following formulas? If the answer is no, write down the correct formula. Do these formulas hold in case A and B commute?
   (a) $(A - B)(A + B) = A^2 - B^2$.
   (b) $(A + B)^2 = A^2 + 2AB + B^2$.
   (c) $(AB)^2 = A^2 B^2$.

5. Suppose that the product ABC is defined. Show that the product $C^T B^T A^T$ is also defined, and $(ABC)^T = C^T B^T A^T$.

6. Let A be a square matrix.
   (a) Show that $(A^2)^T = (A^T)^2$.
   (b) Show that $(A^n)^T = (A^T)^n$, with integer $n \geq 3$.

7. Let $A = \begin{bmatrix} 1 & 0 \\ 1 & 0 \end{bmatrix} \neq O$ and $B = \begin{bmatrix} 0 & 0 \\ 0 & 2 \end{bmatrix} \neq O$. Verify that $AB = O$.

8. Let $A = \begin{bmatrix} 0 & 1 & 0 \\ 0 & 0 & 1 \\ 0 & 0 & 0 \end{bmatrix}$. Show that $A^3 = O$.

9. Let $H = \begin{bmatrix} 3 & 1 & -2 \\ 0 & -4 & 1 \\ 1 & 2 & 0 \end{bmatrix}$.
   (a) Calculate H^T.
   (b) Show that transposition of any square matrix A leaves the diagonal entries unchanged, while interchanging the symmetric off-diagonal entries (a_{ij} and a_{ji}, with $i \neq j$).
   (c) *A square matrix A is called symmetric if $A^T = A$.* Show that then $a_{ij} = a_{ji}$ for all off diagonal entries. Is matrix H symmetric?
   (d) Let B be any $m \times n$ matrix. Show that the matrix $B^T B$ is square and symmetric, and the same is true for BB^T.

10. Let $x \in R^n$.
    (a) Show that x^T is a $1 \times n$ matrix, or a row vector.
    (b) Calculate the product $x^T x$ in terms of the coordinates of x, and show that $x^T x > 0$, provided that $x \neq 0$.

2.2 The inverse of a square matrix

An $n \times n$ matrix A is said to be *invertible* if there is an $n \times n$ matrix C such that

$$CA = I \quad \text{and} \quad AC = I,$$

where I is an $n \times n$ identity matrix. Such a matrix C is called the *inverse of A*, and denoted A^{-1}, so that

$$A^{-1}A = AA^{-1} = I. \tag{2.1}$$

For example, if $A = \begin{bmatrix} 2 & 1 \\ 3 & 2 \end{bmatrix}$, then $A^{-1} = \begin{bmatrix} 2 & -1 \\ -3 & 2 \end{bmatrix}$, because

$$\begin{bmatrix} 2 & 1 \\ 3 & 2 \end{bmatrix} \begin{bmatrix} 2 & -1 \\ -3 & 2 \end{bmatrix} = \begin{bmatrix} 2 & -1 \\ -3 & 2 \end{bmatrix} \begin{bmatrix} 2 & 1 \\ 3 & 2 \end{bmatrix} = \begin{bmatrix} 1 & 0 \\ 0 & 1 \end{bmatrix} = I.$$

Not every square matrix has an inverse. For example, $A = \begin{bmatrix} 0 & 1 \\ 0 & 0 \end{bmatrix}$ is *not invertible* (no inverse exists). Indeed, if we try any $C = \begin{bmatrix} c_{11} & c_{12} \\ c_{21} & c_{22} \end{bmatrix}$, then

$$AC = \begin{bmatrix} 0 & 1 \\ 0 & 0 \end{bmatrix} \begin{bmatrix} c_{11} & c_{12} \\ c_{21} & c_{22} \end{bmatrix} = \begin{bmatrix} c_{21} & c_{22} \\ 0 & 0 \end{bmatrix} \neq \begin{bmatrix} 1 & 0 \\ 0 & 1 \end{bmatrix},$$

for any choice of C. Noninvertible matrices are also called *singular*.

If an $n \times n$ matrix A is invertible, then the system

$$Ax = b$$

has a unique solution $x = A^{-1}b$. Indeed, multiplying both sides of this equation by A^{-1} gives

$$A^{-1}Ax = A^{-1}b,$$

which simplifies to $Ix = A^{-1}b$, or $x = A^{-1}b$. *The corresponding homogeneous system* (when $b = 0$)

$$Ax = 0 \tag{2.2}$$

has *a unique solution* $x = A^{-1}0 = 0$, the trivial solution. The trivial solution is the only solution of (2.2), and that happens when A has n pivots (a pivot in every column).

Conclusion: if an $n \times n$ matrix A is invertible, it has n pivots. It follows that in case A has fewer than n pivots, A is not invertible (singular).

Theorem 2.2.1. *An $n \times n$ matrix A is invertible if and only if A has n pivots.*

Proof. If A is invertible, we just proved that A has n pivots. Conversely assume that A has n pivots. It will be shown later on in this section how to construct the inverse matrix A^{-1}. ☐

Given n vectors in R^n, let us use them as columns of an $n \times n$ matrix, and call this matrix A. These columns are linearly independent if and only if A has n pivots, as we learned previously. We can then restate the preceding theorem.

Theorem 2.2.2. *A square matrix is invertible if and only if its columns are linearly independent.*

Suppose A is a 3×3 matrix. If A is invertible, then A has 3 pivots, and its columns are linearly independent. If A is not invertible, then the number of pivots is either 1 or 2, and the columns of A are linearly dependent.

Elementary matrices

The matrix

$$E_2(-3) = \begin{bmatrix} 1 & 0 & 0 \\ 0 & -3 & 0 \\ 0 & 0 & 1 \end{bmatrix}$$

is obtained by multiplying the second row of I by -3 (or performing $-3R_2$ on the identity matrix I). Calculate the product of this matrix and an arbitrary one

$$\begin{bmatrix} 1 & 0 & 0 \\ 0 & -3 & 0 \\ 0 & 0 & 1 \end{bmatrix} \begin{bmatrix} a_{11} & a_{12} & a_{13} \\ a_{21} & a_{22} & a_{23} \\ a_{31} & a_{32} & a_{33} \end{bmatrix} = \begin{bmatrix} a_{11} & a_{12} & a_{13} \\ -3a_{21} & -3a_{22} & -3a_{23} \\ a_{31} & a_{32} & a_{33} \end{bmatrix},$$

so that multiplying an arbitrary matrix from the left by $E_2(-3)$ is the same as performing an elementary operation $-3R_2$ on that matrix. In general, one defines an *elementary matrix* $E_i(a)$ by multiplying row i of the $n \times n$ identity matrix I by number a. If A is an arbitrary $n \times n$ matrix, then the result of multiplication $E_i(a)A$ is that the elementary operation aR_i is performed on A. We call $E_i(a)$ *an elementary matrix of the first kind.*

The matrix

$$E_{13} = \begin{bmatrix} 0 & 0 & 1 \\ 0 & 1 & 0 \\ 1 & 0 & 0 \end{bmatrix}$$

is obtained by interchanging the first and third rows of I (or performing $R_1 \leftrightarrow R_3$ on I). Calculate the product of E_{13} and an arbitrary matrix to get

$$\begin{bmatrix} 0 & 0 & 1 \\ 0 & 1 & 0 \\ 1 & 0 & 0 \end{bmatrix} \begin{bmatrix} a_{11} & a_{12} & a_{13} \\ a_{21} & a_{22} & a_{23} \\ a_{31} & a_{32} & a_{33} \end{bmatrix} = \begin{bmatrix} a_{31} & a_{32} & a_{33} \\ a_{21} & a_{22} & a_{23} \\ a_{11} & a_{12} & a_{13} \end{bmatrix},$$

so that multiplying an arbitrary matrix from the left by E_{13} is the same as performing an elementary operation $R_1 \leftrightarrow R_3$ on that matrix. In general, one defines an elementary matrix E_{ij} by interchanging rows i and j of the $n \times n$ identity matrix I. If A is an arbitrary $n \times n$ matrix, then the result of multiplication $E_{ij}A$ is that an elementary operation $R_i \leftrightarrow R_j$ is performed on A; E_{ij} is called *an elementary matrix of the second kind.*

The matrix

$$E_{13}(2) = \begin{bmatrix} 1 & 0 & 0 \\ 0 & 1 & 0 \\ 2 & 0 & 1 \end{bmatrix}$$

is obtained from I by adding to its third row the first row multiplied by 2 (or performing $R_3 + 2R_1$ on I). Calculate the product of $E_{13}(2)$ and an arbitrary matrix to obtain

$$\begin{bmatrix} 1 & 0 & 0 \\ 0 & 1 & 0 \\ 2 & 0 & 1 \end{bmatrix} \begin{bmatrix} a_{11} & a_{12} & a_{13} \\ a_{21} & a_{22} & a_{23} \\ a_{31} & a_{32} & a_{33} \end{bmatrix} = \begin{bmatrix} a_{11} & a_{12} & a_{13} \\ a_{21} & a_{22} & a_{23} \\ a_{31} + 2a_{11} & a_{32} + 2a_{12} & a_{33} + 2a_{13} \end{bmatrix},$$

so that multiplying an arbitrary matrix from the left by $E_{13}(2)$ is the same as performing an elementary operation $R_3 + 2R_1$ on that matrix. In general, one defines an elementary matrix $E_{ij}(a)$ by performing $R_j + aR_i$ on the $n \times n$ identity matrix I. If A is an arbitrary $n \times n$ matrix, the result of multiplication $E_{ij}(a)A$ is that an elementary operation $R_j + aR_i$ is performed on A; E_{ij} is called *an elementary matrix of the third kind.*

We summarize. *If a matrix A is multiplied from the left by an elementary matrix, the result is the same as applying the corresponding elementary operation to A.*

Calculating A^{-1}

Given an $n \times n$ matrix A, we wish to determine if A is invertible, and if it is invertible, calculate the inverse A^{-1}.

Let us row reduce A by applying elementary operations, which is the same as multiplying from the left by elementary matrices. Denote by E_1 the first elementary matrix used. (In case one has $a_{11} = 1$ and $a_{21} = 2$, the first elementary operation is $R_2 - 2R_1$, so that $E_1 = E_{12}(-2)$. If it so happens that $a_{11} = 0$ and $a_{21} = 1$, then the first elementary operation is $R_1 \leftrightarrow R_2$, and then $E_1 = E_{12}$.) The first step of row reduction results in the matrix E_1A. Denote by E_2 the second elementary matrix used. After two steps of row reduction, we have $E_2(E_1A) = E_2E_1A$. If A is invertible, it has n pivots, and then we can row reduce A to I by complete forward elimination, after say p steps. In terms of elementary matrices,

$$E_p \cdots E_2 E_1 A = I. \tag{2.3}$$

This implies that the product $E_p \cdots E_2 E_1$ is the inverse of A, $E_p \cdots E_2 E_1 = A^{-1}$, or

$$E_p \cdots E_2 E_1 I = A^{-1}. \tag{2.4}$$

Compare (2.3) with (2.4): *the same sequence of elementary operations that reduces A to I, turns I into A^{-1}.*

The result is a method for computing A^{-1}. Form a long matrix $[A \vdots I]$ of size $n \times 2n$. Apply row operations on the entire long matrix, with the goal of obtaining I is the first position. Once this is achieved, the matrix in the second position is A^{-1}. In short,

$$[A \vdots I] \rightarrow [I \vdots A^{-1}].$$

Example 1. Let $A = \begin{bmatrix} 1 & 2 & -1 \\ 2 & 3 & -2 \\ -1 & -2 & 0 \end{bmatrix}$. Form the matrix $[A \vdots I]$,

$$\left[\begin{array}{ccc|ccc} 1 & 2 & -1 & 1 & 0 & 0 \\ 2 & 3 & -2 & 0 & 1 & 0 \\ -1 & -2 & 0 & 0 & 0 & 1 \end{array}\right].$$

Perform $R_2 - 2R_1$ and $R_3 + R_1$ on the entire matrix to get

$$\left[\begin{array}{ccc|ccc} 1 & 2 & -1 & 1 & 0 & 0 \\ 0 & -1 & 0 & -2 & 1 & 0 \\ 0 & 0 & -1 & 1 & 0 & 1 \end{array}\right].$$

Perform $-R_2$ and $-R_3$ on the entire matrix, to make all pivots equal to 1,

$$\left[\begin{array}{ccc|ccc} 1 & 2 & -1 & 1 & 0 & 0 \\ 0 & 1 & 0 & 2 & -1 & 0 \\ 0 & 0 & 1 & -1 & 0 & -1 \end{array}\right].$$

Perform $R_1 + R_3$ to obtain

$$\left[\begin{array}{ccc|ccc} 1 & 2 & 0 & 0 & 0 & -1 \\ 0 & 1 & 0 & 2 & -1 & 0 \\ 0 & 0 & 1 & -1 & 0 & -1 \end{array}\right].$$

Finally, perform $R_1 - 2R_2$ to have

$$\left[\begin{array}{ccc|ccc} 1 & 0 & 0 & -4 & 2 & -1 \\ 0 & 1 & 0 & 2 & -1 & 0 \\ 0 & 0 & 1 & -1 & 0 & -1 \end{array}\right].$$

The process is complete, $A^{-1} = \begin{bmatrix} -4 & 2 & -1 \\ 2 & -1 & 0 \\ -1 & 0 & -1 \end{bmatrix}$.

Example 2. Let $B = \begin{bmatrix} -1 & 2 & 1 \\ 2 & -4 & -3 \\ 1 & -2 & 1 \end{bmatrix}$. Form the matrix $[B \vdots I]$,

$$\left[\begin{array}{ccc|ccc} -1 & 2 & 1 & 1 & 0 & 0 \\ 2 & -4 & -3 & 0 & 1 & 0 \\ 1 & -2 & 1 & 0 & 0 & 1 \end{array}\right].$$

Perform $R_2 + 2R_1$ and $R_3 + R_1$ on the entire matrix to get

$$\left[\begin{array}{ccc|ccc} ① & 2 & 1 & 1 & 0 & 0 \\ 0 & 0 & ① & 2 & 1 & 0 \\ 0 & 0 & 2 & 1 & 0 & 1 \end{array}\right].$$

Game over! The matrix B does not have a pivot in the second column, so that B has fewer than 3 pivots and is therefore singular (there is no inverse), by Theorem 2.2.1.

For a 2×2 matrix $A = \begin{bmatrix} a & b \\ c & d \end{bmatrix}$ there is an easier way to calculate the inverse. One checks by multiplication of matrices that $A^{-1} = \frac{1}{ad-bc}\begin{bmatrix} d & -b \\ -c & a \end{bmatrix}$, provided that $ad - bc \neq 0$. In case $ad - bc = 0$, the matrix A has no inverse, as will be justified later on.

The inverses of diagonal matrices are also easy to find. For example, if $A = \begin{bmatrix} a & 0 & 0 \\ 0 & b & 0 \\ 0 & 0 & c \end{bmatrix}$, with nonzero a, b, c, then $A^{-1} = \begin{bmatrix} \frac{1}{a} & 0 & 0 \\ 0 & \frac{1}{b} & 0 \\ 0 & 0 & \frac{1}{c} \end{bmatrix}$. If one of the diagonal entries of A is zero, then the matrix A is singular, since it has fewer than three pivots.

Exercises

1. Write down the 3×3 elementary matrix which corresponds to the following elementary operation: to row 3 add 4 times row 2. What is the notation used for this matrix?

   Answer: $E_{23}(4) = \begin{bmatrix} 1 & 0 & 0 \\ 0 & 1 & 0 \\ 0 & 4 & 1 \end{bmatrix}$.

2. Write down the 3×3 elementary matrix which corresponds to the following elementary operation: multiply row 3 by -5.

3. Write down the 4×4 elementary matrix which corresponds to the following elementary operation: interchange rows 1 and 4.

   Answer: $E_{14} = \begin{bmatrix} 0 & 0 & 0 & 1 \\ 0 & 1 & 0 & 0 \\ 0 & 0 & 1 & 0 \\ 1 & 0 & 0 & 0 \end{bmatrix}$.

4. Explain why the following matrices are singular (not invertible):

   (a) $A = \begin{bmatrix} 0 & 0 \\ 4 & 1 \end{bmatrix}$.

   (b) $A = \begin{bmatrix} -3 & 0 \\ 5 & 0 \end{bmatrix}$.

   (c) $A = \begin{bmatrix} 1 & 0 & 0 \\ 0 & -4 & 0 \\ 0 & 0 & 0 \end{bmatrix}$.

   (d) $A = \begin{bmatrix} 0 & 1 & 1 \\ 2 & 4 & 5 \\ 0 & 0 & 0 \end{bmatrix}$.

   Hint. Count the number of pivots.

5. Find the inverses of the following matrices without performing the Gaussian elimination:

   (a) $A = \begin{bmatrix} 1 & 0 & 0 \\ 0 & 1 & 0 \\ 0 & 4 & 1 \end{bmatrix}$.

   Hint. $A = E_{23}(4)$. Observe that $E_{23}(-4)A = I$, since performing $R_3 - 4R_2$ on A gives I. It follows that $A^{-1} = E_{23}(-4)$.

   Answer: $A^{-1} = \begin{bmatrix} 1 & 0 & 0 \\ 0 & 1 & 0 \\ 0 & -4 & 1 \end{bmatrix}$.

   (b) $A = \begin{bmatrix} 0 & 0 & 0 & 1 \\ 0 & 1 & 0 & 0 \\ 0 & 0 & 1 & 0 \\ 1 & 0 & 0 & 0 \end{bmatrix}$.

Hint. $A = E_{14}$. Then $E_{14}A = I$, since switching the first and the fourth rows of A produces I. It follows that $A^{-1} = E_{14}$.

Answer: $A^{-1} = A$.

(c) $A = \begin{bmatrix} 3 & 0 \\ 0 & -2 \end{bmatrix}$.

(d) $A = \begin{bmatrix} \frac{1}{4} & 0 & 0 \\ 0 & -1 & 0 \\ 0 & 0 & 5 \end{bmatrix}$.

(e) $A = \begin{bmatrix} 4 & 0 & 0 \\ 0 & 0 & 0 \\ 0 & 0 & 5 \end{bmatrix}$.

Answer: The matrix is singular.

(f) $A = \begin{bmatrix} 1 & -1 \\ 3 & -2 \end{bmatrix}$.

Answer: $\begin{bmatrix} -2 & 1 \\ -3 & 1 \end{bmatrix}$.

6. Find the inverses of the following matrices by using Gaussian elimination:

(a) $A = \begin{bmatrix} 1 & 2 & 0 \\ 0 & -1 & 1 \\ 1 & -2 & 1 \end{bmatrix}$.

Answer: $A^{-1} = \frac{1}{3} \begin{bmatrix} 1 & -2 & 2 \\ 1 & 1 & -1 \\ 1 & 4 & -1 \end{bmatrix}$.

(b) $A = \begin{bmatrix} 1 & 3 \\ 2 & 6 \end{bmatrix}$.

Answer: The matrix is singular.

(c) $A = \begin{bmatrix} 0 & 0 & 1 \\ 0 & -1 & 1 \\ 1 & -2 & 1 \end{bmatrix}$.

Answer: $A^{-1} = \begin{bmatrix} 1 & -2 & 1 \\ 1 & -1 & 0 \\ 1 & 0 & 0 \end{bmatrix}$.

(d) $A = \begin{bmatrix} 1 & 2 & 3 \\ 2 & 0 & 2 \\ 3 & 2 & 1 \end{bmatrix}$.

Answer: $A^{-1} = \frac{1}{4} \begin{bmatrix} -1 & 1 & 1 \\ 1 & -2 & 1 \\ 1 & 1 & -1 \end{bmatrix}$.

(e) $A = \begin{bmatrix} 1 & 1 & 2 \\ 0 & 0 & 1 \\ 1 & 0 & 1 \end{bmatrix}$.

Answer: $A^{-1} = \begin{bmatrix} 0 & -1 & 1 \\ 1 & -1 & -1 \\ 0 & 1 & 0 \end{bmatrix}$.

(f) $A = \begin{bmatrix} 1 & 0 & 1 \\ 1 & 1 & 1 \\ 2 & 1 & 1 \end{bmatrix}$.

Answer: $A^{-1} = \begin{bmatrix} 0 & -1 & 1 \\ -1 & 1 & 0 \\ 1 & 1 & -1 \end{bmatrix}$.

(g) $A = \begin{bmatrix} 1 & 1 & -1 \\ 1 & 2 & 1 \\ -1 & -1 & 0 \end{bmatrix}$.

Answer: $A^{-1} = \begin{bmatrix} -1 & -1 & -3 \\ 1 & 1 & 2 \\ -1 & 0 & -1 \end{bmatrix}$.

(h) $A = \begin{bmatrix} 1 & -1 \\ 3 & -2 \end{bmatrix}$.

Answer: $A^{-1} = \begin{bmatrix} -2 & 1 \\ -3 & 1 \end{bmatrix}$.

(i) $B = \begin{bmatrix} 1 & -1 & 0 \\ 3 & -2 & 0 \\ 0 & 0 & 5 \end{bmatrix}$.

Answer: $B^{-1} = \begin{bmatrix} -2 & 1 & 0 \\ -3 & 1 & 0 \\ 0 & 0 & \frac{1}{5} \end{bmatrix}$.

Compare with the preceding example. The matrix B is an example of a *block diagonal matrix*.

(h) $C = \begin{bmatrix} 1 & -1 & 0 & 0 \\ 3 & -2 & 0 & 0 \\ 0 & 0 & 5 & 0 \\ 0 & 0 & 0 & -1 \end{bmatrix}$.

Answer: $C^{-1} = \begin{bmatrix} -2 & 1 & 0 & 0 \\ -3 & 1 & 0 & 0 \\ 0 & 0 & \frac{1}{5} & 0 \\ 0 & 0 & 0 & -1 \end{bmatrix}$.

The matrix C is another example of a block diagonal matrix.

7. The third column of a 3×3 matrix is equal to the sum of the first two columns. Is this matrix invertible? Explain.

8. Suppose that A and B are nonsingular $n \times n$ matrices, and $(AB)^2 = A^2B^2$. Show that $AB = BA$.

9. Let E_{13} and E_{24} be 4×4 matrices.
    (a) Calculate $P = E_{13}E_{24}$.

    Answer: $P = \begin{bmatrix} 0 & 0 & 1 & 0 \\ 0 & 0 & 0 & 1 \\ 1 & 0 & 0 & 0 \\ 0 & 1 & 0 & 0 \end{bmatrix}$.

    (b) Let A be any 4×4 matrix. Show that PA is obtained from A by interchanging row 1 with row 3, and row 2 with row 4.

    (If A is given by its rows $A = \begin{bmatrix} R_1 \\ R_2 \\ R_3 \\ R_4 \end{bmatrix}$, then $PA = \begin{bmatrix} R_3 \\ R_4 \\ R_1 \\ R_2 \end{bmatrix}$.)

    (c) Show that $P^2 = I$.

    The matrix P is an example of a *permutation matrix*.

10. (a) Suppose that a square matrix A is invertible. Show that A^T is also invertible, and

$$(A^T)^{-1} = (A^{-1})^T.$$

    Hint. Take the transpose of $AA^{-1} = I$.

    (b) Show that a square matrix is invertible if and only if its rows are linearly independent.

    Hint. Use Theorem 2.2.2.

    (c) Suppose that the third row of a 7×7 matrix is equal to the sum of the first and second rows. Is this matrix invertible?

11. A square matrix A is called *nilpotent* if $A^k = O$, the zero matrix, for some positive integer k.

    (a) Show that $A = \begin{bmatrix} 0 & 1 & 0 & 0 \\ 0 & 0 & 1 & 0 \\ 0 & 0 & 0 & 1 \\ 0 & 0 & 0 & 0 \end{bmatrix}$ is nilpotent.

    Hint. Calculate A^4.

    (b) If A is nilpotent show that $I - A$ is invertible, and calculate $(I - A)^{-1}$.

    Answer: $(I - A)^{-1} = I + A + A^2 + \cdots + A^{k-1}$.

2.3 *LU* decomposition

In this section we study inverses of elementary matrices, and develop $A = LU$ decomposition of any square matrix A, a useful tool.

Examining the definition of the inverse matrix ($A^{-1}A = AA^{-1} = I$), one sees that A plays the role of the inverse matrix for A^{-1}, so that $A = (A^{-1})^{-1}$, or

$$(A^{-1})^{-1} = A.$$

Another property of inverse matrices is

$$(cA)^{-1} = \frac{1}{c}A^{-1}, \quad \text{for any number } c \neq 0,$$

which is true because $(cA)(\frac{1}{c}A^{-1}) = AA^{-1} = I$.

Given two invertible $n \times n$ matrices A and B, we claim that the matrix AB is also invertible, and

$$(AB)^{-1} = B^{-1}A^{-1}. \tag{3.1}$$

Indeed,

$$(B^{-1}A^{-1})(AB) = B^{-1}(A^{-1}A)B = B^{-1}IB = B^{-1}B = I,$$

and one shows similarly that $(AB)(B^{-1}A^{-1}) = I$. Similar rule holds for arbitrary number of invertible matrices. For example,

$$(ABC)^{-1} = C^{-1}B^{-1}A^{-1}.$$

Indeed, apply (3.1) twice to get

$$(ABC)^{-1} = [(AB)C]^{-1} = C^{-1}(AB)^{-1} = C^{-1}B^{-1}A^{-1}.$$

We show next that inverses of elementary matrices are also elementary matrices, of the same type. We have

$$E_i\left(\frac{1}{a}\right)E_i(a) = I,$$

because the elementary matrix $E_i(\frac{1}{a})$ performs an elementary operation $\frac{1}{a}R_i$ on $E_i(a)$, which results in I, so that

$$E_i(a)^{-1} = E_i\left(\frac{1}{a}\right). \tag{3.2}$$

For example, $E_2(-5)^{-1} = E_2(-\frac{1}{5})$, so that in the 3×3 case

$$\begin{bmatrix} 1 & 0 & 0 \\ 0 & -5 & 0 \\ 0 & 0 & 1 \end{bmatrix}^{-1} = \begin{bmatrix} 1 & 0 & 0 \\ 0 & -\frac{1}{5} & 0 \\ 0 & 0 & 1 \end{bmatrix}.$$

Next

$$E_{ij}^{-1} = E_{ij}, \tag{3.3}$$

(the matrix E_{ij} is its own inverse) because

$$E_{ij}E_{ij} = I.$$

Indeed, the matrix E_{ij} on the left switches the rows i and j of the other E_{ij}, putting the rows back in order to give I. Finally,

$$E_{ij}(a)^{-1} = E_{ij}(-a), \tag{3.4}$$

because

$$E_{ij}(-a)E_{ij}(a) = I.$$

Indeed, performing $R_j - aR_i$ on $E_{ij}(a)$ produces I. For example, $E_{13}(4)^{-1} = E_{13}(-4)$, so that in the 3×3 case

$$\begin{bmatrix} 1 & 0 & 0 \\ 0 & 1 & 0 \\ 4 & 0 & 1 \end{bmatrix}^{-1} = \begin{bmatrix} 1 & 0 & 0 \\ 0 & 1 & 0 \\ -4 & 0 & 1 \end{bmatrix}.$$

Some products of elementary matrices can be calculated at a glance, by performing the products from right to left. For example,

$$L = E_{12}(2)E_{13}(-3)E_{23}(4) = E_{12}(2)\left[E_{13}(-3)E_{23}(4)\right]$$

$$= \begin{bmatrix} 1 & 0 & 0 \\ 2 & 1 & 0 \\ 0 & 0 & 1 \end{bmatrix}\left(\begin{bmatrix} 1 & 0 & 0 \\ 0 & 1 & 0 \\ -3 & 0 & 1 \end{bmatrix}\begin{bmatrix} 1 & 0 & 0 \\ 0 & 1 & 0 \\ 0 & 4 & 1 \end{bmatrix}\right) = \begin{bmatrix} 1 & 0 & 0 \\ 2 & 1 & 0 \\ -3 & 4 & 1 \end{bmatrix}. \tag{3.5}$$

Indeed, the product of the last two matrices in (3.5), namely

$$E_{13}(-3)E_{23}(4) = \begin{bmatrix} 1 & 0 & 0 \\ 0 & 1 & 0 \\ -3 & 0 & 1 \end{bmatrix}\begin{bmatrix} 1 & 0 & 0 \\ 0 & 1 & 0 \\ 0 & 4 & 1 \end{bmatrix} = \begin{bmatrix} 1 & 0 & 0 \\ 0 & 1 & 0 \\ -3 & 4 & 1 \end{bmatrix},$$

is obtained by applying $R_3 - 3R_1$ to $E_{23}(4)$. Applying $R_2 + 2R_1$ to the last matrix gives L in (3.5).

This matrix L is an example of a *lower triangular matrix*, defined as a square matrix with all elements above the diagonal equal to 0 (other elements are arbitrary). The matrix $L_1 = \begin{bmatrix} 2 & 0 & 0 \\ 3 & -3 & 0 \\ 0 & -5 & 0 \end{bmatrix}$ gives another example of a lower triangular matrix. All elementary matrices of the type $E_{ij}(a)$ are lower triangular. The matrix $U = \begin{bmatrix} 1 & -1 & 0 \\ 0 & -3 & 4 \\ 0 & 0 & 0 \end{bmatrix}$ is an example of an *upper triangular matrix*, defined as a square matrix with all elements below the diagonal equal to 0 (the elements on the diagonal and above the diagonal are not restricted).

Let us perform row reduction on the matrix $A = \begin{bmatrix} 1 & -1 & 1 \\ 2 & -1 & 2 \\ -3 & 7 & 4 \end{bmatrix}$. Performing $R_2 - 2R_1$, $R_3 + 3R_1$, followed by $R_3 - 4R_2$, produces an upper triangular matrix

$$U = \begin{bmatrix} 1 & -1 & 1 \\ 0 & 1 & 0 \\ 0 & 0 & 7 \end{bmatrix}. \tag{3.6}$$

Rephrasing these elementary operations in terms of the elementary matrices gives

$$E_{23}(-4)E_{13}(3)E_{12}(-2)A = U.$$

To express A, multiply both sides from the left by the inverse of the matrix $E_{23}(-4)E_{13}(3)E_{12}(-2)$:

$$A = \left[E_{23}(-4)E_{13}(3)E_{12}(-2)\right]^{-1}U = E_{12}(-2)^{-1}E_{13}(3)^{-1}E_{23}(-4)^{-1}U$$
$$= E_{12}(2)E_{13}(-3)E_{23}(4)U = LU,$$

where L is the lower triangular matrix calculated in (3.5), and the upper triangular matrix U is shown in (3.6), so that

$$A = \begin{bmatrix} 1 & -1 & 1 \\ 2 & -1 & 2 \\ -3 & 7 & 4 \end{bmatrix} = \begin{bmatrix} 1 & 0 & 0 \\ 2 & 1 & 0 \\ -3 & 4 & 1 \end{bmatrix}\begin{bmatrix} 1 & -1 & 1 \\ 0 & 1 & 0 \\ 0 & 0 & 7 \end{bmatrix}.$$

Matrix A is decomposed as product of a lower triangular matrix L and an upper triangular matrix U.

Similar $A = LU$ *decomposition* can be calculated for any $n \times n$ matrix A, for which forward elimination can be performed without switching the rows. The upper triangular matrix U is the result of row reduction (the row echelon form). The lower triangular matrix L has 1's on the diagonal, and $(L)_{ji} = a$ if the operation $R_j - aR_i$ was used in row reduction (here $(L)_{ji}$ denotes the ji entry of the matrix L). If the operation $R_j - aR_i$ was not used in row reduction, then $(L)_{ji} = 0$. For example, suppose that the elementary operations $R_3 - 3R_1$ followed by $R_3 + 4R_2$ reduced a 3×3 matrix A to an upper triangular matrix U (so that $a_{21} = 0$, and we had a "free zero" in that position). Then $L = \begin{bmatrix} 1 & 0 & 0 \\ 0 & 1 & 0 \\ 3 & -4 & 1 \end{bmatrix}$.

We shall use later the following theorem.

Theorem 2.3.1. *Every invertible matrix A can be written as a product of elementary matrices.*

Proof. By the formula (2.3), developed for computation of A^{-1},

$$E_p \cdots E_2 E_1 A = I,$$

for some elementary matrices $E_1, E_2, \ldots, E_p$. Multiply both sides by $(E_p \cdots E_2 E_1)^{-1}$, to obtain

$$A = (E_p \cdots E_2 E_1)^{-1} I = E_1^{-1} E_2^{-1} \cdots E_p^{-1}.$$

The inverses of elementary matrices are themselves elementary matrices. □

If one keeps the $A = LU$ decomposition of *a large matrix* A on file, then to solve

$$Ax = LUx = b,$$

for some $b \in R^n$, set

$$Ux = y, \tag{3.7}$$

and then

$$Ly = b. \tag{3.8}$$

One can quickly solve (3.8) by "forward-substitution" for $y \in R^n$, and then solve (3.7) by back-substitution to get the solution x. This process is much faster than performing Gaussian elimination for $Ax = b$ "from scratch."

Exercises

1. Assuming that A and B are nonsingular $n \times n$ matrices, simplify:
   (a) $B(AB)^{-1}A$.
       Answer: I.
   (b) $(2A)^{-1}A^2$.
       Answer: $\frac{1}{2}A$.
   (c) $[4(AB)^{-1}A]^{-1}$.
       Answer: $\frac{1}{4}B$.
2. Without using Gaussian elimination find the inverses of the following 3×3 elementary matrices:

(a) $E_{13}(2)$.

Answer: $E_{13}(-2) = \begin{bmatrix} 1 & 0 & 0 \\ 0 & 1 & 0 \\ -2 & 0 & 1 \end{bmatrix}$.

(b) $E_2(5)$.

Answer: $E_2(\frac{1}{5}) = \begin{bmatrix} 1 & 0 & 0 \\ 0 & \frac{1}{5} & 0 \\ 0 & 0 & 1 \end{bmatrix}$.

(c) E_{13}.

Answer: $E_{13} = \begin{bmatrix} 0 & 0 & 1 \\ 0 & 1 & 0 \\ 1 & 0 & 0 \end{bmatrix}$.

3. Identify the following 4×4 matrices as elementary matrices, and then find their inverses:

(a) $A = \begin{bmatrix} 1 & 0 & 0 & 0 \\ 0 & 0 & 0 & 1 \\ 0 & 0 & 1 & 0 \\ 0 & 1 & 0 & 0 \end{bmatrix}$.

Answer: $A = E_{24}$, $A^{-1} = E_{24}$.

(b) $B = \begin{bmatrix} 1 & 0 & 0 & 0 \\ 0 & 1 & 0 & 0 \\ 0 & 0 & 1 & 0 \\ 0 & 0 & -5 & 1 \end{bmatrix}$.

Answer: $B = E_{34}(-5)$, $B^{-1} = E_{34}(5)$.

(c) $C = \begin{bmatrix} 1 & 0 & 0 & 0 \\ 0 & 1 & 0 & 0 \\ 0 & 0 & 1 & 0 \\ 0 & 0 & 0 & 7 \end{bmatrix}$.

Answer: $C = E_4(7)$, $C^{-1} = E_4(\frac{1}{7})$.

4. Calculate the products of the following 3×3 elementary matrices, by performing the multiplication from right to left:

(a) $E_{12}(-3)E_{13}(-1)E_{23}(4)$.

Answer: $\begin{bmatrix} 1 & 0 & 0 \\ -3 & 1 & 0 \\ -1 & 4 & 1 \end{bmatrix}$.

(b) $E_{12}E_{13}(-1)E_{23}(4)$.

Answer: $\begin{bmatrix} 0 & 1 & 0 \\ 1 & 0 & 0 \\ -1 & 4 & 1 \end{bmatrix}$.

(c) $E_{13}E_{13}(-1)E_{23}(4)$.

Answer: $\begin{bmatrix} -1 & 4 & 1 \\ 0 & 1 & 0 \\ 1 & 0 & 0 \end{bmatrix}$.

(d) $E_{12}(2)E_{23}(-1)E_{23}$.

Answer: $\begin{bmatrix} 1 & 0 & 0 \\ 2 & 0 & 1 \\ 0 & 1 & -1 \end{bmatrix}$.

(e) $E_3(3)E_{13}(-1)E_{12}$.

Answer: $\begin{bmatrix} 0 & 1 & 0 \\ 1 & 0 & 0 \\ 0 & -3 & 3 \end{bmatrix}$.

5. Find the LU decomposition of the following matrices:

(a) $\begin{bmatrix} 1 & 2 \\ 3 & 4 \end{bmatrix}$.

Answer: $L = \begin{bmatrix} 1 & 0 \\ 3 & 1 \end{bmatrix}$, $U = \begin{bmatrix} 1 & 2 \\ 0 & -2 \end{bmatrix}$.

(b) $\begin{bmatrix} 1 & 1 & 1 \\ 1 & 2 & 2 \\ 1 & 2 & 3 \end{bmatrix}$.

Answer: $L = \begin{bmatrix} 1 & 0 & 0 \\ 1 & 1 & 0 \\ 1 & 1 & 1 \end{bmatrix}$, $U = \begin{bmatrix} 1 & 1 & 1 \\ 0 & 1 & 1 \\ 0 & 0 & 1 \end{bmatrix}$.

(c) $\begin{bmatrix} 1 & 1 & -1 \\ 1 & 2 & 2 \\ 2 & 3 & 5 \end{bmatrix}$.

(d) $\begin{bmatrix} 1 & 1 & 1 \\ -1 & 1 & 0 \\ 2 & 2 & 3 \end{bmatrix}$.

Answer: $L = \begin{bmatrix} 1 & 0 & 0 \\ -1 & 1 & 0 \\ 2 & 0 & 1 \end{bmatrix}$, $U = \begin{bmatrix} 1 & 1 & 1 \\ 0 & 2 & 1 \\ 0 & 0 & 1 \end{bmatrix}$.

(e) $\begin{bmatrix} 1 & 2 & 1 & 0 \\ 0 & 2 & 1 & -1 \\ 2 & 4 & 3 & 1 \\ 0 & -2 & 0 & 2 \end{bmatrix}$.

Answer: $L = \begin{bmatrix} 1 & 0 & 0 & 0 \\ 0 & 1 & 0 & 0 \\ 2 & 0 & 1 & 0 \\ 0 & -1 & 1 & 1 \end{bmatrix}$, $U = \begin{bmatrix} 1 & 2 & 1 & 0 \\ 0 & 2 & 1 & -1 \\ 0 & 0 & 1 & 1 \\ 0 & 0 & 0 & 0 \end{bmatrix}$.

6. (a) For the matrix $A = \begin{bmatrix} 0 & 1 & -1 \\ 1 & 2 & 2 \\ 2 & 3 & 4 \end{bmatrix}$ the LU decomposition is not possible (explain why).

    Calculate the LU decomposition for the matrix $E_{12}A$.

    (b*) Show that any nonsingular $n \times n$ matrix A admits a decomposition $PA = LU$, where P is a permutation matrix.

    Hint. Choose P to perform all row exchanges needed in the row reduction of A.

7. Assume that $A = E_{12}(3) \, E_3(-2) \, E_{23}$.

    (a) Express the inverse matrix A^{-1} as a product of elementary matrices.

    Answer: $A^{-1} = E_{23} \, E_3(-\frac{1}{2}) \, E_{12}(-3)$.

    (b) In case A is 3×3, write down A^{-1}.

    Answer: $A^{-1} = \begin{bmatrix} 1 & 0 & 0 \\ 0 & 0 & -\frac{1}{2} \\ -3 & 1 & 0 \end{bmatrix}$.

8. Suppose that S is invertible and $A = S^{-1}BS$.

    (a) Show that $B = SAS^{-1}$.

    (b) Suppose that A is also invertible. Show that B is invertible, and express B^{-1}.

9. Assume that A, B, and $A + B$ are nonsingular $n \times n$ matrices. Show that

$$(A^{-1} + B^{-1})^{-1} = A(A + B)^{-1}B.$$

    Hint. Show that the inverses of these matrices are equal.

10. Show that in general

$$(A + B)^{-1} \neq A^{-1} + B^{-1}.$$

    Hint. $A = 3I$, $B = 5I$ provides an easy example (or a *counterexample*).

2.4 Subspaces, bases, and dimension

The space R^3 is a *vector space*, meaning that one can add vectors and multiply vectors by scalars. Vectors of the form $\begin{bmatrix} 1 \\ x_2 \\ x_3 \end{bmatrix}$ form *a subset* (a part) of R^3. Let us call this subset H_1.

For example, the vectors $\begin{bmatrix} 1 \\ -2 \\ 3 \end{bmatrix}$ and $\begin{bmatrix} 1 \\ 3 \\ 0 \end{bmatrix}$ both belong to H_1, but their sum $\begin{bmatrix} 2 \\ 1 \\ 3 \end{bmatrix}$ does not (vectors in H_1 have the first component 1). Vectors of the form $\begin{bmatrix} 0 \\ x_2 \\ x_3 \end{bmatrix}$ form another subset of R^3, which we call H_2. The sum of any two vectors in H_2,

$$\begin{bmatrix} 0 \\ x_2 \\ x_3 \end{bmatrix} + \begin{bmatrix} 0 \\ y_2 \\ y_3 \end{bmatrix} = \begin{bmatrix} 0 \\ x_2 + y_2 \\ x_3 + y_3 \end{bmatrix},$$

belongs to H_2, and also a scalar multiple of any vector in H_2,

$$c \begin{bmatrix} 0 \\ x_2 \\ x_3 \end{bmatrix} = \begin{bmatrix} 0 \\ c\,x_2 \\ c\,x_3 \end{bmatrix},$$

belongs to H_2, for any scalar c.

Definition 2.4.1. A subset H of vectors in R^n is called a subspace if for any vectors u and v in H and any scalar c,
(i) $u + v$ belongs to H (*H is closed under addition*);
(ii) $c\,u$ belongs to H (*H is closed under scalar multiplication*).

So that addition of vectors and multiplication of vectors by scalars do not take us out of H. The set H_2 above is a subspace, while H_1 is not a subspace, because it is not closed under addition, as we discussed above (H_1 is also not closed under scalar multiplication). In simple terms, a subspace H is a part (subset) of R^n, where one can add vectors and multiply vectors by scalars without leaving H.

Using $c = 0$ in part (ii) of Definition 2.4.1, one sees that *any subspace contains the zero vector*. Hence, if a set does not contains the zero vector, it is not a subspace. For example, let H_3 be a subset of vectors $\begin{bmatrix} x_1 \\ x_2 \\ x_3 \\ x_4 \end{bmatrix}$ of R^4, such that $x_1 + x_2 + x_3 + x_4 = 1$. Then H_3 is not a subspace, because the zero vector $\begin{bmatrix} 0 \\ 0 \\ 0 \\ 0 \end{bmatrix}$ does not belong to H_3.

A special subspace, called *the zero subspace* $\{0\}$, consists of only the zero vector in R^n. The space R^n itself also satisfies Definition 2.4.1, and it can be regarded as a subspace of itself.

Given vectors $v_1, v_2, \ldots, v_p$ in R^n, their span, $S = \text{Span}\{v_1, v_2, \ldots, v_p\}$, is a subspace of R^n. Indeed, suppose $x \in S$ and $y \in S$ ($\in$ is a mathematical symbol meaning "belongs"). Then $x = x_1 v_1 + x_2 v_2 + \cdots + x_p v_p$ and $y = y_1 v_1 + y_2 v_2 + \cdots + y_p v_p$ for some numbers x_i and y_i. Calculate $x + y = (x_1 + y_1)v_1 + (x_2 + y_2)v_2 + \cdots + (x_p + y_p)v_p \in S$, and $c\,x = (cx_1)v_1 + (cx_2)v_2 + \cdots + (cx_p)v_p \in S$, to verify that S is a subspace.

Definition 2.4.2. Given a subspace H, we say that the vectors $\{u_1, u_2, \ldots, u_q\}$ in H form a basis of H if they are linearly independent and span H so that $H = \text{Span}\{u_1, u_2, \ldots, u_q\}$.

Theorem 2.4.1. *Suppose that q vectors* $U = \{u_1, u_2, \ldots, u_q\}$ *form a basis of H, and let* $r \geq q + 1$. *Then any r vectors in H are linearly dependent.*

Proof. Let $v_1, v_2, \ldots, v_r$ be some vectors in H, with $r > q$. We wish to show that the relation

$$x_1 v_1 + x_2 v_2 + \cdots + x_r v_r = 0 \tag{4.1}$$

has a nontrivial solution (not all x_i are zero). Express v_i's through the basis U:

$$v_1 = a_{11} u_1 + a_{21} u_2 + \cdots + a_{q1} u_q,$$
$$v_2 = a_{12} u_1 + a_{22} u_2 + \cdots + a_{q2} u_q,$$
$$\vdots$$
$$v_r = a_{1r} u_1 + a_{2r} u_2 + \cdots + a_{qr} u_q,$$

with some numbers a_{ij}, and use them in (4.1). Rearranging we obtain

$$(a_{11} x_1 + a_{12} x_2 + \cdots + a_{1r} x_r) u_1 + (a_{21} x_1 + a_{22} x_2 + \cdots + a_{2r} x_r) u_2$$
$$+ \cdots + (a_{q1} x_1 + a_{q2} x_2 + \cdots + a_{qr} x_r) u_q = 0.$$

To satisfy the latter equation, it is sufficient to make all of the coefficients equal to zero:

$$a_{11} x_1 + a_{12} x_2 + \cdots + a_{1r} x_r = 0,$$
$$a_{21} x_1 + a_{22} x_2 + \cdots + a_{2r} x_r = 0,$$
$$\vdots$$
$$a_{q1} x_1 + a_{q2} x_2 + \cdots + a_{qr} x_r = 0.$$

We have a homogeneous system with more unknowns than equations. By Theorem 1.4.1, it has nontrivial solutions. □

It follows that *any two bases of a subspace have the same number of vectors*. Indeed, if two bases with different number of vectors existed, then vectors in the larger basis would have to be linearly dependent, which is not possible by the definition of a basis. *The common number of vectors in any basis of H is called the dimension of H*, denoted by dim H.

It is intuitively clear that the space R^2 is two-dimensional, R^3 is three-dimensional, etc. To justify rigorously that R^2 is two-dimensional, let us exhibit a basis with two elements in R^2, by considering *the standard basis*, consisting of $e_1 = \begin{bmatrix} 1 \\ 0 \end{bmatrix}$ and $e_2 = \begin{bmatrix} 0 \\ 1 \end{bmatrix}$. These vectors are linearly independent and they span R^2, because any vector $x = \begin{bmatrix} x_1 \\ x_2 \end{bmatrix} \in R^2$ can be written as $x = x_1 e_1 + x_2 e_2$. In R^3 the standard basis consists of $e_1 = \begin{bmatrix} 1 \\ 0 \\ 0 \end{bmatrix}$, $e_2 = \begin{bmatrix} 0 \\ 1 \\ 0 \end{bmatrix}$, and $e_3 = \begin{bmatrix} 0 \\ 0 \\ 1 \end{bmatrix}$, and similarly for other R^n.

Theorem 2.4.2. *If the dimension of a subspace H is p, then any p linearly independent vectors of H form a basis of H.*

Proof. Let $u_1, u_2, \ldots, u_p$ be any p linearly independent vectors of H. We only need to show that they span H. Suppose, on the contrary, that we can find a vector w in H which is not in their span. By Theorem 1.5.1, the $p + 1$ vectors $u_1, u_2, \ldots, u_p, w$ are linearly independent. But that contradicts Theorem 2.4.1.   □

It follows that in R^2 any two noncollinear vectors form a basis. In R^3 any three vectors that do not lie in the same plane form a basis.

Suppose that vectors $B = \{b_1, b_2, \ldots, b_p\}$ form a basis in some subspace H. Then any vector $v \in H$ can be represented through the basis elements,

$$v = x_1 b_1 + x_2 b_2 + \cdots + x_p b_p,$$

with some numbers $x_1, x_2, \ldots, x_p$. This representation is unique, because if there was another representation $v = y_1 b_1 + y_2 b_2 + \cdots + y_p b_p$, then subtraction would give

$$0 = (x_1 - y_1)b_1 + (x_2 - y_2)b_2 + \cdots + (x_p - y_p)b_p,$$

and then $x_1 = y_1, x_2 = y_2, \ldots, x_p = y_p$, by linear independence of vectors in the basis B. The coefficients $x_1, x_2, \ldots, x_p$ are called *the coordinates of v with respect to the basis B,* with the notation

$$[v]_B = \begin{bmatrix} x_1 \\ x_2 \\ \vdots \\ x_p \end{bmatrix}.$$

Example 1. Two linearly independent vectors $b_1 = \begin{bmatrix} 1 \\ -1 \end{bmatrix}$ and $b_2 = \begin{bmatrix} 2 \\ 0 \end{bmatrix}$ form a basis of R^2, $B = \{b_1, b_2\}$. The vector $v = \begin{bmatrix} 5 \\ -3 \end{bmatrix}$ can be decomposed as $v = 3b_1 + b_2$. It follows that the coordinates $[v]_B = \begin{bmatrix} 3 \\ 1 \end{bmatrix}$.

Example 2. The vectors $b_1 = \begin{bmatrix} 1 \\ -1 \end{bmatrix}$, $b_2 = \begin{bmatrix} 2 \\ 0 \end{bmatrix}$, and $b_3 = \begin{bmatrix} 4 \\ -2 \end{bmatrix}$ do not form a basis of R^2, because any three vectors in R^2 are linearly dependent, and, in fact, $b_3 = 2b_1 + b_2$. As in Example 1, b_1 and b_2 form a basis of R^2, $B = \{b_1, b_2\}$, and $[b_3]_B = \begin{bmatrix} 2 \\ 1 \end{bmatrix}$.

Example 3. Let us verify that the vectors $b_1 = \begin{bmatrix} 1 \\ 0 \\ 1 \end{bmatrix}$, $b_2 = \begin{bmatrix} 0 \\ -1 \\ 1 \end{bmatrix}$, and $b_3 = \begin{bmatrix} 1 \\ 2 \\ 3 \end{bmatrix}$ form a basis of R^3, and then find the coordinates of the vector $v = \begin{bmatrix} 3 \\ 3 \\ 4 \end{bmatrix}$ with respect to this basis, $B = \{b_1, b_2, b_3\}$.

To justify that the three vectors b_1, b_2, b_3 form a basis of R^3, we only need to show that they are linearly independent. That involves showing that the matrix $A = [b_1 \ b_2 \ b_3]$ has three pivots. Let us go straight to finding the coordinates of v, representing

$$v = x_1 b_1 + x_2 b_2 + x_3 b_3,$$

and in the process it will be clear that the matrix A has three pivots. We need to solve a 3×3 system with the augmented matrix

$$[b_1 \ b_2 \ b_3 : v] = \begin{bmatrix} 1 & 0 & 1 & | & 3 \\ 0 & -1 & 2 & | & 3 \\ 1 & 1 & 3 & | & 4 \end{bmatrix}.$$

The matrix of this system is precisely A. Perform $R_3 - R_1$, followed by $R_3 + R_2$, to obtain

$$\begin{bmatrix} ① & 0 & 1 & | & 3 \\ 0 & ① & 2 & | & 3 \\ 0 & 0 & ④ & | & 4 \end{bmatrix}.$$

The matrix A has three pivots, therefore the vectors b_1, b_2, b_3 are linearly independent, and hence they form a basis of R^3. Restoring the system, we obtain $x_3 = 1$, $x_2 = -1$, $x_1 = 2$, by back-substitution.

Answer: $[v]_B = \begin{bmatrix} 2 \\ -1 \\ 1 \end{bmatrix}$.

Exercises

1. Do the following subsets form subspaces of the corresponding spaces?
   (a) Vectors in R^3 with $x_1 + x_2 \geq 1$.
   Answer: No, the zero vector is not included in this subset.
   (b) Vectors in R^3 with $x_1^2 + x_2^2 + x_3^2 \leq 1$.
   Answer: No, the subset is not closed under both addition and scalar multiplication.
   (c) Vectors in R^5 with $x_1 + x_4 = 0$.
   Answer: Yes.
   (d) Vectors in R^4 with $x_2 = 0$.
   Answer: Yes.
   (e) Vectors in R^2 with $x_1 x_2 = 1$.
   Answer: No, not closed under addition (also not closed under scalar multiplication).
   (f) Vectors in R^2 with $x_1 x_2 = 0$.
   Answer: No, not closed under addition (it is closed under scalar multiplication).
   (g) Vectors in R^3 with $x_1 = 2x_2 = -3x_3$.
   Answer: Yes, these vectors lie on a line through the origin.
   (h) Vectors in R^3 of the form $\begin{bmatrix} 0 \\ x_2 \\ x_2^2 \end{bmatrix}$.
   Does this subset contain the zero vector?
   Answer: Not a subspace, even though this subset contains the zero vector.

2. Show that all vectors lying on any line through the origin in R^2 form a subspace.
3. (a) Show that all vectors lying on any line through the origin in R^3 form a subspace.
    (b) Show that all vectors lying on any plane through the origin in R^3 form a subspace.
4. (a) Explain why the vectors $b_1 = \begin{bmatrix} 1 \\ 2 \end{bmatrix}$ and $b_2 = \begin{bmatrix} -1 \\ 1 \end{bmatrix}$ form a basis of R^2, and then find the coordinates of the vector e_1 from the standard basis with respect to this basis, $B = \{b_1, b_2\}$.
    Answer: $[e_1]_B = \begin{bmatrix} 1/3 \\ -2/3 \end{bmatrix}$.
    (b) What is the vector $v \in R^2$ if $[v]_B = \begin{bmatrix} 1 \\ 3 \end{bmatrix}$?
    Answer: $v = \begin{bmatrix} -2 \\ 5 \end{bmatrix}$.
    (c) For each of the following vectors $v_1 = \begin{bmatrix} 2 \\ 1 \end{bmatrix}$, $v_2 = \begin{bmatrix} 0 \\ 2 \end{bmatrix}$, and $v_3 = \begin{bmatrix} -2 \\ 2 \end{bmatrix}$ find their coordinates with respect to this basis, $B = \{b_1, b_2\}$.
    Hint. Calculations can be performed simultaneously (in parallel) by considering the augmented matrix $\begin{bmatrix} 1 & -1 & | & 2 & 0 & | & -2 \\ 2 & 1 & | & 1 & 2 & | & 2 \end{bmatrix}$. Perform $R_2 - 2R_1$ on the entire matrix, then restore each system.
    Answer: $[v_1]_B = \begin{bmatrix} 1 \\ -1 \end{bmatrix}$, $[v_2]_B = \begin{bmatrix} 2/3 \\ 2/3 \end{bmatrix}$, $[v_3]_B = \begin{bmatrix} 0 \\ 2 \end{bmatrix}$.
5. Verify that the vectors $b_1 = \begin{bmatrix} 1 \\ 0 \\ 1 \end{bmatrix}$, $b_2 = \begin{bmatrix} 0 \\ -1 \\ 1 \end{bmatrix}$, $b_3 = \begin{bmatrix} 1 \\ 2 \\ 3 \end{bmatrix}$ form a basis of R^3, and then find the coordinates of the vectors $v_1 = \begin{bmatrix} 1 \\ 0 \\ 4 \end{bmatrix}$ and $v_2 = \begin{bmatrix} 2 \\ 1 \\ 5 \end{bmatrix}$ with respect to this basis, $B = \{b_1, b_2, b_3\}$.
6. (a) Show that the vectors $b_1 = \begin{bmatrix} 1 \\ 2 \\ 0 \\ 3 \end{bmatrix}$, $b_2 = \begin{bmatrix} 1 \\ 1 \\ 1 \\ 1 \end{bmatrix}$, $b_3 = \begin{bmatrix} 0 \\ -1 \\ 1 \\ -2 \end{bmatrix}$ are linearly dependent, and express b_3 as a linear combination of b_1 and b_2.
    Answer: $b_3 = -b_1 + b_2$.
    (b) Let $V = \text{Span}\{b_1, b_2, b_3\}$. Find a basis of V, and dimension of V.
    Answer: $B = \{b_1, b_2\}$ is a basis of V; dimension of V is 2.
    (c) Find the coordinates of b_1, b_2, b_3 with respect to the basis in part (b).
    Answer: $[b_1]_B = \begin{bmatrix} 1 \\ 0 \end{bmatrix}$, $[b_2]_B = \begin{bmatrix} 0 \\ 1 \end{bmatrix}$, $[b_3]_B = \begin{bmatrix} -1 \\ 1 \end{bmatrix}$.
7. Let $E = \{e_1, e_2, e_3\}$ be the standard basis in R^3, and $x = \begin{bmatrix} x_1 \\ x_2 \\ x_3 \end{bmatrix}$. Find the coordinates $[x]_E$.
    Answer: $[x]_E = \begin{bmatrix} x_1 \\ x_2 \\ x_3 \end{bmatrix}$.

2.5 Null and column spaces

We now study two important subspaces associated with any $m \times n$ matrix A.

Definition 2.5.1. The null space of A is the set of all vectors $x \in R^n$ satisfying $Ax = 0$. It is denoted by $N(A)$.

Let us justify that the null space is a subspace of R^n. (Recall that the terms "subspace" and "space" are used interchangeably.) Assume that two vectors x_1 and x_2 belong to $N(A)$, meaning that $Ax_1 = 0$ and $Ax_2 = 0$. Then

$$A(x_1 + x_2) = Ax_1 + Ax_2 = 0,$$

so that $x_1 + x_2 \in N(A)$. Similarly, $A(cx_1) = cAx_1 = 0$, so that $cx_1 \in N(A)$, for any number c, justifying that $N(A)$ is a subspace.

Finding the null space of A requires solving the homogeneous system $Ax = 0$, which was studied previously. We can now interpret the answer in terms of dimension and basis of $N(A)$.

Example 1. $A = \begin{bmatrix} -1 & 2 & 0 & 1 \\ 2 & -4 & 1 & -1 \\ 3 & -6 & 1 & -2 \end{bmatrix}$. The augmented matrix of the system $Ax = 0$ is

$$\left[\begin{array}{cccc|c} -1 & 2 & 0 & 1 & 0 \\ 2 & -4 & 1 & -1 & 0 \\ 3 & -6 & 1 & -2 & 0 \end{array}\right].$$

Perform $R_2 + 2R_1$, $R_3 + 3R_1$ to get

$$\left[\begin{array}{cccc|c} ① & 2 & 0 & 1 & 0 \\ 0 & 0 & ① & 1 & 0 \\ 0 & 0 & 1 & 1 & 0 \end{array}\right].$$

The second column does not have a pivot, but the third column does. Forward elimination is completed by performing $R_3 - R_2$:

$$\left[\begin{array}{cccc|c} ① & 2 & 0 & 1 & 0 \\ 0 & 0 & ① & 1 & 0 \\ 0 & 0 & 0 & 0 & 0 \end{array}\right].$$

Restore the system, take the free variables x_2 and x_4 to the right, and solve for the basis variables x_1 and x_3. Obtain $x_1 = 2x_2 + x_4$, $x_3 = -x_4$, where x_2 and x_4 are arbitrary numbers. Putting the answer in the vector form, we then obtain

$$\begin{bmatrix} 2x_2 + x_4 \\ x_2 \\ -x_4 \\ x_4 \end{bmatrix} = x_2 \begin{bmatrix} 2 \\ 1 \\ 0 \\ 0 \end{bmatrix} + x_4 \begin{bmatrix} 1 \\ 0 \\ -1 \\ 1 \end{bmatrix},$$

so that $N(A)$ is span of the vectors $u = \begin{bmatrix} 2 \\ 1 \\ 0 \\ 0 \end{bmatrix}$ and $v = \begin{bmatrix} 1 \\ 0 \\ -1 \\ 1 \end{bmatrix}$, $N(A) = \text{Span}\{u, v\}$. *Conclusions:* the null space $N(A)$ is a subspace of R^4 of dimension 2, $\dim N(A) = 2$, the vectors u and v form a basis of $N(A)$.

For an arbitrary matrix A, the dimension of the null space $N(A)$ is equal to the number of free variables in the row echelon form of A.

If the system $Ax = 0$ has only the trivial solution $x = 0$, then the null space of A is the zero subspace, or $N(A) = \{0\}$, consisting only of the zero vector.

Definition 2.5.2. The column space of a matrix A is the span (the set of all possible linear combinations) of its column vectors. It is denoted by $C(A)$.

If $A = [a_1\, a_2 \ldots a_n]$ is an $m \times n$ matrix given by its columns, the column space $C(A) =$ Span$\{a_1, a_2, \ldots, a_n\}$ consists of all vectors of the form

$$x_1 a_1 + x_2 a_2 + \cdots + x_n a_n = Ax, \tag{5.1}$$

with arbitrary numbers $x_1, x_2, \ldots, x_n$. Columns of A are vectors in R^m, so that $C(A)$ is a subset of R^m. In fact, the column space is *a subspace of R^m*, because any span is a subspace. Formula (5.1) shows that the column space $C(A)$ can be viewed as *the range of the function Ax*.

The rank of a matrix A, denoted by rank A, is the dimension of the column space of A, rank $A = \dim C(A)$.

Example 2. Determine the basis of the column space of the following two matrices. Express the columns that are not in the basis through those in the basis.

(i) $A = \begin{bmatrix} ② & 1 & 3 & 0 & 3 \\ 0 & ① & 1 & 1 & 0 \\ 0 & 0 & 0 & ① & 1 \\ 0 & 0 & 0 & 0 & 0 \end{bmatrix} = [a_1\, a_2\, a_3\, a_4\, a_5]$,

where a_i's denote the columns of A. The matrix A is already in row echelon form, with the pivots circled. The pivot columns a_1, a_2, a_4 are linearly independent. Indeed, the matrix $[a_1\, a_2\, a_4]$ has three pivots. We show next that the other columns, a_3 and a_5, are linear combinations of the pivot columns a_1, a_2, a_4. Indeed, to express a_5 through the pivot columns, we need to find numbers x_1, x_2, x_3 so that

$$x_1 a_1 + x_2 a_2 + x_3 a_4 = a_5.$$

The augmented matrix of this system is

$$[a_1\, a_2\, a_4 \,\vdots\, a_5] = \begin{bmatrix} ② & 1 & 0 & \vdots & 3 \\ 0 & ① & 1 & \vdots & 0 \\ 0 & 0 & ① & \vdots & 1 \\ 0 & 0 & 0 & \vdots & 0 \end{bmatrix}.$$

Back-substitution gives $x_1 = x_2 = x_3 = 1$, so that

$$a_5 = a_1 + a_2 + a_4. \tag{5.2}$$

To express a_3 through the pivot columns, we need to find new numbers x_1, x_2, x_3 so that

$$x_1 a_1 + x_2 a_2 + x_3 a_4 = a_3.$$

The augmented matrix of this system is

$$[a_1\ a_2\ a_4 : a_3] = \begin{bmatrix} ② & 1 & 0 & | & 3 \\ 0 & ④ & 1 & | & 1 \\ 0 & 0 & ① & | & 0 \\ 0 & 0 & 0 & | & 0 \end{bmatrix}.$$

Back-substitution gives $x_3 = 0$, $x_2 = -1$, $x_1 = 2$, so that

$$a_3 = 2a_1 - a_2. \tag{5.3}$$

We claim that *the pivot columns a_1, a_2, a_4 form a basis of $C(A)$*, so that $\dim C(A) =$ $\operatorname{rank} A = 3$. We already know that these vectors are linearly independent, so that it remains to show that they span $C(A)$. The column space $C(A)$ consists of vectors in the form $v = c_1 a_1 + c_2 a_2 + c_3 a_3 + c_4 a_4 + c_5 a_5$ for some numbers c_1, c_2, c_3, c_4, c_5. Using (5.2) and (5.3), any vector $v \in C(A)$ can be expressed as

$$v = c_1 a_1 + c_2 a_2 + c_3(2a_1 - a_2) + c_4 a_4 + c_5(a_1 + a_2 + a_4)$$
$$= (c_1 + 2c_3 + c_5)a_1 + (c_2 - c_3 + c_5)a_2 + c_5 a_4,$$

which is a linear combination of a_1, a_2, a_4.

(ii) $B = \begin{bmatrix} 2 & 1 & 3 & 0 & 3 \\ 0 & -1 & 1 & 1 & 0 \\ 2 & 0 & 4 & 2 & 4 \\ -2 & -2 & -2 & 1 & -3 \end{bmatrix} = [b_1\ b_2\ b_3\ b_4\ b_5],$

where b_i's denote the columns of B.

Calculation shows that the row echelon form of B is the matrix A from the part (i) just discussed. It turns out that the same conclusions as for A hold for B: b_1, b_2, b_4 form a basis of $C(B)$, while $b_5 = b_1 + b_2 + b_4$ and $b_3 = 2b_1 - b_2$, similarly to (5.2) and (5.3). Indeed, to see that b_1, b_2, b_4 are linearly independent, one forms the matrix $[b_1\ b_2\ b_4]$ and row reduces it to the matrix $[a_1\ a_2\ a_4]$ with three pivots. To express b_5 through b_1, b_2, b_4, one forms the augmented matrix $[b_1\ b_2\ b_4 : b_5]$ and row reduces it to the matrix $[a_1\ a_2\ a_4 : a_5]$, which leads to $b_5 = b_1 + b_2 + b_4$. Similar reasoning shows that in any matrix, columns with pivots form a basis of the column space.

Caution: $C(B)$ is not the same as $C(A)$. Indeed, vectors in $C(A)$ have the last component equal to zero, while vectors in $C(B)$ do not.

We summarize

To obtain a basis for the column space $C(B)$, reduce B to its row echelon form. Then the columns with pivots (from the original matrix B) form a basis for $C(B)$. Other columns are expressed through the pivot ones by forming the corresponding augmented matrices, and performing Gaussian elimination. The dimension of $C(B)$, or rank B, is equal to the number of pivot columns.

Recall that the dimension of the null space $N(B)$ is equal to the number of columns without pivots (or the number of free variables). The sum of the dimensions of the column space and of the null space is equal to the total number of columns, which for an $m \times n$ matrix B reads:

$$\operatorname{rank} B + \dim N(B) = n,$$

and is known as *the rank theorem*.

Exercises

1. Find the null space of the given matrix. Identify its basis and dimension.
    (a) $\begin{bmatrix} 1 & 2 \\ 3 & 4 \end{bmatrix}$.
        Answer: The zero subspace of R^2, of dimension 0.
    (b) $A = \begin{bmatrix} 1 & -2 \\ 3 & -6 \end{bmatrix}$.
        Answer: $N(A)$ is spanned by $\begin{bmatrix} 2 \\ 1 \end{bmatrix}$, and its dimension is 1.
    (c) $O = \begin{bmatrix} 0 & 0 \\ 0 & 0 \end{bmatrix}$.
        Answer: $N(O) = R^2$, dimension is 2.
    (d) $\begin{bmatrix} 0 & 1 & -2 \\ 4 & 3 & -6 \\ -4 & -2 & 4 \end{bmatrix}$.
    (e) $E = \begin{bmatrix} 1 & -1 & -2 \\ 2 & -2 & -4 \\ 3 & -3 & -6 \end{bmatrix}$.
        Answer: $N(E) = \operatorname{Span} \left\{ \begin{bmatrix} 2 \\ 0 \\ 1 \end{bmatrix}, \begin{bmatrix} 1 \\ 1 \\ 0 \end{bmatrix} \right\}$, dimension is 2.
    (f) $F = \begin{bmatrix} 1 & 0 & 0 \\ 2 & -2 & 0 \\ 3 & -3 & -6 \end{bmatrix}$.
        Answer: $N(F) = \{0\}$, the zero subspace, of dimension zero.
    (g) $\begin{bmatrix} 2 & 1 & 3 & 0 \\ 2 & 0 & 4 & 1 \\ -2 & -1 & -3 & 1 \end{bmatrix}$.
        Answer: $N(A)$ is spanned by $\begin{bmatrix} -2 \\ 1 \\ 1 \\ 0 \end{bmatrix}$, and $\dim N(A) = 1$.
    (h) $\begin{bmatrix} 2 & 1 & 3 & 0 \\ 2 & 0 & 4 & 1 \\ -2 & -2 & -2 & 1 \end{bmatrix}$.
    (i) $H = \begin{bmatrix} -1 & 1 & 3 & 0 \end{bmatrix}$. Hint: The null space is a subspace of R^4.
        Answer: $N(H) = \operatorname{Span} \left\{ \begin{bmatrix} 1 \\ 1 \\ 0 \\ 0 \end{bmatrix}, \begin{bmatrix} 3 \\ 0 \\ 1 \\ 0 \end{bmatrix}, \begin{bmatrix} 0 \\ 0 \\ 0 \\ 1 \end{bmatrix} \right\}$, its dimension is 3.
2. A 4×5 matrix has two pivots. What is the dimension of its null space?
3. The rank of a 9×7 matrix is 3. What is the dimension of its null space? What is the number of pivots?
4. The rank of a 4×4 matrix is 4.

(a) Describe the null space.

(b) Describe the column space.

5. The rank of a 3×3 matrix is 2. Explain why its null space is a line through the origin, while its column space is a plane through the origin.

6. Assume that matrix A is of size 3×5. Explain why $\dim N(A) \geq 2$.

7. For a 4×4 matrix A the dimension of $N(A)$ is 4. Describe A.
   Answer: $A = O$.

8. Find the basis of the column space for the following matrices, and determine their rank. Express the columns that are not in the basis through those in the basis.

(a) $\begin{bmatrix} -1 & 1 & -1 \\ 0 & 2 & 4 \end{bmatrix}$.

(b) $\begin{bmatrix} -1 & 1 & -1 \\ 1 & 2 & 10 \end{bmatrix}$.
   Answer: $C_3 = 4C_1 + 3C_2$, rank $= 2$.

(c) $\begin{bmatrix} 1 & 1 & 2 \\ -3 & -3 & -6 \end{bmatrix}$.
   Answer: rank $= 1$.

(d) $A = \begin{bmatrix} -1 & 2 & 5 \\ -1 & 2 & 5 \\ 2 & 0 & -2 \end{bmatrix}$.

   Answer: $C(A) = \text{Span}\left\{ \begin{bmatrix} -1 \\ -1 \\ 2 \end{bmatrix}, \begin{bmatrix} 2 \\ 2 \\ 0 \end{bmatrix} \right\}$, rank $= 2$, and $C_3 = -C_1 + 2C_2$.

(e) $A = \begin{bmatrix} 0 & 0 & 1 \\ 0 & 2 & 5 \\ -1 & 0 & -3 \end{bmatrix}$.
   Answer: $C(A) = R^3$.

(f) $\begin{bmatrix} 2 & 1 & 3 & 0 \\ 2 & 0 & 4 & 1 \\ -2 & -1 & -3 & 1 \end{bmatrix}$.
   Answer: Column space is spanned by C_1, C_2 and C_4, rank is 3, and $C_3 = 2C_1 - C_2$.

(g) $B = \begin{bmatrix} 1 & -1 & 0 & 1 & 1 \\ 2 & -1 & 1 & 1 & -3 \\ 0 & 1 & 1 & -1 & -5 \end{bmatrix}$.

   Answer: $C(B) = \text{Span}\left\{ \begin{bmatrix} 1 \\ 2 \\ 0 \end{bmatrix}, \begin{bmatrix} -1 \\ -1 \\ 1 \end{bmatrix} \right\}$, rank $= 2$, and $C_3 = C_1 + C_2$, $C_4 = -C_2$, $C_5 = -4C_1 - 5C_2$.

9. Consider the following subspace of R^3:

$$V = \text{Span}\left\{ \begin{bmatrix} 2 \\ 0 \\ 1 \end{bmatrix}, \begin{bmatrix} 1 \\ 1 \\ 0 \end{bmatrix}, \begin{bmatrix} -2 \\ -4 \\ -6 \end{bmatrix} \right\}.$$

Find a basis of V and dim V.

Hint. Use these vectors as columns of a matrix.

10. Let $A = \begin{bmatrix} -1 & -1 \\ 1 & 1 \end{bmatrix}$.

(a) Show that the vector $\begin{bmatrix} 1 \\ -1 \end{bmatrix}$ belongs to both the null space $N(A)$ and the column space $C(A)$.

(b) Show that $N(A) = C(A)$.

(c) Show that $N(A^2) = R^2$.

11. Let A be an arbitrary $n \times n$ matrix.

(a) Show that any vector in $N(A)$ belongs to $N(A^2)$.

(b) Show that the converse statement is false.

Hint. Try $A = \begin{bmatrix} -1 & -1 \\ 1 & 1 \end{bmatrix}$.

12. Let A be an $m \times n$ matrix with linearly independent columns.

(a) Show that the system $Ax = b$ has at most one solution for any vector b.

Hint. If $C_1, C_2, \ldots, C_n$ are the columns of A, and $x_1, x_2, \ldots, x_n$ are the components of x, then $x_1 C_1 + x_2 C_2 + \cdots + x_n C_n = b$.

(b) Suppose that $b \in C(A)$. Show that the system $Ax = b$ has exactly one solution.

3 Determinants

A 4×4 matrix involves 16 numbers. Its determinant is just one number, but it carries significant information about the matrix.

3.1 Cofactor expansion

To each *square matrix A*, one associates a number called *the determinant of A*, and denoted by either $\det A$ or $|A|$. For 2×2 matrices,

$$\begin{vmatrix} a & b \\ c & d \end{vmatrix} = ad - bc.$$

For 3×3 matrices, the formula is

$$\begin{vmatrix} a_{11} & a_{12} & a_{13} \\ a_{21} & a_{22} & a_{23} \\ a_{31} & a_{32} & a_{33} \end{vmatrix} = a_{11}a_{22}a_{33} + a_{12}a_{23}a_{31} + a_{13}a_{21}a_{32} - a_{11}a_{23}a_{32} - a_{12}a_{21}a_{33} - a_{13}a_{22}a_{31}.$$

$$(1.1)$$

It seems impossible to memorize this formula, but we shall learn how to produce it.

For an $n \times n$ matrix A, define *the minor M_{ij}* as the $(n-1) \times (n-1)$ determinant obtained by removing row i and column j in A. For example, for the matrix

$$A = \begin{bmatrix} 1 & 0 & -3 \\ -1 & 6 & 2 \\ 3 & 2 & 1 \end{bmatrix},$$

the minors are $M_{11} = \begin{vmatrix} 6 & 2 \\ 2 & 1 \end{vmatrix} = 2$, $M_{12} = \begin{vmatrix} -1 & 2 \\ 3 & 1 \end{vmatrix} = -7$, $M_{13} = \begin{vmatrix} -1 & 6 \\ 3 & 2 \end{vmatrix} = -20$, and so on. Define also *the cofactor*

$$C_{ij} = (-1)^{i+j} M_{ij}.$$

For the above matrix, $C_{11} = (-1)^{1+1} M_{11} = 2$, $C_{12} = (-1)^{1+2} M_{12} = 7$, $C_{13} = (-1)^{1+3} M_{13} = -20$, and so on.

Cofactor expansion will allow us to define 3×3 determinants through 2×2 ones, then 4×4 determinants through 3×3 ones, and so on. For an $n \times n$ matrix, the cofactor expansion in row i is

$$|A| = a_{i1}C_{i1} + a_{i2}C_{i2} + \cdots + a_{in}C_{in}.$$

The cofactor expansion in column j is

https://doi.org/10.1515/9783111086507-003

$$|A| = a_{1j}C_{1j} + a_{2j}C_{2j} + \cdots + a_{nj}C_{nj}.$$

For 3×3 determinants, there are 6 cofactor expansions (in 3 rows, and in 3 columns), but all of them lead to the same formula (1.1). Similarly, for $n \times n$ determinants all cofactor expansions lead to the same number, $|A|$. For the above matrix, cofactor expansion in the first row gives

$$|A| = 1 \cdot C_{11} + 0 \cdot C_{12} + (-3) \cdot C_{13} = 62.$$

In practice one does not calculate $(-1)^{i+j}$, but uses *the checker-board pattern*

$$\begin{bmatrix} + & - & + \\ - & + & - \\ + & - & + \end{bmatrix}$$

to get the right signs of the cofactors (and similarly for larger matrices). Let us expand the same determinant in the second row:

$$\begin{vmatrix} 1 & 0 & -3 \\ -1 & 6 & 2 \\ 3 & 2 & 1 \end{vmatrix} = -(-1)\begin{vmatrix} 0 & -3 \\ 2 & 1 \end{vmatrix} + 6\begin{vmatrix} 1 & -3 \\ 3 & 1 \end{vmatrix} - 2\begin{vmatrix} 1 & 0 \\ 3 & 2 \end{vmatrix} = 62.$$

One tries to pick a row (or column) with many zeroes to perform a cofactor expansion. Indeed, if $a_{ij} = 0$, there is no need to calculate C_{ij}, because $a_{ij}C_{ij} = 0$ anyway. If all entries of some row are zero, then $|A| = 0$.

Example. Expanding in the first column gives

$$\begin{vmatrix} 2 & 0 & 3 & -4 \\ 0 & 3 & 8 & 1 \\ 0 & 0 & 4 & -2 \\ 0 & 0 & 0 & 5 \end{vmatrix} = 2 \cdot \begin{vmatrix} 3 & 8 & 1 \\ 0 & 4 & -2 \\ 0 & 0 & 5 \end{vmatrix} = 2 \cdot 3 \cdot \begin{vmatrix} 4 & -2 \\ 0 & 5 \end{vmatrix} = 2 \cdot 3 \cdot 4 \cdot 5 = 120.$$

(The 3×3 determinant on the second step was also expanded in the first column.)

The matrix in the last example was upper triangular. Similar reasoning shows that *the determinant of any upper triangular matrix equals to the product of its diagonal entries*. For a lower triangular matrix, like

$$\begin{vmatrix} 2 & 0 & 0 & 0 \\ 12 & -3 & 0 & 0 \\ 2 & \frac{1}{3} & 4 & 0 \\ -1 & 2 & 7 & 0 \end{vmatrix} = 2 \cdot (-3) \cdot 4 \cdot 0 = 0,$$

the expansion was performed in the first row on each step. In general, *the determinant of any lower triangular matrix equals to the product of its diagonal entries*. Diagonal matrices can be viewed as either upper triangular or lower triangular. Therefore, *the determinant of any diagonal matrix equals to the product of its diagonal entries.* For example, if I is the $n \times n$ identity matrix, then

$$|-2I| = (-2) \cdot (-2) \cdots (-2) = (-2)^n.$$

Cofactor expansions are not practical for computing $n \times n$ determinants for $n \geq 5$. Let us count the number of multiplications such an expansion takes. For a 2×2 matrix, it takes 2 multiplications. For a 3×3 matrix, one needs to calculate three 2×2 determinants, which takes $3 \cdot 2 = 3!$ multiplications, plus 3 more multiplications in the cofactor expansion, for a total of $3! + 3$. For an $n \times n$ matrix, it takes $n! + n$ multiplications. If $n = 20$, this number is 2432902008176640020, and computations would take many thousands of years on the fastest computers. An efficient way for computing determinants, based on Gaussian elimination, is developed in the next section.

Exercises

1. Find x so that $\left|\begin{smallmatrix} x & 3 \\ -1 & 2 \end{smallmatrix}\right| = \left|\begin{smallmatrix} 0 & x \\ 1 & 5 \end{smallmatrix}\right|$.

    Answer: $x = -1$.

2. Let $A = \left[\begin{smallmatrix} 1 & 1 & -1 \\ -1 & 1 & 2 \\ 0 & 2 & 3 \end{smallmatrix}\right]$. Calculate $\det A$

    (a) By expanding in the second row.
    (b) By expanding in the second column.
    (c) By expanding in the third row.

    Answer: $|A| = 4$.

3. Calculate the determinants of the following matrices:

    (a) $\left[\begin{smallmatrix} 1 & 0 & 0 \\ 0 & 2 & 0 \\ 0 & 0 & 3 \end{smallmatrix}\right]$.

    Answer: $3!$

    (b) $\left[\begin{smallmatrix} 1 & 0 & 0 & 0 \\ 0 & -2 & 0 & 0 \\ 0 & 0 & -3 & 0 \\ 0 & 0 & 0 & -4 \end{smallmatrix}\right]$.

    Answer: $-4!$

    (c) Any diagonal matrix.

    (d) $\left[\begin{smallmatrix} 1 & 0 \\ -5 & 2 \end{smallmatrix}\right]$.

    (e) $\left[\begin{smallmatrix} 1 & 0 & 0 \\ -5 & 2 & 0 \\ 6 & 12 & 3 \end{smallmatrix}\right]$.

    Answer: 6.

    (f) Any lower triangular matrix.
    (g) Any upper triangular matrix.

    (h) $\left[\begin{smallmatrix} 0 & 0 & a \\ 0 & b & 5 \\ c & -2 & 3 \end{smallmatrix}\right]$.

    Answer: $-abc$.

(i) $\begin{bmatrix} 1 & -1 & 0 & 3 \\ 0 & 2 & -2 & 1 \\ -1 & -2 & 0 & 2 \\ 1 & 1 & 1 & 2 \end{bmatrix}$.

Answer: -27.

(j) $\begin{bmatrix} 2 & 0 & 0 & 0 \\ 0 & a & b & 0 \\ 0 & c & d & 0 \\ 0 & 0 & 0 & 3 \end{bmatrix}$. (A block diagonal matrix.)

Answer: $2 \cdot \left| \begin{smallmatrix} a & b \\ c & d \end{smallmatrix} \right| \cdot 3 = 6\,(ad - bc)$.

(k) $\begin{vmatrix} a & b & 0 & 0 \\ c & d & 0 & 0 \\ 0 & 0 & e & f \\ 0 & 0 & g & h \end{vmatrix}$. (A block diagonal matrix.)

Answer: $\left| \begin{smallmatrix} a & b \\ c & d \end{smallmatrix} \right| \cdot \left| \begin{smallmatrix} e & f \\ g & h \end{smallmatrix} \right| = (ad - bc)\,(eh - fg)$.

(l) $\begin{vmatrix} 2 & -1 & 0 & 5 \\ 4 & -2 & 0 & -3 \\ 1 & 3 & 0 & 1 \\ 0 & -7 & 0 & 8 \end{vmatrix}$.

Answer: 0.

(m) A matrix with a row of zeroes.

Answer: The determinant is 0.

4. Calculate $|A^2|$ and relate it to $|A|$ for the following matrices:

   (a) $A = \begin{bmatrix} 2 & -4 \\ 0 & 3 \end{bmatrix}$.

   (b) $A = \begin{bmatrix} 1 & -1 \\ 1 & 1 \end{bmatrix}$.

5. Let $A = \begin{bmatrix} 0 & 0 & \cdots & 0 & 1 \\ 0 & 1 & \cdots & 0 & 0 \\ \vdots & \vdots & \ddots & \vdots & \vdots \\ 0 & 0 & \cdots & 1 & 0 \\ 1 & 0 & \cdots & 0 & 0 \end{bmatrix}$, an $n \times n$ matrix. Show that $|A| = -1$.

   Hint. Expand in the first row, then expand in the last row.

6. Calculate the $n \times n$ determinant $D_n = \begin{vmatrix} 2 & 1 & 0 & \cdots & 0 & 0 \\ 1 & 2 & 1 & \cdots & 0 & 0 \\ 0 & 1 & 2 & \cdots & 0 & 0 \\ \vdots & \vdots & \vdots & \ddots & \vdots & \vdots \\ 0 & 0 & 0 & \cdots & 2 & 1 \\ 0 & 0 & 0 & \cdots & 1 & 2 \end{vmatrix}$.

   Hint. Expanding in the first row, obtain the recurrence relation $D_n = 2D_{n-1} - D_{n-2}$. Beginning with $D_2 = 3$ and $D_3 = 4$, use this recurrence relation to calculate $D_4 = 5$ and $D_5 = 6$, and so on.

   Answer: $D_n = n + 1$.

7. Let A be a 5×5 matrix, with $a_{ij} = (i - 3)j$. Show that $|A| = 0$.

   Hint. What is the third row of A?

8. Suppose that a square matrix has integer entries. Show that its determinant is an integer. Prove that the converse statement is not true, by considering, for example,

   $\begin{vmatrix} \frac{3}{2} & \frac{1}{2} \\ \frac{5}{2} & -\frac{1}{2} \end{vmatrix}$.

3.2 Properties of determinants

An $n \times n$ matrix A can be listed by its rows $A = \begin{bmatrix} R_1 \\ R_2 \\ \vdots \\ R_n \end{bmatrix}$, which are n-dimensional row

vectors. Let us highlight R_i (row i) in A:

$$A = \begin{bmatrix} a_{11} & a_{12} & \cdots & a_{1n} \\ \vdots & \vdots & & \vdots \\ a_{i1} & a_{i2} & \cdots & a_{in} \\ \vdots & \vdots & & \vdots \\ a_{n1} & a_{n2} & \cdots & a_{nn} \end{bmatrix}.$$

Using the summation notation, the cofactor expansion in row i takes the form

$$|A| = a_{i1}C_{i1} + a_{i2}C_{i2} + \cdots + a_{in}C_{in} = \sum_{s=1}^{n} a_{is}C_{is}.$$

The first three properties deal with the elementary row operations.

Property 1. If some row of A is multiplied by a number k to produce B, then $\det B = k \det A$.

Indeed, assume that row i of A is multiplied by k. We need to show that

$$|B| = \begin{vmatrix} R_1 \\ \vdots \\ kR_i \\ \vdots \\ R_n \end{vmatrix} = k \begin{vmatrix} R_1 \\ \vdots \\ R_i \\ \vdots \\ R_n \end{vmatrix} = k|A|. \tag{2.1}$$

Expand $|B|$ in row i, and use the summation notation to obtain

$$|B| = \sum_{s=1}^{n} (ka_{is})\, C_{is} = k \sum_{s=1}^{n} a_{is}C_{is} = k|A|,$$

justifying Property 1. (In row i, cofactors are the same for B and A, since row i is removed in both matrices when calculating cofactors.) In (2.1), the number k is "factored out" of row i.

If $B = kA$, then all n rows of A are multiplied by k to produce B. It follows that $\det B = k^n \det A$ (by factoring k out of each row), or

$$|kA| = k^n|A|.$$

Property 2. If any two rows of A are interchanged to produce B, then $\det B = -\det A$.

Indeed, for 2×2 matrices this property is immediately verified. Suppose that A is a 3×3 matrix, $A = \begin{bmatrix} R_1 \\ R_2 \\ R_3 \end{bmatrix}$, and $B = \begin{bmatrix} R_3 \\ R_2 \\ R_1 \end{bmatrix}$ is obtained from A by switching rows 1 and 3. Expand both $|B|$ and $|A|$ in the second row. In the expansion of $|B|$, one will encounter

2×2 determinants with the rows switched, compared with the expansion of $|A|$, giving $|B| = -|A|$. Then one justifies this property for 4×4 matrices, and so on.

It follows that if a matrix has two identical rows, its determinant is zero. Indeed, interchange the identical rows, to get a matrix B. By Property 2, $|B| = -|A|$. On the other hand, $B = A$, so that $|B| = |A|$. It follows that $|A| = -|A|$, giving $|A| = 0$. If two rows are proportional, the determinant is again zero. For example, using Property 1,

$$\begin{vmatrix} R_1 \\ kR_1 \\ R_3 \end{vmatrix} = k \begin{vmatrix} R_1 \\ R_1 \\ R_3 \end{vmatrix} = 0.$$

Assume that row j in A is replaced by R_i, so that $R_j = R_i$. The resulting matrix has zero determinant,

$$\begin{vmatrix} R_1 \\ \vdots \\ R_i \\ \vdots \\ R_i \\ \vdots \\ R_n \end{vmatrix} = 0.$$

Let us expand this determinant in the jth row,

$$a_{i1}C_{j1} + a_{i2}C_{j2} + \cdots + a_{in}C_{jn} = 0.$$

(Once row j is removed, the cofactors are the same as in matrix A.) Comparing that with the cofactor expansion of $|A|$ in row i,

$$a_{i1}C_{i1} + a_{i2}C_{i2} + \cdots + a_{in}C_{in} = |A|,$$

we conclude the following theorem.

Theorem 3.2.1. *If all elements of row i are multiplied by the cofactors of another row j and added, the result is zero. If all elements of row i are multiplied by their own cofactors and added, the result is $|A|$. In short,*

$$\sum_{s=1}^{n} a_{is}C_{js} = \begin{cases} 0 & \text{if } j \neq i, \\ |A| & \text{if } j = i. \end{cases}$$

Property 3. If a multiple of one row of A is added to another row to produce a matrix B, then $\det B = \det A$. (In other words, elementary operations of type $R_j + kR_i$ leave the value of the determinant unchanged.)

Indeed, assume that B was obtained from A by using $R_j + kR_i$. Expand $|B|$ in row j, use the summation convention and the preceding Theorem 3.2.1:

$$|B| = \begin{vmatrix} R_1 \\ \vdots \\ R_j + kR_i \\ \vdots \\ R_n \end{vmatrix} = \sum_{s=1}^{n} (a_{js} + ka_{is})\, C_{js} = \sum_{s=1}^{n} a_{js} C_{js} + k \sum_{s=1}^{n} a_{is} C_{js} = |A|.$$

Using Properties 1, 2, 3, one row reduces any determinant to that of an upper triangular matrix (which is the product if its diagonal entries). This method (based on Gaussian elimination) is very efficient, allowing computation of 20×20 determinants on basic laptops. (Entering a 20×20 determinant is likely to take longer than its computation.)

Example. To evaluate the following 4×4 determinant, perform $R_1 \leftrightarrow R_2$, and then factor 2 out of the (new) first row to obtain

$$\begin{vmatrix} 0 & 1 & 2 & 3 \\ 2 & -2 & 0 & -6 \\ 1 & 1 & 0 & 1 \\ 2 & -2 & 4 & 4 \end{vmatrix} = - \begin{vmatrix} 2 & -2 & 0 & -6 \\ 0 & 1 & 2 & 3 \\ 1 & 1 & 0 & 1 \\ 2 & -2 & 4 & 4 \end{vmatrix} = -2 \begin{vmatrix} 1 & -1 & 0 & -3 \\ 0 & 1 & 2 & 3 \\ 1 & 1 & 0 & 1 \\ 2 & -2 & 4 & 4 \end{vmatrix}.$$

Performing $R_3 - R_1$, $R_4 - 2R_1$ for the resulting determinant (dropping the factor of -2, for now), followed by $R_3 - 2R_2$, and finally $R_4 + R_3$, gives

$$\begin{vmatrix} 1 & -1 & 0 & -3 \\ 0 & 1 & 2 & 3 \\ 1 & 1 & 0 & 1 \\ 2 & -2 & 4 & 4 \end{vmatrix} = \begin{vmatrix} 1 & -1 & 0 & -3 \\ 0 & 1 & 2 & 3 \\ 0 & 2 & 0 & 4 \\ 0 & 0 & 4 & 10 \end{vmatrix} = \begin{vmatrix} 1 & -1 & 0 & -3 \\ 0 & 1 & 2 & 3 \\ 0 & 0 & -4 & -2 \\ 0 & 0 & 4 & 10 \end{vmatrix}$$

$$= \begin{vmatrix} 1 & -1 & 0 & -3 \\ 0 & 1 & 2 & 3 \\ 0 & 0 & -4 & -2 \\ 0 & 0 & 0 & 8 \end{vmatrix} = 1 \cdot 1 \cdot (-4) \cdot 8 = -32.$$

The original determinant is then $(-2) \cdot (-32) = 64$.

In practice one combines row reduction with cofactor expansion. For example, after performing $R_2 + R_1$ and $R_3 - R_1$,

$$\begin{vmatrix} 1 & 0 & 2 \\ -1 & 1 & -1 \\ 1 & 1 & 5 \end{vmatrix} = \begin{vmatrix} 1 & 0 & 2 \\ 0 & 1 & 1 \\ 0 & 1 & 3 \end{vmatrix} = 1 \cdot \begin{vmatrix} 1 & 1 \\ 1 & 3 \end{vmatrix} = 2,$$

the determinant is evaluated by expanding in the first column.

If Gaussian elimination for A does not involve row exchanges, $|A|$ is equal to the product of the diagonal entries in the resulting upper triangular matrix, otherwise $|A|$ is $\pm$ the product of the diagonal entries in the row echelon form. It follows that $|A| \neq 0$ is equivalent to all of these diagonal entries being nonzero, so that A has n pivots, which in turn is equivalent to A being invertible. We conclude that A *is invertible if and only if* $|A| \neq 0$.

Determinants of elementary matrices are easy to calculate. Indeed, $|E_i(k)| = k$ (a diagonal matrix), $|E_{ij}| = -1$ (by Property 2), and $|E_{ij}(k)| = 1$ (a lower triangular matrix). We can then restate Property 1 as

$$\left|E_i(k)A\right| = k|A| = \left|E_i(k)\right|\left|A\right|,$$

Property 2 as

$$\left|E_{ij}A\right| = -|A| = \left|E_{ij}\right|\left|A\right|,$$

and Property 3 as

$$\left|E_{ij}(k)A\right| = |A| = \left|E_{ij}(k)\right|\left|A\right|.$$

Summarizing,

$$|EA| = |E||A|, \tag{2.2}$$

where E is an elementary matrix of any kind.

Property 4. For any two $n \times n$ matrices,

$$|AB| = |A||B|. \tag{2.3}$$

Proof. Case (i) $|A| = 0$. Then A is not invertible. We claim that AB is also not invertible. Indeed, if the inverse $(AB)^{-1}$ existed, we would have $AB(AB)^{-1} = I$, which means that $B(AB)^{-1}$ is the inverse of A, but A has no inverse. Since AB is not invertible, $|AB| = 0$, and (2.3) holds.

Case (ii) $|A| \neq 0$. By Theorem 2.3.1, a nonsingular matrix A can be written as a product of elementary matrices (of various kinds)

$$A = E_1 E_2 \cdots E_p.$$

Applying (2.2) to products of two matrices at a time,

$$|A| = |E_1|\,|E_2 \cdots E_p| = |E_1|\,|E_2| \cdots |E_p|. \tag{2.4}$$

Similarly,

$$|AB| = |E_1E_2 \cdots E_p B| = |E_1| \, |E_2 \cdots E_p \, B| = |E_1| \, |E_2| \cdots |E_p| \, |B| = |A| \, |B|,$$

using (2.4) on the last step. □

Recall that powers of a square matrix A are defined as follows: $A^2 = AA$, $A^3 = A^2 A$, etc. Then $|A^2| = |A| \, |A| = |A|^2$, and in general

$$|A^k| = |A|^k, \quad \text{for any positive integer } k.$$

Property 5. If A is invertible, then $|A| \neq 0$, and

$$|A^{-1}| = \frac{1}{|A|}.$$

Indeed,

$$
\begin{aligned}
|AA^{-1}| &= |I| = 1, \\
|A||A^{-1}| &= 1,
\end{aligned}
\tag{2.5}
$$

by Property 4. Then $|A| \neq 0$, and $|A^{-1}| = \frac{1}{|A|}$.

We conclude again that in case $|A| = 0$, the matrix A is not invertible (existence of A^{-1} would produce a contradiction in (2.5)).

Property 6. $|A^T| = |A|$.

Indeed, the transpose A^T has the rows and columns of A interchanged, while cofactor expansion works equally well for rows and columns.

The last property implies that all of the facts stated above for rows are also true for columns. For example, if two columns of A are proportional, then $|A| = 0$. If a multiple of column i is subtracted from column j, the determinant remains unchanged. If a column of A is the zero vector, then $|A| = 0$.

Exercises

1. Calculate the following determinants by combining row reduction and cofactor expansion:

(a) $\begin{vmatrix} 1 & 2 & 0 \\ 3 & -1 & 1 \\ 1 & -2 & 1 \end{vmatrix}$.

(b) $\begin{vmatrix} 0 & -2 & 3 \\ 3 & -1 & 1 \\ 1 & -1 & 1 \end{vmatrix}$.

Hint. Perform $R_1 \leftrightarrow R_3$.

(c) $\begin{vmatrix} 0 & -2 & 3 & 1 \\ -1 & -1 & 1 & 0 \\ 2 & -1 & -1 & 2 \\ 1 & -4 & 3 & 3 \end{vmatrix}$.

Answer: 0.

(d) $\begin{vmatrix} 1 & 0 & -1 & 1 \\ 1 & 1 & 2 & -1 \\ 0 & 1 & 2 & 3 \\ 2 & 1 & -2 & 3 \end{vmatrix}$.

Answer: 12.

(e) $\begin{vmatrix} 1 & 1 & -1 & 1 \\ 1 & 1 & 2 & -1 \\ -1 & -1 & 2 & 3 \\ 2 & 1 & -2 & 3 \end{vmatrix}$.

Answer: −14.

(f) $\begin{vmatrix} 1 & 1 & -1 & 1 \\ 1 & 2 & 2 & -1 \\ -1 & -2 & 2 & 3 \\ 2 & 1 & -2 & 3 \end{vmatrix}$.

Answer: −10.

(g) $\begin{vmatrix} 1 & 1 & 1 \\ a & b & c \\ a^2 & b^2 & c^2 \end{vmatrix}$. (*Vandermonde determinant*)

Hint. Perform $R_2 - aR_1$, $R_3 - a^2R_1$, then expand in the first column.

Answer: $(b - a)(c - a)(c - b)$.

2. Assuming that $\begin{vmatrix} a & b & c \\ d & e & f \\ g & h & k \end{vmatrix} = 5$, find the following determinants:

(a) $\begin{vmatrix} a & b & c \\ d+3a & e+3b & f+3c \\ g & h & k \end{vmatrix}$.

Answer: 5.

(b) $\begin{vmatrix} a & b & c \\ 2d & 2e & 2f \\ g & h & k \end{vmatrix}$.

Answer: 10.

(c) $\begin{vmatrix} 3a & 3b & 3c \\ 2d & 2e & 2f \\ g & h & k \end{vmatrix}$.

Answer: 30.

(d) $\begin{vmatrix} a & b & c \\ 2d+3a & 2e+3b & 2f+3c \\ g & h & k \end{vmatrix}$.

(e) $\begin{vmatrix} d & e & f \\ a & b & c \\ g & h & k \end{vmatrix}$.

Answer: −5.

(f) $\begin{vmatrix} d & e & f \\ g & h & k \\ a & b & c \end{vmatrix}$.

Answer: 5.

(g) $\begin{vmatrix} a & b & -c \\ d & e & -f \\ g & h & -k \end{vmatrix}$.

Answer: −5.

(h) $\begin{vmatrix} a & b & 0 \\ d & e & 0 \\ g & h & 0 \end{vmatrix}$.

3. (a) If every column of A adds to zero, show that $|A| = 0$.
   (b) If every row of A adds to zero, what is $|A|$?

4. Let A and B be 4×4 matrices such that $|A| = 3$ and $|B| = \frac{1}{2}$. Find the following determinants:

(a) $|A^T|$.

(b) $|2A|$.

Answer: 48.

(c) $|B^2|$.

(d) $|BA|$.

(e) $|A^{-1}B|$.

Answer: $\frac{1}{6}$.

(f) $|2AB^{-1}|$

Answer: 96.

(g) $|A^2(-B)^T|$.

Answer: $\frac{9}{2}$.

5. Let A be a 7×7 matrix such that $|-A| = |A|$. Show that $|A| = 0$.

6. True or false?

(a) $|BA| = |AB|$.

(b) $|-A| = |A|$.

Answer: False.

(c) If A^3 is invertible, then $|A| \neq 0$.

Answer: True.

(d) $|A + B| = |A| + |B|$.

Answer: False.

(e) $|(A^2)^{-1}| = |(A^{-1})^2| = \frac{1}{|A|^2}$, provided that $|A| \neq 0$.

Answer: True.

7. Show that

$$\begin{vmatrix} 1 & 1 & 1 \\ x & a & c \\ y & b & d \end{vmatrix} = 0$$

is an equation of the straight line through the points (a, b) and (c, d) in the xy-plane.

Hint. The graph of a linear equation is a straight line.

8. Show that

$$\begin{vmatrix} 1 & 1 & 1 & 1 \\ x & a_1 & b_1 & c_1 \\ y & a_2 & b_2 & c_2 \\ z & a_3 & b_3 & c_3 \end{vmatrix} = 0$$

is an equation of the plane passing through the points (a_1, a_2, a_3), (b_1, b_2, b_3) and (c_1, c_2, c_3).

Hint. Expanding in the first column, obtain a linear equation in x, y, z.

9. Let $A = \begin{bmatrix} 1 & 2 & 0 \\ 0 & -1 & 1 \\ 1 & -2 & 1 \end{bmatrix}$ and $B = \begin{bmatrix} 1 & -2 & 1 \\ 2 & -4 & 2 \\ 1 & -3 & 1 \end{bmatrix}$. Calculate $\det\left(A^3 B\right)$.

Hint. What is $\det B$?

10. Calculate the $n \times n$ determinant $\begin{vmatrix} 1 & 1 & 1 & \cdots & 1 & 1 \\ 2 & 3 & 2 & \cdots & 2 & 2 \\ 2 & 2 & 4 & \cdots & 2 & 2 \\ \vdots & \vdots & \vdots & \ddots & \vdots & \vdots \\ 2 & 2 & 2 & \cdots & n & 2 \\ 2 & 2 & 2 & \cdots & 2 & n+1 \end{vmatrix}$.

Hint. Apply $R_2 - 2R_1$, $R_3 - 2R_1$, and so on.

Answer: $(n - 1)!$.

11. Let A be an $n \times n$ matrix, and the matrix B is obtained by writing the rows of A in the reverse order. Show that $|B| = (-1)^{\frac{n(n-1)}{2}} |A|$.

Hint. $1 + 2 + 3 + \cdots + (n - 1) = \frac{n(n-1)}{2}$.

12. Let A be an $n \times n$ *skew-symmetric matrix*, defined by the relation $A^T = -A$.

    (a) Show that $a_{ij} = -a_{ji}$.

    (b) Show that all diagonal entries are zero ($a_{ii} = 0$ for all i).

    (c) Let n be odd. Show that $|A| = 0$.

13. Let A be an $n \times n$ matrix, with $a_{ij} = \min(i, j)$.

    (a) If $n = 4$, show that $A = \begin{bmatrix} 1 & 1 & 1 & 1 \\ 1 & 2 & 2 & 2 \\ 1 & 2 & 3 & 3 \\ 1 & 2 & 3 & 4 \end{bmatrix}$ and find its determinant.

    (b) Show that $|A| = 1$ for any n.

    Hint. From column n subtract column $(n-1)$, then from column $(n-1)$ subtract column $(n-2)$, and so on.

14. Let n be odd. Show that there is no $n \times n$ matrix A with real entries, such that $A^2 = -I$.

15. If the rows of A (or the columns of A) are linearly dependent, show that $|A| = 0$.

    Hint. One of the rows is a linear combination of the others. Use elementary operations to produce a row of zeros.

3.3 Cramer's rule

Determinants provide an alternative way for calculation of inverse matrices, and for solving linear systems with a square matrix.

Let

$$A = \begin{bmatrix} a_{11} & a_{12} & \cdots & a_{1n} \\ a_{21} & a_{22} & \cdots & a_{2n} \\ \vdots & \vdots & \ddots & \vdots \\ a_{n1} & a_{n2} & \cdots & a_{nn} \end{bmatrix} \tag{3.1}$$

be an $n \times n$ matrix, with $|A| \neq 0$. Form *the adjugate matrix*

$$\text{Adj}\, A = \begin{bmatrix} C_{11} & C_{21} & \cdots & C_{n1} \\ C_{12} & C_{22} & \cdots & C_{n2} \\ \vdots & \vdots & \ddots & \vdots \\ C_{1n} & C_{2n} & \cdots & C_{nn} \end{bmatrix}$$

consisting of cofactors of A, in transposed order. Theorem 3.2.1 implies that the product of A and Adj A

$$A\,\text{Adj}\, A = \begin{bmatrix} |A| & 0 & \cdots & 0 \\ 0 & |A| & \cdots & 0 \\ \vdots & \vdots & & \vdots \\ 0 & 0 & \cdots & |A| \end{bmatrix} = |A|I,$$

where I is the $n \times n$ identity matrix. Indeed, the diagonal elements of the product matrix are computed by multiplying elements of rows of A by their own cofactors and adding (which gives $|A|$), while the off-diagonal elements of the product matrix are computed by multiplying rows of A by cofactors of other rows and adding (which gives 0). It follows that $A(\frac{1}{|A|} \text{Adj}\, A) = I$, producing a formula for the inverse matrix,

$$A^{-1} = \frac{1}{|A|} \text{Adj}\, A = \frac{1}{|A|} \begin{bmatrix} C_{11} & C_{21} & \cdots & C_{n1} \\ C_{12} & C_{22} & \cdots & C_{n2} \\ \vdots & \vdots & \ddots & \vdots \\ C_{1n} & C_{2n} & \cdots & C_{nn} \end{bmatrix}. \tag{3.2}$$

Example 1. $A = \begin{bmatrix} a & b \\ c & d \end{bmatrix}$. Then $|A| = ad - bc$, $C_{11} = d$, $C_{12} = -c$, $C_{21} = -b$, $C_{22} = a$, giving

$$A^{-1} = \frac{1}{ad - bc} \begin{bmatrix} d & -b \\ -c & a \end{bmatrix},$$

provided that $ad - bc \neq 0$. What happens if $|A| = ad - bc = 0$? Then A has no inverse, as a consequence of the following theorem, proved in the preceding section.

Theorem 3.3.1. *An $n \times n$ matrix A is invertible if and only if $|A| \neq 0$.*

Example 2. Find the inverse of $A = \begin{bmatrix} 1 & 1 & 0 \\ 0 & 0 & -1 \\ 1 & 2 & 0 \end{bmatrix}$.

Calculate $|A| = 1$, $C_{11} = \begin{vmatrix} 0 & -1 \\ 2 & 0 \end{vmatrix} = 2$, $C_{12} = -\begin{vmatrix} 0 & -1 \\ 1 & 0 \end{vmatrix} = -1$, $C_{13} = \begin{vmatrix} 0 & 0 \\ 1 & 2 \end{vmatrix} = 0$, $C_{21} = -\begin{vmatrix} 1 & 0 \\ 2 & 0 \end{vmatrix} = 0$,
$C_{22} = \begin{vmatrix} 1 & 0 \\ 1 & 0 \end{vmatrix} = 0$, $C_{23} = -\begin{vmatrix} 1 & 1 \\ 1 & 2 \end{vmatrix} = -1$, $C_{31} = \begin{vmatrix} 1 & 0 \\ 0 & -1 \end{vmatrix} = -1$, $C_{32} = -\begin{vmatrix} 1 & 0 \\ 0 & -1 \end{vmatrix} = 1$, $C_{33} = \begin{vmatrix} 1 & 1 \\ 0 & 0 \end{vmatrix} = 0$ to obtain

$$A^{-1} = \begin{bmatrix} C_{11} & C_{21} & C_{31} \\ C_{12} & C_{22} & C_{32} \\ C_{13} & C_{23} & C_{33} \end{bmatrix} = \begin{bmatrix} 2 & 0 & -1 \\ -1 & 0 & 1 \\ 0 & -1 & 0 \end{bmatrix}.$$

We now turn to an $n \times n$ system of equations $Ax = b$, with the matrix

$$A = \begin{bmatrix} a_{11} & a_{12} & \cdots & a_{1n} \\ a_{21} & a_{22} & \cdots & a_{2n} \\ \vdots & \vdots & \ddots & \vdots \\ a_{n1} & a_{n2} & \cdots & a_{nn} \end{bmatrix}, \quad \text{the vector of right-hand sides } b = \begin{bmatrix} b_1 \\ b_2 \\ \vdots \\ b_n \end{bmatrix},$$

and the vector of unknowns $x = \begin{bmatrix} x_1 \\ x_2 \\ \vdots \\ x_n \end{bmatrix}$, or in components

$$a_{11}x_1 + a_{12}x_2 + \cdots + a_{1n}x_n = b_1,$$
$$a_{21}x_1 + a_{22}x_2 + \cdots + a_{2n}x_n = b_2,$$
$$\vdots$$
$$a_{n1}x_1 + a_{n2}x_2 + \cdots + a_{nn}x_n = b_n. \tag{3.3}$$

Define the matrix

$$A_1 = \begin{bmatrix} b_1 & a_{12} & \cdots & a_{1n} \\ b_2 & a_{22} & \cdots & a_{2n} \\ \vdots & \vdots & \ddots & \vdots \\ b_n & a_{n2} & \cdots & a_{nn} \end{bmatrix},$$

obtained by replacing the first column of A by the vector of the right-hand sides. Similarly, define

$$A_2 = \begin{bmatrix} a_{11} & b_1 & \cdots & a_{1n} \\ a_{21} & b_2 & \cdots & a_{2n} \\ \vdots & \vdots & \ddots & \vdots \\ a_{n1} & b_n & \cdots & a_{nn} \end{bmatrix}, \ldots, A_n = \begin{bmatrix} a_{11} & a_{12} & \cdots & b_1 \\ a_{21} & a_{22} & \cdots & b_2 \\ \vdots & \vdots & \ddots & \vdots \\ a_{n1} & a_{n2} & \cdots & b_n \end{bmatrix}.$$

By expanding in the first column, calculate

$$|A_1| = b_1 C_{11} + b_2 C_{21} + \cdots + b_n C_{n1}, \tag{3.4}$$

where C_{ij} are cofactors of the original matrix A. One shows similarly that

$$|A_i| = b_1 C_{1i} + b_2 C_{2i} + \cdots + b_n C_{ni},$$

for all i.

Theorem 3.3.2 (Cramer's rule). *Assume that $|A| \neq 0$. Then the unique solution of the system* (3.3) *is given by*

$$x_1 = \frac{|A_1|}{|A|}, \quad x_2 = \frac{|A_2|}{|A|}, \ldots, x_n = \frac{|A_n|}{|A|}.$$

Proof. By the preceding Theorem 3.3.1, A^{-1} exists. Then the unique solution of the system (3.3) is $x = A^{-1}b$. Using the expression of A^{-1} from (3.2) gives

$$X = \begin{bmatrix} x_1 \\ x_2 \\ \vdots \\ x_n \end{bmatrix} = \frac{1}{|A|} \begin{bmatrix} C_{11} & C_{21} & \cdots & C_{n1} \\ C_{12} & C_{22} & \cdots & C_{n2} \\ \vdots & \vdots & \ddots & \vdots \\ C_{1n} & C_{2n} & \cdots & C_{nn} \end{bmatrix} \begin{bmatrix} b_1 \\ b_2 \\ \vdots \\ b_n \end{bmatrix}.$$

Now compare the first components on the left and on the right. Using (3.4),

$$x_1 = \frac{1}{|A|} (b_1 C_{11} + b_2 C_{21} + \cdots + b_n C_{n1}) = \frac{|A_1|}{|A|}.$$

One shows similarly that $x_i = \frac{|A_i|}{|A|}$ for all i. □

Cramer's rule calculates each component of solution separately, without having to calculate the other components.

Example 3. Solve the system

$$2x - y = 3,$$
$$-x + 5y = 4.$$

Solution: $x = \dfrac{\begin{vmatrix} 3 & -1 \\ 4 & 5 \end{vmatrix}}{\begin{vmatrix} 2 & -1 \\ -1 & 5 \end{vmatrix}} = \dfrac{19}{9}, y = \dfrac{\begin{vmatrix} 2 & 3 \\ -1 & 4 \end{vmatrix}}{\begin{vmatrix} 2 & -1 \\ -1 & 5 \end{vmatrix}} = \dfrac{11}{9}.$

Cramer's rule is very convenient for 2×2 systems. For 3×3 systems, it requires a tedious evaluation of four 3×3 determinants (Gaussian elimination is preferable).

For $n \times n$ homogeneous systems

$$Ax = 0, \tag{3.5}$$

we shall use the following theorem, which is just a logical consequence of Theorem 3.3.1.

Theorem 3.3.3. *The system* (3.5) *has nontrivial solutions if and only if* $|A| = 0$.

Proof. Assume that nontrivial solutions exist. We claim that $|A| = 0$. Indeed, if $|A| \neq 0$, then, by Theorem 3.3.1, A^{-1} exists, so that (3.5) has only the trivial solution ($x = A^{-1}0 = 0$), a contradiction. Conversely, assume that $|A| = 0$. Then by Theorem 3.3.1, the matrix A is not invertible, hence the system (3.5) has free variables, resulting in nontrivial solutions. □

3.3.1 Vector product

In Calculus, a common notation for the coordinate vectors in R^3 is $\mathbf{i} = e_1, \mathbf{j} = e_2$, and $\mathbf{k} = e_3$. Given two vectors $\mathbf{a} = a_1\mathbf{i} + a_2\mathbf{j} + a_3\mathbf{k}$ and $\mathbf{b} = b_1\mathbf{i} + b_2\mathbf{j} + b_3\mathbf{k}$, *the vector product* of $\mathbf{a}$ and $\mathbf{b}$ is defined to be the vector

$$\mathbf{a} \times \mathbf{b} = (a_2 b_3 - a_3 b_2)\,\mathbf{i} + (a_3 b_1 - a_1 b_3)\,\mathbf{j} + (a_1 b_2 - a_2 b_1)\,\mathbf{k}. \tag{3.6}$$

Perhaps it is not easy to memorize this formula, but determinants come to the rescue:

$$\mathbf{a} \times \mathbf{b} = \begin{vmatrix} \mathbf{i} & \mathbf{j} & \mathbf{k} \\ a_1 & a_2 & a_3 \\ b_1 & b_2 & b_3 \end{vmatrix}.$$

Indeed, expanding this determinant in the first row gives the formula (3.6). By the properties of determinants, it follows that for any vector **a**,

$$\mathbf{a} \times \mathbf{a} = \begin{vmatrix} \mathbf{i} & \mathbf{j} & \mathbf{k} \\ a_1 & a_2 & a_3 \\ a_1 & a_2 & a_3 \end{vmatrix} = \mathbf{0},$$

where **0** is the zero vector, and similarly

$$\mathbf{a} \times \mathbf{b} = -\mathbf{b} \times \mathbf{a},$$

for any vectors **a** and **b**. Recall also the notion of *the scalar product*

$$\mathbf{a} \cdot \mathbf{b} = a_1 b_1 + a_2 b_2 + a_3 b_3.$$

If $\mathbf{c} = c_1\mathbf{i} + c_2\mathbf{j} + c_3\mathbf{k}$, then *the triple product* is defined as $\mathbf{a} \cdot (\mathbf{b} \times \mathbf{c})$. We obtain (using expansion in the first row)

$$\mathbf{a} \cdot (\mathbf{b} \times \mathbf{c}) = a_1(b_2 c_3 - b_3 c_2) + a_2(b_3 c_1 - b_1 c_3) + a_3(b_1 c_2 - b_2 c_1)$$
$$= \begin{vmatrix} a_1 & a_2 & a_3 \\ b_1 & b_2 & b_3 \\ c_1 & c_2 & c_3 \end{vmatrix}.$$

If V denotes the volume of the parallelepiped determined by vectors **a**, **b**, **c**, it is known from Calculus that

$$V = |\mathbf{a} \cdot (\mathbf{b} \times \mathbf{c})| = \left| \begin{vmatrix} a_1 & a_2 & a_3 \\ b_1 & b_2 & b_3 \\ c_1 & c_2 & c_3 \end{vmatrix} \right|.$$

If vectors **a**, **b**, **c** are linearly dependent, then this determinant is zero. Geometrically, linearly dependent vectors lie in the same plane, and hence the volume $V = 0$.

Since $|A^T| = |A|$, it follows that the absolute value of the determinant

$$\begin{vmatrix} a_1 & b_1 & c_1 \\ a_2 & b_2 & c_2 \\ a_3 & b_3 & c_3 \end{vmatrix}$$

also gives the volume of the parallelepiped determined by vectors **a**, **b**, **c**.

There are a number of useful *vector identities* involving vector and scalar products. For example,

$$\mathbf{a} \times (\mathbf{b} \times \mathbf{c}) = \mathbf{b}(\mathbf{a} \cdot \mathbf{c}) - \mathbf{c}(\mathbf{a} \cdot \mathbf{b}),$$

which is memorized as a "bac minus cab" identity. The proof involves a straightforward calculation of both sides in components.

3.3.2 Block matrices

Assume that a 4×4 matrix A is partitioned into four *submatrices*

$$A = \left[\begin{array}{cc|cc} a_{11} & a_{12} & a_{13} & a_{14} \\ a_{21} & a_{22} & a_{23} & a_{24} \\ \hline a_{31} & a_{32} & a_{33} & a_{34} \\ a_{41} & a_{42} & a_{43} & a_{44} \end{array}\right] = \left[\begin{array}{c|c} A_1 & A_2 \\ \hline A_3 & A_4 \end{array}\right],$$

with 2×2 matrices $A_1 = \left[\begin{smallmatrix} a_{11} & a_{12} \\ a_{21} & a_{22} \end{smallmatrix}\right], A_2 = \left[\begin{smallmatrix} a_{13} & a_{14} \\ a_{23} & a_{24} \end{smallmatrix}\right], A_3 = \left[\begin{smallmatrix} a_{31} & a_{32} \\ a_{41} & a_{42} \end{smallmatrix}\right]$, and $A_4 = \left[\begin{smallmatrix} a_{33} & a_{34} \\ a_{43} & a_{44} \end{smallmatrix}\right]$. Suppose that a 4×4 matrix B is partitioned similarly,

$$B = \left[\begin{array}{cc|cc} b_{11} & b_{12} & b_{13} & b_{14} \\ b_{21} & b_{22} & b_{23} & b_{24} \\ \hline b_{31} & b_{32} & b_{33} & b_{34} \\ b_{41} & b_{42} & b_{43} & b_{44} \end{array}\right] = \left[\begin{array}{c|c} B_1 & B_2 \\ \hline B_3 & B_4 \end{array}\right],$$

with 2×2 matrices B_1, B_2, B_3, B_4. It follows from the definition of matrix multiplication that the product AB can be evaluated by regarding A and B as 2×2 (block) matrices,

$$AB = \left[\begin{array}{c|c} A_1 & A_2 \\ \hline A_3 & A_4 \end{array}\right] \left[\begin{array}{c|c} B_1 & B_2 \\ \hline B_3 & B_4 \end{array}\right] = \left[\begin{array}{c|c} A_1B_1 + A_2B_3 & A_1B_2 + A_2B_4 \\ \hline A_3B_1 + A_4B_3 & A_3B_2 + A_4B_4 \end{array}\right], \tag{3.7}$$

where A_1B_1 and the other terms are themselves products of 2×2 matrices. In other words, we treat the 2×2 blocks as numbers, until the last step.

Using the expansion of determinants $|A| = \sum \pm a_{1i_1} a_{2i_2} a_{3i_3} a_{4i_4}$, it is possible to show that for the 4×4 matrix A, partitioned as above,

$$|A| = |A_1| |A_4| - |A_2| |A_3|,$$

where again we treat blocks as numbers, and $|A_i|$ are 2×2 determinants.

In particular, for 4×4 *block diagonal matrices* $A = \left[\begin{array}{c|c} A_1 & O \\ \hline O & A_4 \end{array}\right]$, where O is the 2×2 zero matrix, one has

$$|A| = |A_1| |A_4|.$$

The last formula can be also justified by Gaussian elimination. Indeed, the row echelon form of A is an upper triangular matrix, and the product of its diagonal entries gives $|A|$. That product splits into $|A_1|$ and $|A_4|$.

If, similarly, $B = \begin{bmatrix} B_1 & 0 \\ \hline 0 & B_4 \end{bmatrix}$, where B_1, B_4, and O are 2×2 matrices, then by (3.7)

$$\begin{bmatrix} A_1 & 0 \\ \hline 0 & A_4 \end{bmatrix} \begin{bmatrix} B_1 & 0 \\ \hline 0 & B_4 \end{bmatrix} = \begin{bmatrix} A_1 B_1 & 0 \\ \hline 0 & A_4 B_4 \end{bmatrix}.$$

It follows that

$$\begin{bmatrix} A_1 & 0 \\ \hline 0 & A_4 \end{bmatrix}^{-1} = \begin{bmatrix} A_1^{-1} & 0 \\ \hline 0 & A_4^{-1} \end{bmatrix},$$

provided that A_1^{-1} and A_4^{-1} exist.

Similar formulas apply to other types of block matrices, where the blocks are not necessarily square matrices. For example, let us partition a 3×3 matrix A into four submatrices as follows:

$$A = \begin{bmatrix} a_{11} & a_{12} & a_{13} \\ a_{21} & a_{22} & a_{23} \\ \hline a_{31} & a_{32} & a_{33} \end{bmatrix} = \begin{bmatrix} A_1 & A_2 \\ \hline A_3 & A_4 \end{bmatrix},$$

where $A_1 = \begin{bmatrix} a_{11} & a_{12} \\ a_{21} & a_{22} \end{bmatrix}$, $A_2 = \begin{bmatrix} a_{13} \\ a_{23} \end{bmatrix}$ of size 2×1, $A_3 = \begin{bmatrix} a_{31} & a_{32} \end{bmatrix}$ of size 1×2, and a scalar $A_4 = a_{33}$ of size 1×1. If a 3×3 matrix B is partitioned similarly, $B = \begin{bmatrix} B_1 & B_2 \\ B_3 & B_4 \end{bmatrix}$, then it is straightforward to check that the product AB can be calculated by treating blocks as numbers:

$$AB = \begin{bmatrix} a_{11} & a_{12} & a_{13} \\ a_{21} & a_{22} & a_{23} \\ \hline a_{31} & a_{32} & a_{33} \end{bmatrix} \begin{bmatrix} b_{11} & b_{12} & b_{13} \\ b_{21} & b_{22} & b_{23} \\ \hline b_{31} & b_{32} & b_{33} \end{bmatrix}$$

$$= \begin{bmatrix} A_1 & A_2 \\ \hline A_3 & A_4 \end{bmatrix} \begin{bmatrix} B_1 & B_2 \\ \hline B_3 & B_4 \end{bmatrix} = \begin{bmatrix} C_1 & C_2 \\ \hline C_3 & C_4 \end{bmatrix},$$

where $C_1 = A_1 B_1 + A_2 B_3$ is of size 2×2, $C_2 = A_1 B_2 + A_2 B_4$ is of size 2×1, $C_3 = A_3 B_1 + A_4 B_3$ is of size 1×2, and a scalar $C_4 = A_3 B_2 + A_4 B_4$ (all matrix products are defined). So that the block structure of AB is the same as that for A and B. In case $A_2 = O$ and $A_3 = O$, the matrix $A = \begin{bmatrix} a_{11} & a_{12} & 0 \\ a_{21} & a_{22} & 0 \\ \hline 0 & 0 & a_{33} \end{bmatrix} = \begin{bmatrix} A_1 & 0 \\ \hline 0 & a_{33} \end{bmatrix}$ is *block-diagonal*, with the inverse

$$A^{-1} = \begin{bmatrix} a_{11} & a_{12} & 0 \\ a_{21} & a_{22} & 0 \\ \hline 0 & 0 & a_{33} \end{bmatrix}^{-1} = \begin{bmatrix} A_1^{-1} & 0 \\ \hline 0 & \frac{1}{a_{33}} \end{bmatrix},$$

provided that A_1^{-1} exists, and $a_{33} \neq 0$. For the determinant, one has

$$|A| = \begin{vmatrix} a_{11} & a_{12} & 0 \\ a_{21} & a_{22} & 0 \\ 0 & 0 & a_{33} \end{vmatrix} = |A_1| \, a_{33} = (a_{11}a_{22} - a_{12}a_{21}) \, a_{33}.$$

Exercises

1. Use the adjugate matrix to calculate the inverse for the following matrices:

   (a) $\begin{bmatrix} 1 & 2 \\ 1 & 1 \end{bmatrix}$.

   (b) $\begin{bmatrix} 1 & -2 \\ -2 & 4 \end{bmatrix}$.

   Answer: The matrix is singular.

   (c) $C = \begin{bmatrix} 1 & 2 & 0 \\ 0 & -1 & 1 \\ 1 & -2 & 1 \end{bmatrix}$.

   Answer: $C^{-1} = \frac{1}{3} \begin{bmatrix} 1 & -2 & 2 \\ 1 & 1 & -1 \\ 1 & 4 & -1 \end{bmatrix}$.

   (d) $D = \begin{bmatrix} 0 & -1 & 0 \\ 1 & 0 & 0 \\ 0 & 0 & 5 \end{bmatrix}$.

   Answer: $D^{-1} = \begin{bmatrix} 0 & 1 & 0 \\ -1 & 0 & 0 \\ 0 & 0 & \frac{1}{5} \end{bmatrix}$.

   (e) $\begin{bmatrix} 1 & 2 & 3 \\ 4 & 5 & 6 \\ 7 & 8 & 9 \end{bmatrix}$.

   Answer: The matrix is singular.

   (f) $\begin{bmatrix} 1 & 0 & 0 \\ 0 & -5 & 0 \\ 0 & 0 & 9 \end{bmatrix}$.

   (g) $G = \begin{bmatrix} 1 & 1 & 1 & 0 \\ 1 & 0 & 0 & -1 \\ -1 & 0 & 0 & 0 \\ 0 & 0 & 1 & 0 \end{bmatrix}$.

   Answer: $G^{-1} = \begin{bmatrix} 0 & 0 & -1 & 0 \\ 1 & 0 & 1 & -1 \\ 0 & 0 & 0 & 1 \\ 0 & -1 & -1 & 0 \end{bmatrix}$.

   (h) $H = \begin{bmatrix} 1 & 1 & 1 & 0 \\ 1 & 1 & 0 & 1 \\ 1 & 0 & 1 & 1 \\ 0 & 1 & 1 & 1 \end{bmatrix}$.

   Answer: $H^{-1} = \frac{1}{3} \begin{bmatrix} 1 & 1 & 1 & -2 \\ 1 & 1 & -2 & 1 \\ 1 & -2 & 1 & 1 \\ -2 & 1 & 1 & 1 \end{bmatrix}$.

   (i) $R = \begin{bmatrix} \cos\theta & -\sin\theta \\ \sin\theta & \cos\theta \end{bmatrix}$.

   Answer: $R^{-1} = \begin{bmatrix} \cos\theta & \sin\theta \\ -\sin\theta & \cos\theta \end{bmatrix}$.

2. Use Cramer's rule to solve the following systems. In case Cramer's rule does not work, apply Gaussian elimination.

   (a)
   $$x_1 - x_2 = 2,$$
   $$2x_1 + x_2 = -3.$$

   (b)
   $$5x_1 - x_2 = 0,$$

$$2x_1 + x_2 = 0.$$

(c)

$$4x_1 - 2x_2 = 5,$$
$$-2x_1 + x_2 = -1.$$

Answer: The system is inconsistent.

(d)

$$2x_1 - x_2 = 1,$$
$$-2x_1 + x_2 = -1.$$

Answer: $x_1 = \frac{1}{2}t + \frac{1}{2}, x_2 = t, t$ is arbitrary.

(e)

$$x_1 - x_3 = 1,$$
$$x_1 + 3x_2 + x_3 = 0,$$
$$x_1 + x_2 + x_3 = 1.$$

Answer: $x_1 = \frac{5}{4}, x_2 = -\frac{1}{2}, x_3 = \frac{1}{4}.$

(f)

$$x_2 - x_3 = 1,$$
$$x_1 + 3x_2 + x_3 = 0,$$
$$x_1 + x_2 + x_3 = 1.$$

Answer: $x_1 = 3, x_2 = -\frac{1}{2}, x_3 = -\frac{3}{2}.$

(g)

$$x_1 + x_2 - x_3 = 1,$$
$$x_1 + 3x_2 + 2x_3 = 2,$$
$$x_1 + x_2 - 3x_3 = 1.$$

Answer: $x_1 = \frac{1}{2}, x_2 = \frac{1}{2}, x_3 = 0.$

(h)

$$x_1 + 3x_2 + 2x_3 = 2,$$
$$x_1 + x_2 - 3x_3 = 1,$$
$$2x_2 + 5x_3 = -1.$$

Answer: The system has no solution.

3. Let A be an $n \times n$ matrix.

(a) Show that

$$|\text{Adj}\, A| = |A|^{n-1}.$$

Hint. Recall that $A\,\text{Adj}\, A = |A|I$, so that $|A\,\text{Adj}\, A| = |A|\,|\text{Adj}\, A| = \det(|A|I) = |A|^n$.

(b) Show that $\text{Adj}\, A$ is singular if and only if A is singular.

4. (a) Show that a lower triangular matrix is invertible if and only if all of its diagonal entries are nonzero.

   (b) Show that the inverse of a nonsingular lower triangular matrix is also lower triangular.

5. Let A be a nonsingular matrix with integer entries. Show that the inverse matrix A^{-1} contains only integer entries if and only if $|A| = \pm 1$.

   Hint. If $|A| = \pm 1$, then, by (3.2), $A^{-1} = \pm\,\text{Adj}\, A$ has integer entries. Conversely, suppose that every entry of the inverse matrix A^{-1} is an integer. It follows that $|A|$ and $|A^{-1}|$ are both integers. Since we have

$$|A|\,|A^{-1}| = |AA^{-1}| = |I| = 1,$$

   it follows that $|A| = \pm 1$.

6. For an $n\times n$ system $Ax = b$ assume that the determinant of A is zero (so that Cramer's rule does not work). Show that either there is no solution, or else there are infinitely many solutions.

7. Justify the following identities, for any vectors in R^3:

   (a) $\mathbf{a} \cdot (\mathbf{b} \times \mathbf{c}) = (\mathbf{a} \times \mathbf{b}) \cdot \mathbf{c}$.

   (b) $\mathbf{a} \times (\mathbf{b} \times \mathbf{c}) = \mathbf{b}(\mathbf{a} \cdot \mathbf{c}) - \mathbf{c}(\mathbf{a} \cdot \mathbf{b})$.

   (c) $\|\mathbf{a} \times \mathbf{b}\| = \|\mathbf{a}\|\,\|\mathbf{b}\| \sin \theta$, where θ is the angle between $\mathbf{a}$ and $\mathbf{b}$.

   (d) $(\mathbf{a} \times \mathbf{b}) \cdot (\mathbf{c} \times \mathbf{d}) = (\mathbf{a} \cdot \mathbf{c})(\mathbf{b} \cdot \mathbf{d}) - (\mathbf{a} \cdot \mathbf{d})(\mathbf{b} \cdot \mathbf{c})$.

   Hint. Write each vector in components. Part (d) is tedious.

8. (a) Find the inverse and the determinant of the following 5×5 block diagonal matrix:

$$A = \begin{bmatrix} 1 & -3 & 0 & 0 & 0 \\ -1 & 4 & 0 & 0 & 0 \\ 0 & 0 & \cos\theta & -\sin\theta & 0 \\ 0 & 0 & \sin\theta & \cos\theta & 0 \\ 0 & 0 & 0 & 0 & 4 \end{bmatrix}.$$

Answer: $A^{-1} = \begin{bmatrix} 4 & 3 & 0 & 0 & 0 \\ 1 & 1 & 0 & 0 & 0 \\ 0 & 0 & \cos\theta & \sin\theta & 0 \\ 0 & 0 & -\sin\theta & \cos\theta & 0 \\ 0 & 0 & 0 & 0 & \frac{1}{4} \end{bmatrix}$, $|A| = 4$.

(b) Let $x = \begin{bmatrix} x_1 \\ x_2 \\ x_3 \\ x_4 \\ x_5 \end{bmatrix}$, $y = \begin{bmatrix} x_1 \\ x_2 \\ 0 \\ 0 \\ 0 \end{bmatrix}$, $z = \begin{bmatrix} 0 \\ 0 \\ x_3 \\ x_4 \\ 0 \end{bmatrix}$, and $w = \begin{bmatrix} 0 \\ 0 \\ 0 \\ 0 \\ x_5 \end{bmatrix}$.

Evaluate Ay, Az, Aw, and compare with Ax.

4 Eigenvectors and eigenvalues

4.1 Characteristic equation

The vector $z = \begin{bmatrix} 1 \\ -1 \end{bmatrix}$ is very special for the matrix $A = \begin{bmatrix} 3 & 1 \\ 1 & 3 \end{bmatrix}$. Calculate

$$Az = \begin{bmatrix} 3 & 1 \\ 1 & 3 \end{bmatrix} \begin{bmatrix} 1 \\ -1 \end{bmatrix} = \begin{bmatrix} 2 \\ -2 \end{bmatrix} = 2 \begin{bmatrix} 1 \\ -1 \end{bmatrix} = 2z,$$

so that $Az = 2z$, and *the vectors z and Az go along the same line*. We say that z is *an eigenvector of A corresponding to an eigenvalue 2*.

In general, we say that *a vector $x \in R^n$ is an eigenvector of an $n \times n$ matrix A, corresponding to an eigenvalue λ* if

$$Ax = \lambda x, \quad x \neq 0. \tag{1.1}$$

(Eigenvalue is a number denoted by a Greek letter lambda.) Notice that the zero vector is not eligible to be an eigenvector. If A is 2×2, then an eigenvector must satisfy $x = \begin{bmatrix} x_1 \\ x_2 \end{bmatrix} \neq \begin{bmatrix} 0 \\ 0 \end{bmatrix}$.

If $c \neq 0$ is any scalar, and (1.1) holds, then

$$A(c\,x) = cAx = c\lambda x = \lambda(c\,x),$$

which implies that $c\,x$ is also an eigenvector of the matrix A, corresponding to *the same eigenvalue λ*. In particular, $c\begin{bmatrix} 1 \\ -1 \end{bmatrix}$ gives us infinitely many eigenvectors of the 2×2 matrix A above, corresponding to the eigenvalue $\lambda = 2$.

Let us rewrite (1.1) as $Ax = \lambda Ix$, or $Ax - \lambda Ix = 0$, and then in the form

$$(A - \lambda I)x = 0, \tag{1.2}$$

where I is the identity matrix. To find x, one needs to solve a homogeneous system of linear equations, with the matrix $A - \lambda I$. To have nonzero solutions $x \neq 0$, this matrix must be singular, with determinant zero,

$$|A - \lambda I| = 0. \tag{1.3}$$

Expanding this determinant gives a polynomial equation for λ, called *the characteristic equation*, and its roots are the eigenvalues. (The polynomial itself is called *the characteristic polynomial*.) If the matrix A is 2×2, we obtain a quadratic equation, which has two roots λ_1 and λ_2 (possibly equal). In case A is 3×3, one needs to solve a cubic equation, with three roots λ_1, λ_2, and λ_3 (possibly repeated). An $n \times n$ matrix has n eigenvalues $\lambda_1, \lambda_2, \ldots, \lambda_n$, some possibly repeated. To calculate the eigenvectors corresponding to λ_1, we solve the system

$$(A - \lambda_1 I)x = 0,$$

and proceed similarly for other eigenvalues.

https://doi.org/10.1515/9783111086507-004

Example 1. Consider $A = \begin{bmatrix} 3 & 1 \\ 1 & 3 \end{bmatrix}$. Calculate

$$A - \lambda I = \begin{bmatrix} 3 & 1 \\ 1 & 3 \end{bmatrix} - \lambda \begin{bmatrix} 1 & 0 \\ 0 & 1 \end{bmatrix} = \begin{bmatrix} 3 & 1 \\ 1 & 3 \end{bmatrix} - \begin{bmatrix} \lambda & 0 \\ 0 & \lambda \end{bmatrix} = \begin{bmatrix} 3 - \lambda & 1 \\ 1 & 3 - \lambda \end{bmatrix}.$$

(To calculate $A - \lambda I$, subtract λ from each of the diagonal entries of A.)
The characteristic equation

$$|A - \lambda I| = \begin{vmatrix} 3 - \lambda & 1 \\ 1 & 3 - \lambda \end{vmatrix} = (3 - \lambda)^2 - 1 = 0$$

has the roots $\lambda_1 = 2$ and $\lambda_2 = 4$, the eigenvalues of A (writing $3 - \lambda = \pm 1$ gives the eigenvalues quickly).
(i) To find the eigenvectors corresponding to $\lambda_1 = 2$, we need to solve the system $(A - 2I)x = 0$ for $x = \begin{bmatrix} x_1 \\ x_2 \end{bmatrix}$, which is

$$x_1 + x_2 = 0,$$
$$x_1 + x_2 = 0.$$

(The matrix $A - 2I = \begin{bmatrix} 1 & 1 \\ 1 & 1 \end{bmatrix}$ is obtained from $A - \lambda I$ by setting $\lambda = 2$.) Discarding the second equation, setting the free variable $x_2 = c$, an arbitrary number, and solving for $x_1 = -c$, we obtain that $x = \begin{bmatrix} -c \\ c \end{bmatrix} = c \begin{bmatrix} -1 \\ 1 \end{bmatrix}$ are the eigenvectors corresponding to $\lambda_1 = 2$.
(ii) To find the eigenvectors corresponding to $\lambda_2 = 4$, one solves the system $(A - 4I)x = 0$, or

$$-x_1 + x_2 = 0,$$
$$x_1 - x_2 = 0,$$

because $A - 4I = \begin{bmatrix} -1 & 1 \\ 1 & -1 \end{bmatrix}$. Discard the second equation, set $x_2 = c$, and solve for $x_1 = c$.
Conclusion: $x = c \begin{bmatrix} 1 \\ 1 \end{bmatrix}$ are the eigenvectors corresponding to $\lambda_2 = 4$.

Example 2. Let $A = \begin{bmatrix} 2 & 1 & 1 \\ 0 & 2 & 0 \\ 1 & 5 & 2 \end{bmatrix}$.
The characteristic equation is

$$|A - \lambda I| = \begin{vmatrix} 2 - \lambda & 1 & 1 \\ 0 & 2 - \lambda & 0 \\ 1 & 5 & 2 - \lambda \end{vmatrix} = 0.$$

(Subtract λ from the diagonal entries of A to obtain $A - \lambda I$.) Expand the determinant in the second row, then simplify to get

$$(2 - \lambda)[(2 - \lambda)^2 - 1] = 0,$$

or

$$(2 - \lambda)(\lambda^2 - 4\lambda + 3) = 0.$$

Setting the first factor to zero gives the first eigenvalue $\lambda_1 = 2$. Setting the second factor to zero, $\lambda^2 - 4\lambda + 3 = 0$, gives $\lambda_2 = 1$ and $\lambda_3 = 3$.

Next, for each eigenvalue we calculate the corresponding eigenvectors.

(i) $\lambda_1 = 2$. The corresponding eigenvectors are solutions of $(A - 2I)x = 0$. Calculate $A - 2I = \begin{bmatrix} 0 & 1 & 1 \\ 0 & 0 & 0 \\ 1 & 5 & 0 \end{bmatrix}$. (In future calculations this step will be performed mentally.) Restore the system $(A - 2I)x = 0$, and discard the second equation consisting of all zeroes to obtain

$$x_2 + x_3 = 0,$$
$$x_1 + 5x_2 = 0.$$

We expect to get infinitely many eigenvectors. So let us calculate one of them, and multiply the resulting vector by c. To this end, set $x_3 = 1$. Then $x_2 = -1$, and $x_1 = 5$. We thus obtain $c \begin{bmatrix} 5 \\ -1 \\ 1 \end{bmatrix}$. (Alternatively, set the free variable $x_3 = c$, an arbitrary number. Then $x_2 = -c$ and $x_1 = 5c$, giving again $c \begin{bmatrix} 5 \\ -1 \\ 1 \end{bmatrix}$.)

(ii) $\lambda_2 = 1$. The corresponding eigenvectors are nontrivial solutions of $(A - I)x = 0$. Restore this system to get

$$x_1 + x_2 + x_3 = 0,$$
$$x_2 = 0,$$
$$x_1 + 5x_2 + x_3 = 0.$$

From the second equation $x_2 = 0$, and then both the first and third equations simplify to $x_1 + x_3 = 0$. Set $x_3 = 1$, then $x_1 = -1$ to obtain $c \begin{bmatrix} -1 \\ 0 \\ 1 \end{bmatrix}$. (Alternatively, set the free variable $x_3 = c$, an arbitrary number. Then $x_2 = 0$ and $x_1 = -c$, giving again $c \begin{bmatrix} -1 \\ 0 \\ 1 \end{bmatrix}$.)

(iii) $\lambda_3 = 3$. The corresponding eigenvectors are nontrivial solutions of $(A - 3I)x = 0$. Restore this system to get

$$-x_1 + x_2 + x_3 = 0,$$
$$-x_2 = 0,$$
$$x_1 + 5x_2 - x_3 = 0.$$

From the second equation $x_2 = 0$, and then both the first and third equations simplify to $x_1 - x_3 = 0$. Set $x_3 = c$, then $x_1 = c$ to obtain $c \begin{bmatrix} 1 \\ 0 \\ 1 \end{bmatrix}$. One can present an eigenvector corresponding to $\lambda_3 = 3$ as $\begin{bmatrix} 1 \\ 0 \\ 1 \end{bmatrix}$, with implied arbitrary multiple of c.

4.1.1 Properties of eigenvectors and eigenvalues

A square matrix is called triangular if it is either upper triangular, lower triangular, or diagonal.

Property 1. The diagonal entries of a triangular matrix are its eigenvalues.

For example, for $A = \begin{bmatrix} 2 & 0 & 0 \\ -1 & 3 & 0 \\ 3 & 0 & 4 \end{bmatrix}$ the characteristic equation is

$$|A - \lambda I| = \begin{vmatrix} 2 - \lambda & 0 & 0 \\ -1 & 3 - \lambda & 0 \\ 3 & 0 & 4 - \lambda \end{vmatrix} = 0,$$

giving

$$(2 - \lambda)(3 - \lambda)(4 - \lambda) = 0.$$

The eigenvalues are $\lambda_1 = 2$, $\lambda_2 = 3$, and $\lambda_3 = 4$. In general, the determinant of any triangular matrix equals to the product of its diagonal entries, and the same reasoning applies.

For an $n \times n$ matrix A define its trace to be the sum of all diagonal elements

$$\operatorname{tr} A = a_{11} + a_{22} + \cdots + a_{nn}.$$

Property 2. Let $\lambda_1, \lambda_2, \ldots, \lambda_n$ be the eigenvalues of an $n \times n$ matrix A, possibly repeated. Then

$$\lambda_1 + \lambda_2 + \cdots + \lambda_n = \operatorname{tr} A,$$
$$\lambda_1 \cdot \lambda_2 \cdots \lambda_n = |A|.$$

These formulas are clearly true for triangular matrices. For example, if

$$A = \begin{bmatrix} 2 & 0 & 0 \\ -1 & 3 & 0 \\ 5 & -4 & 3 \end{bmatrix},$$

then $\lambda_1 = 2$, $\lambda_2 = 3$, $\lambda_3 = 3$, so that $\lambda_1 + \lambda_2 + \lambda_3 = \operatorname{tr} A = 8$, and $\lambda_1 \cdot \lambda_2 \cdot \lambda_3 = |A| = 18$.

Let us justify Property 2 for any 2×2 matrix $A = \begin{bmatrix} a_{11} & a_{12} \\ a_{21} & a_{22} \end{bmatrix}$. The characteristic equation

$$\begin{vmatrix} a_{11} - \lambda & a_{12} \\ a_{21} & a_{22} - \lambda \end{vmatrix} = (a_{11} - \lambda)(a_{22} - \lambda) - a_{12}a_{21} = 0$$

can be expanded to

$$\lambda^2 - (a_{11} + a_{22})\lambda + a_{11}a_{22} - a_{12}a_{21} = 0,$$

or

$$\lambda^2 - (\operatorname{tr} A)\lambda + |A| = 0. \tag{1.4}$$

The eigenvalues λ_1 and λ_2 are the roots of this equation, so that we can factor (1.4) as

$$(\lambda - \lambda_1)(\lambda - \lambda_2) = 0.$$

Expanding

$$\lambda^2 - (\lambda_1 + \lambda_2)\lambda + \lambda_1\lambda_2 = 0. \tag{1.5}$$

Comparing (1.4) with (1.5), which are two versions of the same equation, we conclude that $\lambda_1 + \lambda_2 = \operatorname{tr} A$, and $\lambda_1\lambda_2 = |A|$, as claimed.

For example, if

$$A = \begin{bmatrix} -4 & 6 \\ -1 & 3 \end{bmatrix},$$

then $\lambda_1 + \lambda_2 = -1$, $\lambda_1\lambda_2 = -6$. We can now obtain the eigenvalues $\lambda_1 = -3$ and $\lambda_2 = 2$ without evaluating the characteristic polynomial.

Property 3. A square matrix A is invertible if and only if all of its eigenvalues are different from zero.

Proof. Matrix A is invertible if and only if $|A| \neq 0$. But having $|A| = \lambda_1 \cdot \lambda_2 \cdots \lambda_n \neq 0$ requires all eigenvalues to be different from zero. □

It follows that a matrix with the zero eigenvalue $\lambda = 0$ is singular.

Property 4. Let λ be an eigenvalue of an invertible matrix A. Then $\frac{1}{\lambda}$ is an eigenvalue of A^{-1}, corresponding to the same eigenvector.

Proof. By Property 3, $\lambda \neq 0$. Multiplying $Ax = \lambda x$ by A^{-1} from the left gives $x = \lambda A^{-1}x$, or $A^{-1}x = \frac{1}{\lambda}x$. □

For example, if A has eigenvalues $-2, 1, 4$, then A^{-1} has eigenvalues $-\frac{1}{2}, 1, \frac{1}{4}$.

We say that *two matrices A and B are similar if there is an invertible matrix P such that $B = P^{-1}AP$* (one can then express $A = PBP^{-1}$).

Property 5. Two similar matrices A and B share the same characteristic polynomial, and therefore they have the same set of eigenvalues.

Proof. The characteristic polynomial of B,

$$|B - \lambda I| = |P^{-1}AP - \lambda I| = |P^{-1}AP - \lambda P^{-1}IP|$$
$$= |P^{-1}(A - \lambda I)P| = |P^{-1}||A - \lambda I||P| = |A - \lambda I|,$$

is the same as the characteristic polynomial of A, by using properties of determinants (on the last step we used that $|P^{-1}| = \frac{1}{|P|}$). □

Property 6. Let λ be an eigenvalue of A. Then λ^2 is an eigenvalue of A^2, corresponding to the same eigenvector.

Indeed, multiplying the relation $Ax = \lambda x$ by matrix A from the left gives

$$A^2x = A(Ax) = A(\lambda x) = \lambda Ax = \lambda \lambda x = \lambda^2 x.$$

One shows similarly that λ^k is an eigenvalue of A^k, for any positive integer k. For example, if A has eigenvalues $-2, 1, 4$, then A^3 has eigenvalues $-8, 1, 64$.

Exercises

1. Verify that the vector $\begin{bmatrix} 1 \\ 0 \\ 1 \end{bmatrix}$ is an eigenvector of the matrix $\begin{bmatrix} 2 & -4 & 1 \\ 0 & 2 & 0 \\ 1 & -3 & 2 \end{bmatrix}$ corresponding to the eigenvalue $\lambda = 3$.

2. Determine the eigenvalues of the following matrices. Verify that the sum of the eigenvalues is equal to the trace, while the product of the eigenvalues is equal to the determinant.

   (a) $A = \begin{bmatrix} 1 & 2 \\ 0 & -1 \end{bmatrix}$.

   Answer: $\lambda_1 = 1, \lambda_2 = -1$, $\operatorname{tr} A = \lambda_1 + \lambda_2 = 0$, $|A| = \lambda_1\lambda_2 = -1$.

   (b) $\begin{bmatrix} 3 & 0 \\ 0 & -4 \end{bmatrix}$.

   (c) $\begin{bmatrix} 3 & 0 \\ -4 & 5 \end{bmatrix}$.

   (d) $\begin{bmatrix} 3 & 1 & -2 \\ 0 & 0 & 4 \\ 0 & 0 & -7 \end{bmatrix}$.

   Answer: $\lambda_1 = 3, \lambda_2 = 0, \lambda_3 = -7$, $\operatorname{tr} A = \lambda_1 + \lambda_2 + \lambda_3 = -4$, $|A| = 0$.

   (e) $A = \begin{bmatrix} 3 & 2 \\ 4 & 1 \end{bmatrix}$.

   Answer: $\lambda_1 = -1, \lambda_2 = 5$.

   (f) $A = \begin{bmatrix} -2 & 0 & 0 \\ 4 & 2 & 1 \\ 3 & 1 & 2 \end{bmatrix}$.

   Answer: $\lambda_1 = -2, \lambda_2 = 1, \lambda_3 = 3$.

   (g) $A = \begin{bmatrix} -2 & -1 & 4 \\ 3 & 2 & -5 \\ 0 & 0 & 1 \end{bmatrix}$.

   Answer: $\lambda_1 = -1, \lambda_2 = 1, \lambda_3 = 1$.

   (h) $A = \begin{bmatrix} -1 & 1 & 0 \\ 1 & -2 & 1 \\ 0 & 1 & -1 \end{bmatrix}$.

   Answer: $\lambda_1 = -3, \lambda_2 = -1, \lambda_3 = 0$.

   (i) $A = \begin{bmatrix} 0 & -1 \\ 1 & 0 \end{bmatrix}$.

   Answer: $\lambda_1 = -i, \lambda_2 = i$, $\operatorname{tr} A = \lambda_1 + \lambda_2 = 0$, $\det A = \lambda_1\lambda_2 = 1$.

3. Calculate the eigenvalues and corresponding eigenvectors for the following matrices:

(a) $\begin{bmatrix} 2 & 1 \\ 5 & -2 \end{bmatrix}$.

Answer: $\lambda_1 = -3$ with $\begin{bmatrix} -1 \\ 5 \end{bmatrix}$, $\lambda_2 = 3$ with $\begin{bmatrix} 1 \\ 1 \end{bmatrix}$.

(b) $\begin{bmatrix} 3 & 0 \\ 0 & -5 \end{bmatrix}$.

Answer: $\lambda_1 = 3$ with $\begin{bmatrix} 1 \\ 0 \end{bmatrix}$, $\lambda_2 = -5$ with $\begin{bmatrix} 0 \\ 1 \end{bmatrix}$.

(c) $\begin{bmatrix} 4 & 6 \\ -1 & -1 \end{bmatrix}$.

Answer: $\lambda_1 = 1$ with $\begin{bmatrix} -2 \\ 1 \end{bmatrix}$, $\lambda_2 = 2$ with $\begin{bmatrix} -3 \\ 1 \end{bmatrix}$.

(d) $\begin{bmatrix} 0 & 4 \\ 1 & 0 \end{bmatrix}$.

Answer: $\lambda_1 = -2$ with $\begin{bmatrix} -2 \\ 1 \end{bmatrix}$, $\lambda_2 = 2$ with $\begin{bmatrix} 2 \\ 1 \end{bmatrix}$.

(e) $\begin{bmatrix} 2 & 0 & 0 & 0 \\ 0 & -3 & 0 & 0 \\ 0 & 0 & 0 & 0 \\ 0 & 0 & 0 & 5 \end{bmatrix}$.

(f) Any $n \times n$ diagonal matrix.

(g) $\begin{bmatrix} 2 & 1 & 1 \\ -1 & -2 & 1 \\ 3 & 3 & 0 \end{bmatrix}$. Hint. Factor the characteristic equation.

Answer: $\lambda_1 = -3$ with $\begin{bmatrix} 0 \\ -1 \\ 1 \end{bmatrix}$, $\lambda_2 = 0$ with $\begin{bmatrix} -1 \\ 1 \\ 1 \end{bmatrix}$, and $\lambda_3 = 3$ with $\begin{bmatrix} 1 \\ 0 \\ 1 \end{bmatrix}$.

(h) $\begin{bmatrix} 2 & -4 & 1 \\ 0 & 2 & 0 \\ 1 & -3 & 2 \end{bmatrix}$. Hint. Expand in the second row.

Answer: $\lambda_1 = 1$ with $\begin{bmatrix} -1 \\ 0 \\ 1 \end{bmatrix}$, $\lambda_2 = 2$ with $\begin{bmatrix} 3 \\ 1 \\ 4 \end{bmatrix}$, and $\lambda_3 = 3$ with $\begin{bmatrix} 1 \\ 0 \\ 1 \end{bmatrix}$.

(i) $\begin{bmatrix} 1 & 2 & 1 \\ 2 & -2 & 1 \\ 0 & 0 & 5 \end{bmatrix}$.

Answer: $\lambda_1 = -3$ with $\begin{bmatrix} -1 \\ 2 \\ 0 \end{bmatrix}$, $\lambda_2 = 2$ with $\begin{bmatrix} 2 \\ 1 \\ 0 \end{bmatrix}$, and $\lambda_3 = 5$ with $\begin{bmatrix} 3 \\ 2 \\ 8 \end{bmatrix}$.

4. Let A be a 2×2 matrix, with trace 6, and one of the eigenvalues equal to -1. What is the determinant $|A|$?

Answer: $|A| = -7$.

5. (a) Write down two different 2×2 matrices with trace equal to 5 and determinant equal to 4.

   (b) What are the eigenvalues of any such matrix?

   Answer: 1 and 4.

6. Let A be a 3×3 matrix with the eigenvalues $-2, 1, \frac{1}{4}$.

   (a) Find $|A^3|$.

   Answer: $-\frac{1}{8}$.

   (b) Find $|A^{-1}|$.

   Answer: -2.

7. Let A be an invertible matrix. Show that zero cannot be an eigenvalue of A^{-1}.

8. Assume that the matrix A has an eigenvalue zero. Show that the matrix AB is not invertible, for any matrix B.

9. Let λ be an eigenvalue of A, corresponding to an eigenvector x, and k is any number. Show that $k\lambda$ is an eigenvalue of kA, corresponding to the same eigenvector x.

10 (a) Show that the matrix A^T has the same eigenvalues as A.

   Hint. $|A^T - \lambda I| = |(A - \lambda I)^T| = |A - \lambda I|$.

   (b) Show that the eigenvectors of A and A^T are in general different.

   Hint. Consider, for example, $A = \begin{bmatrix} 1 & 1 \\ 0 & 2 \end{bmatrix}$.

11. Let λ be an eigenvalue of A, corresponding to an eigenvector x.

   (a) Show that $\lambda^2 + 5$ is an eigenvalue of $A^2 + 5I$, corresponding to the same eigenvector x.

   (b) Show that $3\lambda^2 + 5$ is an eigenvalue of $3A^2 + 5I$, corresponding to the same eigenvector x.

   (c) Consider a quadratic polynomial $p(x) = 3x^2 - 7x + 5$. Define *a polynomial of matrix A* as $p(A) = 3A^2 - 7A + 5I$. Show that $p(\lambda)$ is an eigenvalue of $p(A)$, corresponding to the same eigenvector x.

12. Let A and B be any two $n \times n$ matrices, and c_1, c_2 two arbitrary numbers.

   (a) Show that $\text{tr}(A + B) = \text{tr}\, A + \text{tr}\, B$, and more generally,

$$\text{tr}(c_1 A + c_2 B) = c_1 \,\text{tr}\, A + c_2 \,\text{tr}\, B.$$

   (b) Show that $\text{tr}(AB) = \text{tr}(BA)$.

   Hint. $\text{tr}(AB) = \sum_{i,j=1}^{n} a_{ij} b_{ji} = \sum_{i,j=1}^{n} b_{ji} a_{ij} = \text{tr}(BA)$.

   (c) Show that it is impossible to find two $n \times n$ matrices A and B, so that

$$AB - BA = I.$$

   (d*) Show that it is impossible to find two $n \times n$ matrices A and B, with A invertible, so that

$$AB - BA = A.$$

   Hint. Multiply both sides by A^{-1}, to obtain

$$A(A^{-1}B) - (A^{-1}B)A = I.$$

13. Show that similar matrices have the same trace.

14. Suppose that two $n \times n$ matrices A and B have a common eigenvector x. Show that $\det(AB - BA) = 0$.

   Hint. Show that x is an eigenvector of $AB - BA$, and determine the corresponding eigenvalue.

15. Assume that all columns of a square matrix A add up to the same number b. Show that $\lambda = b$ is an eigenvalue of A.

   Hint. All columns of $A - bI$ add up to zero, and then $|A - bI| = 0$.

4.2 A complete set of eigenvectors

Throughout this section A will denote an arbitrary $n \times n$ matrix. Eigenvectors of A are vectors in R^n. Recall that the maximal number of linearly independent vectors in R^n is n, and any n linearly independent vectors in R^n form a basis of R^n. We say that an $n \times n$ matrix A has *a complete set of eigenvectors if A has n linearly independent eigenvectors.* For a 2×2 matrix, one needs two linearly independent eigenvectors for a complete set, for a 3×3 matrix it takes three, and so on. A complete set of eigenvectors forms a basis of R^n. Such *eigenvector bases* will play a prominent role in the next section. The following theorem provides a condition for A to have a complete set of eigenvectors.

Theorem 4.2.1. *Eigenvectors of A corresponding to distinct eigenvalues form a linearly independent set.*

Proof. We begin with the case of two eigenvectors u_1 and u_2 of A, corresponding to the eigenvalues λ_1 and λ_2 respectively, so that $Au_1 = \lambda_1 u_1$, $Au_2 = \lambda_2 u_2$, and $\lambda_2 \neq \lambda_1$. We need to show that u_1 and u_2 are linearly independent. Assume that the opposite is true. Then $u_2 = au_1$ for some number $a \neq 0$ (if $a = 0$, then $u_2 = 0$, while eigenvectors are nonzero vectors). Evaluate

$$Au_2 = A(au_1) = a\lambda_1 u_1 = \lambda_1 u_2 \neq \lambda_2 u_2,$$

contradicting the definition of u_2. Therefore u_1 and u_2 are linearly independent.

Next, consider the case of three eigenvectors u_1, u_2, u_3 of A, corresponding to the eigenvalues λ_1, λ_2, λ_3, respectively, so that $Au_1 = \lambda_1 u_1$, $Au_2 = \lambda_2 u_2$, $Au_3 = \lambda_3 u_3$, and λ_1, λ_2, λ_3 are three different (distinct) numbers. We just proved that u_1 and u_2 are linearly independent. To prove that u_1, u_2, u_3 are linearly independent, assume that the opposite is true. Then one of these vectors, say u_3, is a linear combination of the other two, so that

$$u_3 = au_1 + \beta u_2, \tag{2.1}$$

with some numbers a and β. Observe that a and β cannot be both zero, because otherwise $u_3 = 0$, contradicting the fact that u_3 is an eigenvector. Multiply both sides of (2.1) by A to get

$$Au_3 = aAu_1 + \beta Au_2,$$

which yields

$$\lambda_3 u_3 = a\lambda_1 u_1 + \beta\lambda_2 u_2. \tag{2.2}$$

From equation (2.2) subtract equation (2.1) multiplied by λ_3 to obtain

$$a(\lambda_1 - \lambda_3)u_1 + \beta(\lambda_2 - \lambda_3)u_2 = 0.$$

The coefficients $a(\lambda_1 - \lambda_3)$ and $\beta(\lambda_2 - \lambda_3)$ cannot be both zero, which implies that u_1 and u_2 are linearly dependent, a contradiction, proving linear independence of u_1, u_2, u_3. By a similar argument, we show that any set of four eigenvectors corresponding to distinct eigenvalues are linearly independent, and so on. $\square$

If an $n \times n$ matrix A has n distinct eigenvalues $\lambda_1, \lambda_2, \ldots, \lambda_n$, then the corresponding eigenvectors $u_1, u_2, \ldots, u_n$ are linearly independent according to this theorem, and form a complete set. If some of the eigenvalues $\lambda_1, \lambda_2, \ldots, \lambda_n$ are repeated, then A has fewer than n distinct eigenvalues. The next example shows that some matrices with repeated eigenvalues still have a complete set of eigenvectors.

Example 1. $A = \begin{bmatrix} 2 & 1 & 1 \\ 1 & 2 & 1 \\ 1 & 1 & 2 \end{bmatrix}$. Expanding the characteristic equation

$$|A - \lambda I| = \begin{vmatrix} 2 - \lambda & 1 & 1 \\ 1 & 2 - \lambda & 1 \\ 1 & 1 & 2 - \lambda \end{vmatrix} = 0$$

in, say, the first row, produces a cubic equation

$$\lambda^3 - 6\lambda^2 + 9\lambda - 4 = 0.$$

To solve it, we need to guess a root: $\lambda_1 = 1$ is a root, which implies that the cubic polynomial has a factor $\lambda - 1$. The second factor is found by division of the polynomials, giving

$$(\lambda - 1)(\lambda^2 - 5\lambda + 4) = 0.$$

Setting the second factor to zero, $\lambda^2 - 5\lambda + 4 = 0$, gives the other two roots $\lambda_2 = 1$ and $\lambda_3 = 4$. The eigenvalues are $1, 1, 4$. The eigenvalue $\lambda_1 = 1$ is repeated, while the eigenvalue $\lambda_3 = 4$ is simple.

To find the eigenvectors of the double eigenvalue $\lambda_1 = 1$, one needs to solve the system $(A - I)x = 0$, which is

$$x_1 + x_2 + x_3 = 0,$$
$$x_1 + x_2 + x_3 = 0,$$
$$x_1 + x_2 + x_3 = 0.$$

Discarding both the second and third equations leaves

$$x_1 + x_2 + x_3 = 0.$$

Here x_2 and x_3 are free variables. Letting $x_3 = t$ and $x_2 = s$, two arbitrary numbers, calculate $x_1 = -t - s$. The solution set is then

$$\begin{bmatrix} -t-s \\ s \\ t \end{bmatrix} = t \begin{bmatrix} -1 \\ 0 \\ 1 \end{bmatrix} + s \begin{bmatrix} -1 \\ 1 \\ 0 \end{bmatrix} = tu_1 + su_2,$$

where $u_1 = \begin{bmatrix} -1 \\ 0 \\ 1 \end{bmatrix}$ and $u_2 = \begin{bmatrix} -1 \\ 1 \\ 0 \end{bmatrix}$. Conclusion: the linear combinations with arbitrary coefficients, or *the span*, of two linearly independent eigenvectors u_1 and u_2 gives the space of all eigenvectors corresponding to $\lambda_1 = 1$, also known as *the eigenspace of* $\lambda_1 = 1$.

The eigenvectors corresponding to the eigenvalue $\lambda_3 = 4$ are solutions of the system $(A - 4I)x = 0$, which is

$$-2x_1 + x_2 + x_3 = 0,$$
$$x_1 - 2x_2 + x_3 = 0,$$
$$x_1 + x_2 - 2x_3 = 0.$$

Discard the third equation as superfluous, because adding the first two equations gives the negative of the third. In the remaining equations,

$$-2x_1 + x_2 + x_3 = 0,$$
$$x_1 - 2x_2 + x_3 = 0,$$

set $x_3 = 1$, then solve the resulting system

$$-2x_1 + x_2 = -1,$$
$$x_1 - 2x_2 = -1,$$

obtaining $x_1 = 1$ and $x_2 = 1$. Conclusion: $c \begin{bmatrix} 1 \\ 1 \\ 1 \end{bmatrix}$ are the eigenvectors corresponding to $\lambda_3 = 4$, with c arbitrary. The answer can also be written as cu_3, where $u_3 = \begin{bmatrix} 1 \\ 1 \\ 1 \end{bmatrix}$ is an eigenvector corresponding to $\lambda_3 = 4$.

Observe that u_3 is not in the span of u_1 and u_2 (because vectors in that span are eigenvectors corresponding to λ_1). By Theorem 1.5.1, the vectors u_1, u_2, u_3 are linearly independent, so that they form a complete set of eigenvectors.

Example 2. Let $A = \begin{bmatrix} 3 & -2 \\ 0 & 3 \end{bmatrix}$. Here $\lambda_1 = \lambda_2 = 3$ is a repeated eigenvalue. The system $(A - 3I)x = 0$ reduces to

$$-2x_2 = 0,$$

so that $x_2 = 0$, while x_1 is arbitrary. There is only one linearly independent eigenvector $\begin{bmatrix} x_1 \\ 0 \end{bmatrix} = x_1 \begin{bmatrix} 1 \\ 0 \end{bmatrix}$. This matrix does not have a complete set of eigenvectors.

4.2.1 Complex eigenvalues

For the matrix $A = \begin{bmatrix} 0 & -1 \\ 1 & 0 \end{bmatrix}$, the characteristic equation is

$$|A - \lambda I| = \begin{vmatrix} -\lambda & -1 \\ 1 & -\lambda \end{vmatrix} = \lambda^2 + 1 = 0.$$

Its roots are $\lambda_1 = i$, and $\lambda_2 = -i$. The corresponding eigenvectors will also have complex valued entries, although the procedure for finding eigenvectors remains the same.

(i) $\lambda_1 = i$. The corresponding eigenvectors satisfy the system $(A - iI)x = 0$, or in components

$$-ix_1 - x_2 = 0,$$
$$x_1 - ix_2 = 0.$$

Discard the second equation, because it can be obtained multiplying the first equation by i. In the first equation,

$$-ix_1 - x_2 = 0,$$

set $x_2 = c$, then $x_1 = -\frac{c}{i} = c\,i$. Thus we obtain the eigenvectors $c\begin{bmatrix} i \\ 1 \end{bmatrix}$, where c is *any complex number*.

(ii) $\lambda_2 = -i$. The corresponding eigenvectors satisfy the system $(A + iI)x = 0$, or in components

$$ix_1 - x_2 = 0,$$
$$x_1 + ix_2 = 0.$$

Discard the second equation, because it can be obtained multiplying the first equation by $-i$. In the first equation,

$$ix_1 - x_2 = 0,$$

set $x_2 = c$, then $x_1 = \frac{c}{i} = -c\,i$. Thus we obtain the eigenvectors $c\begin{bmatrix} -i \\ 1 \end{bmatrix}$, where c is any complex number.

Recall that given a complex number $z = x + iy$, with real x and y, one defines *the complex conjugate* as $\bar{z} = x - iy$. If $z = x$, a real number, then $\bar{z} = x = z$. One has $z\bar{z} = x^2 + y^2 = |z|^2$, where $|z| = \sqrt{x^2 + y^2}$ is called *the modulus of z*. Given complex numbers $z_1, z_2, \ldots, z_n$, one has

$$\overline{z_1 + z_2 + \cdots + z_n} = \bar{z}_1 + \bar{z}_2 + \cdots + \bar{z}_n,$$
$$\overline{z_1 \cdot z_2 \cdots z_n} = \bar{z}_1 \cdot \bar{z}_2 \cdots \bar{z}_n.$$

Given a vector $z = \begin{bmatrix} z_1 \\ z_2 \\ \vdots \\ z_n \end{bmatrix}$ with complex entries, one defines its complex conjugate as $\bar{z} = \begin{bmatrix} \bar{z}_1 \\ \bar{z}_2 \\ \vdots \\ \bar{z}_n \end{bmatrix}$. The eigenvalues of the matrix A above were complex conjugates of one another, as well as the corresponding eigenvectors. The same is true in general, as the following theorem shows.

Theorem 4.2.2. *Let A be a square matrix with real entries. Let λ be a complex (not real) eigenvalue, and z a corresponding complex eigenvector. Then $\bar{\lambda}$ is also an eigenvalue, and $\bar{z}$ a corresponding eigenvector.*

Proof. We are given that

$$Az = \lambda z.$$

Take complex conjugates of both sides (elements of A are real numbers) to get

$$A\bar{z} = \bar{\lambda}\bar{z},$$

which implies that $\bar{\lambda}$ is an eigenvalue, and $\bar{z}$ a corresponding eigenvector. (The i-th component of Az is $\sum_{k=1}^{n} a_{ik}z_k$, and $\overline{\sum_{k=1}^{n} a_{ik}z_k} = \sum_{k=1}^{n} a_{ik}\bar{z}_k$.) □

Exercises

1. Find the eigenvectors of the following matrices, and determine if they form a complete set:

   (a) $\begin{bmatrix} 1 & 2 \\ 0 & -1 \end{bmatrix}$.

   Answer: $\begin{bmatrix} -1 \\ 1 \end{bmatrix}$ with $\lambda_1 = -1$ and $\begin{bmatrix} 1 \\ 0 \end{bmatrix}$ with $\lambda_1 = 1$, a complete set.

   (b) $\begin{bmatrix} 1 & 2 \\ 0 & 1 \end{bmatrix}$.

   Answer: $\begin{bmatrix} 1 \\ 0 \end{bmatrix}$ corresponding to $\lambda_1 = \lambda_2 = 1$, not a complete set.

   (c) $\begin{bmatrix} 1 & 0 \\ 0 & 1 \end{bmatrix}$.

   Answer: $\begin{bmatrix} 1 \\ 0 \end{bmatrix}$ and $\begin{bmatrix} 0 \\ 1 \end{bmatrix}$ corresponding to $\lambda_1 = \lambda_2 = 1$, a complete set.

   (d) $\begin{bmatrix} 1 & 3 & 6 \\ -3 & -5 & -6 \\ 3 & 3 & 4 \end{bmatrix}$.

   Hint. Observe that $\lambda_1 = -2$ is a root of the characteristic equation

$$\lambda^3 - 12\lambda - 16 = 0,$$

   then obtain the other two roots $\lambda_2 = -2$ and $\lambda_1 = 4$ by factoring.

   Answer: $\begin{bmatrix} -2 \\ 0 \\ 1 \end{bmatrix}$ and $\begin{bmatrix} -1 \\ 1 \\ 0 \end{bmatrix}$ corresponding to $\lambda_1 = \lambda_2 = -2$, and $\begin{bmatrix} 1 \\ -1 \\ 1 \end{bmatrix}$ corresponding to $\lambda_3 = 4$, a complete set.

(e) $\begin{bmatrix} 0 & 1 & 1 \\ 0 & 0 & 1 \\ 0 & 0 & 1 \end{bmatrix}$.

Answer: $\begin{bmatrix} 1 \\ 0 \\ 0 \end{bmatrix}$ corresponding to $\lambda_1 = \lambda_2 = 0$, and $\begin{bmatrix} 2 \\ 1 \\ 1 \end{bmatrix}$ corresponding to $\lambda_3 = 1$, not a complete set.

(f) $\begin{bmatrix} -1 & 1 & 1 \\ 1 & -1 & 1 \\ 1 & 0 & 0 \end{bmatrix}$.

Answer: $\begin{bmatrix} -2 \\ 1 \\ 1 \end{bmatrix}$ corresponding to $\lambda_1 = -2$, $\begin{bmatrix} -1 \\ -1 \\ 1 \end{bmatrix}$ corresponding to $\lambda_2 = -1$, and $\begin{bmatrix} 1 \\ 1 \\ 1 \end{bmatrix}$ corresponding to $\lambda_3 = 1$, a complete set.

(g) $\begin{bmatrix} 0 & 1 & 2 \\ -5 & -3 & -7 \\ 1 & 0 & 0 \end{bmatrix}$.

Answer: $\begin{bmatrix} -1 \\ -1 \\ 1 \end{bmatrix}$ corresponding to $\lambda_1 = \lambda_2 = \lambda_3 = -1$, not a complete set.

2. Find the eigenvalues and corresponding eigenvectors.

(a) $\begin{bmatrix} 1 & 1 \\ -1 & 1 \end{bmatrix}$.

Answer: $\lambda_1 = 1 - i$ with $\begin{bmatrix} i \\ 1 \end{bmatrix}$, and $\lambda_2 = 1 + i$ with $\begin{bmatrix} -i \\ 1 \end{bmatrix}$.

(b) $\begin{bmatrix} 3 & 3 & 2 \\ 1 & 1 & -2 \\ -3 & -1 & 0 \end{bmatrix}$.

Answer: $\lambda_1 = -2i$ with $\begin{bmatrix} i \\ -i \\ 1 \end{bmatrix}$, $\lambda_2 = 2i$ with $\begin{bmatrix} -i \\ i \\ 1 \end{bmatrix}$, and $\lambda_3 = 4$ with $\begin{bmatrix} -1 \\ -1 \\ 1 \end{bmatrix}$.

(c) $\begin{bmatrix} 1 & 2 & -1 \\ -2 & -1 & 1 \\ -1 & 1 & 0 \end{bmatrix}$.

Answer: $\lambda_1 = -i$ with $\begin{bmatrix} 1+i \\ 1-i \\ 2 \end{bmatrix}$, $\lambda_2 = i$ with $\begin{bmatrix} 1-i \\ 1+i \\ 2 \end{bmatrix}$, and $\lambda_3 = 0$ with $\begin{bmatrix} 1 \\ 1 \\ 3 \end{bmatrix}$.

(d) $\begin{bmatrix} \cos\theta & -\sin\theta \\ \sin\theta & \cos\theta \end{bmatrix}$, θ is a real number.

Hint. $\lambda_1 = \cos\theta - i\sin\theta$, $\lambda_2 = \cos\theta + i\sin\theta$.

3. Let A be an $n \times n$ matrix with real entries, and suppose n is odd. Show that A has at least one real eigenvalue.

Hint. The characteristic equation is a polynomial equation of odd degree.

4. Find the complex conjugate $\bar{z}$ and the modulus $|z|$ for the following numbers: (a) $3 - 4i$; (b) $5i$; (c) -7; (d) $\cos\frac{\pi}{5} + i\sin\frac{\pi}{5}$; (e) $e^{i\theta}$, θ is real.

5. Let A be a 2×2 matrix with $\operatorname{tr} A = 2$ and $\det(A) = 2$. What are the eigenvalues of A?

6. A matrix A^2 has eigenvalues -1 and -4. What is the smallest possible size of the matrix A (with real entries)?

Answer: 4×4.

4.3 Diagonalization

Any $n \times n$ matrix A,

$$A = \begin{bmatrix} a_{11} & a_{12} & \cdots & a_{1n} \\ a_{21} & a_{22} & \cdots & a_{2n} \\ \vdots & \vdots & \ddots & \vdots \\ a_{n1} & a_{n2} & \cdots & a_{nn} \end{bmatrix} = [C_1 \, C_2 \, \ldots \, C_n],$$

can be written through its column vectors, where

$$C_1 = \begin{bmatrix} a_{11} \\ a_{21} \\ \vdots \\ a_{n1} \end{bmatrix}, \quad C_2 = \begin{bmatrix} a_{12} \\ a_{22} \\ \vdots \\ a_{n2} \end{bmatrix}, \dots, C_n = \begin{bmatrix} a_{1n} \\ a_{2n} \\ \vdots \\ a_{nn} \end{bmatrix}.$$

Recall that given a vector $x = \begin{bmatrix} x_1 \\ x_2 \\ \vdots \\ x_n \end{bmatrix}$, the product Ax was defined as the vector

$$Ax = x_1 C_1 + x_2 C_2 + \cdots + x_n C_n. \tag{3.1}$$

If $B = [K_1 K_2 \dots K_n]$ is another $n \times n$ matrix, with the column vectors $K_1, K_2, \dots, K_n$, then the product AB was defined as follows:

$$AB = A[K_1 K_2 \dots K_n] = [AK_1 AK_2 \dots AK_n],$$

where the products $AK_1, AK_2, \dots, AK_n$ are calculated using (3.1).

Let D be a diagonal matrix

$$D = \begin{bmatrix} \lambda_1 & 0 & \cdots & 0 \\ 0 & \lambda_2 & \cdots & 0 \\ \vdots & \vdots & \ddots & \vdots \\ 0 & 0 & \cdots & \lambda_n \end{bmatrix}. \tag{3.2}$$

Calculate the product

$$AD = \begin{bmatrix} A \begin{bmatrix} \lambda_1 \\ 0 \\ \vdots \\ 0 \end{bmatrix} & A \begin{bmatrix} 0 \\ \lambda_2 \\ \vdots \\ 0 \end{bmatrix} & \cdots & A \begin{bmatrix} 0 \\ 0 \\ \vdots \\ \lambda_n \end{bmatrix} \end{bmatrix} = [\lambda_1 C_1 \ \lambda_2 C_2 \ \cdots \ \lambda_n C_n].$$

Conclusion: *multiplying a matrix A from the right by a diagonal matrix D, results in the columns of A being multiplied by the corresponding entries of D.* In particular, to multiply two diagonal matrices (in either order), one multiplies the corresponding diagonal entries. For example, let $D_1 = \begin{bmatrix} a & 0 & 0 \\ 0 & b & 0 \\ 0 & 0 & c \end{bmatrix}$ and $D_2 = \begin{bmatrix} 2 & 0 & 0 \\ 0 & 3 & 0 \\ 0 & 0 & 4 \end{bmatrix}$, then

$$D_1 D_2 = D_2 D_1 = \begin{bmatrix} 2a & 0 & 0 \\ 0 & 3b & 0 \\ 0 & 0 & 4c \end{bmatrix}.$$

Another example is

$$
\begin{bmatrix} a_{11} & a_{12} & a_{13} \\ a_{21} & a_{22} & a_{23} \\ a_{31} & a_{32} & a_{33} \end{bmatrix} \begin{bmatrix} 2 & 0 & 0 \\ 0 & 3 & 0 \\ 0 & 0 & 4 \end{bmatrix} = \begin{bmatrix} 2a_{11} & 3a_{12} & 4a_{13} \\ 2a_{21} & 3a_{22} & 4a_{23} \\ 2a_{31} & 3a_{32} & 4a_{33} \end{bmatrix}.
$$

Suppose now that an $n \times n$ matrix A has a complete set of n linearly independent eigenvectors $u_1, u_2, \ldots, u_n$, so that $Au_1 = \lambda_1 u_1, Au_2 = \lambda_2 u_2, \ldots, Au_n = \lambda_n u_n$ (the eigenvalues $\lambda_1, \lambda_2, \ldots, \lambda_n$ are not necessarily different). Form a matrix $P = [u_1 \ u_2 \ \ldots \ u_n]$, using the eigenvectors as columns. Observe that P has an inverse matrix P^{-1}, because the columns of P are linearly independent. Calculate

$$
AP = [Au_1 \ Au_2 \ \ldots \ Au_n] = [\lambda_1 u_1 \ \lambda_2 u_2 \ldots \lambda_n u_n] = PD, \tag{3.3}
$$

where D is a diagonal matrix, shown in (3.2), with the eigenvalues of A on the diagonal. Multiplying both sides of (3.3) from the left by P^{-1}, obtain

$$
P^{-1}AP = D. \tag{3.4}
$$

Similarly, multiplying (3.3) by P^{-1} from the right yields

$$
A = PDP^{-1}. \tag{3.5}
$$

One refers to the formulas (3.4) and (3.5) as giving *the diagonalization of matrix A*, and matrix A is called *diagonalizable*. Diagonalizable matrices are *similar to diagonal ones*. The matrix P is called *the diagonalizing matrix*. There are infinitely many choices of the diagonalizing matrix P, because eigenvectors (the columns of P) may be multiplied by arbitrary numbers. If A has some complex (not real) eigenvalues, formulas (3.4) and (3.5) still hold, although some of the entries of P and D are complex.

Example 1. The matrix $A = \begin{bmatrix} 1 & 4 \\ 1 & -2 \end{bmatrix}$ has eigenvalues $\lambda_1 = -3$ with a corresponding eigenvector $u_1 = \begin{bmatrix} -1 \\ 1 \end{bmatrix}$, and $\lambda_2 = 2$ with a corresponding eigenvector $u_2 = \begin{bmatrix} 4 \\ 1 \end{bmatrix}$. Here $P = \begin{bmatrix} -1 & 4 \\ 1 & 1 \end{bmatrix}$ and $D = \begin{bmatrix} -3 & 0 \\ 0 & 2 \end{bmatrix}$. Calculate $P^{-1} = \frac{1}{5} \begin{bmatrix} -1 & 4 \\ 1 & 1 \end{bmatrix}$. Formula (3.4) becomes

$$
\frac{1}{5} \begin{bmatrix} -1 & 4 \\ 1 & 1 \end{bmatrix} \begin{bmatrix} 1 & 4 \\ 1 & -2 \end{bmatrix} \begin{bmatrix} -1 & 4 \\ 1 & 1 \end{bmatrix} = \begin{bmatrix} -3 & 0 \\ 0 & 2 \end{bmatrix}.
$$

Not every matrix can be diagonalized. It follows from (3.3) that the columns of diagonalizing matrix P are eigenvectors of A (since $Au_i = \lambda_i u_i$), and these eigenvectors must be linearly independent in order for P^{-1} to exist. We conclude that *a matrix A is diagonalizible if and only if it has a complete set of eigenvectors.*

Example 2. The matrix $B = \begin{bmatrix} 2 & 1 \\ -1 & 0 \end{bmatrix}$ has a repeated eigenvalue $\lambda_1 = \lambda_2 = 1$, but only one linearly independent eigenvector $u = \begin{bmatrix} -1 \\ 1 \end{bmatrix}$. The matrix B is not diagonalizable.

Example 3. Recall the matrix

$$A = \begin{bmatrix} 2 & 1 & 1 \\ 1 & 2 & 1 \\ 1 & 1 & 2 \end{bmatrix}$$

from the preceding section. It has a repeated eigenvalue $\lambda_1 = \lambda_2 = 1$, together with $\lambda_3 = 4$, and a complete set of eigenvectors $u_1 = \begin{bmatrix} -1 \\ 0 \\ 1 \end{bmatrix}$, and $u_2 = \begin{bmatrix} -1 \\ 1 \\ 0 \end{bmatrix}$ corresponding to $\lambda_1 = \lambda_2 = 1$, and $u_3 = \begin{bmatrix} 1 \\ 1 \\ 1 \end{bmatrix}$ corresponding to $\lambda_3 = 4$. This matrix is diagonalizable, with

$$P = \begin{bmatrix} -1 & -1 & 1 \\ 0 & 1 & 1 \\ 1 & 0 & 1 \end{bmatrix}, \quad P^{-1} = \frac{1}{3}\begin{bmatrix} -1 & -1 & 2 \\ -1 & 2 & -1 \\ 1 & 1 & 1 \end{bmatrix}, \quad D = \begin{bmatrix} 1 & 0 & 0 \\ 0 & 1 & 0 \\ 0 & 0 & 4 \end{bmatrix}.$$

Recall that any n linearly independent vectors form a basis of R^n. If a matrix A has a complete set of eigenvectors, we can use *the eigenvector basis* $B = \{u_1, u_2, \ldots, u_n\}$. Any vector $x \in R^n$ can be decomposed as $x = x_1 u_1 + x_2 u_2 + \cdots + x_n u_n$, by using its coordinates $[x]_B = \begin{bmatrix} x_1 \\ x_2 \\ \vdots \\ x_n \end{bmatrix}$ with respect to this basis B. Calculate

$$Ax = x_1 A u_1 + x_2 A u_2 + \cdots + x_n A u_n = x_1 \lambda_1 u_1 + x_2 \lambda_2 u_2 + \cdots + x_n \lambda_n u_n.$$

It follows that $[Ax]_B = \begin{bmatrix} \lambda_1 x_1 \\ \lambda_2 x_2 \\ \vdots \\ \lambda_n x_n \end{bmatrix}$, and then $[Ax]_B = D[x]_B$. *Conclusion: if one uses the eigenvector basis B in R^n, then the function Ax (or the transformation Ax) is represented by a diagonal matrix D, consisting of eigenvalues of A.*

We discuss some applications of diagonalization next. For any two diagonal matrices of the same size,

$$D_1 D_2 = D_2 D_1,$$

since both products are calculated by multiplying the diagonal entries. For general $n \times n$ matrices A and B, the relation

$$AB = BA \tag{3.6}$$

is rare. The following theorem explains why. If $AB = BA$, one says that *the matrices A and B commute.* Any two diagonal matrices commute.

Theorem 4.3.1. *Two diagonalizable matrices commute if and only if they share the same set of eigenvectors.*

Proof. If two diagonalizable matrices A and B share the same set of eigenvectors, they share the same diagonalizing matrix P, so that $A = PD_1P^{-1}$ and $B = PD_2P^{-1}$, with two diagonal matrices D_1 and D_2. It follows that

$$AB = PD_1P^{-1}PD_2P^{-1} = PD_1(P^{-1}P)D_2P^{-1} = PD_1D_2P^{-1}$$
$$= PD_2D_1P^{-1} = PD_2P^{-1}PD_1P^{-1} = BA.$$

The proof of the converse statement is not included. □

If A is diagonalizable, then

$$A = PDP^{-1},$$

where D is a diagonal matrix with the eigenvalues of A on the diagonal. Calculate

$$A^2 = AA = PDP^{-1}PDP^{-1} = PDDP^{-1} = PD^2P^{-1},$$

and similarly for other powers

$$A^k = PD^kP^{-1} = P\begin{bmatrix} \lambda_1^k & 0 & \cdots & 0 \\ 0 & \lambda_2^k & \cdots & 0 \\ \vdots & \vdots & \ddots & \vdots \\ 0 & 0 & \cdots & \lambda_n^k \end{bmatrix}P^{-1}.$$

Define the limit $\lim_{k\to\infty} A^k$ by taking the limits of each component of A^k. If the eigenvalues of A have modulus $|\lambda_i| < 1$ for all i, then $\lim_{k\to\infty} A^k = O$, *the zero matrix*. Indeed, D^k tends to the zero matrix, while P and P^{-1} are fixed.

Example 4. Let $A = \begin{bmatrix} 1 & 8 \\ 0 & -1 \end{bmatrix}$. Calculate A^{57}.

The eigenvalues of this upper triangular matrix A are $\lambda_1 = 1$ and $\lambda_2 = -1$. Since $\lambda_1 \neq \lambda_2$, the corresponding eigenvectors are linearly independent, and A is diagonalizable, so that

$$A = P\begin{bmatrix} 1 & 0 \\ 0 & -1 \end{bmatrix}P^{-1},$$

with an appropriate diagonalizing matrix P, and the corresponding P^{-1}. Then

$$A^{57} = P\begin{bmatrix} 1^{57} & 0 \\ 0 & (-1)^{57} \end{bmatrix}P^{-1} = P\begin{bmatrix} 1 & 0 \\ 0 & -1 \end{bmatrix}P^{-1} = A = \begin{bmatrix} 1 & 8 \\ 0 & -1 \end{bmatrix}.$$

Similarly, $A^k = A$ if k is an odd integer, while $A^k = I$ if k is an even integer.

Exercises

1. If the matrix A is diagonalizable, determine the diagonalizing matrix P and the diagonal matrix D, and verify that $AP = PD$.

   (a) $A = \begin{bmatrix} 4 & -2 \\ 1 & 1 \end{bmatrix}$.

   Answer: $P = \begin{bmatrix} 2 & 1 \\ 1 & 1 \end{bmatrix}$, $D = \begin{bmatrix} 3 & 0 \\ 0 & 2 \end{bmatrix}$.

   (b) $A = \begin{bmatrix} 2 & -1 \\ 0 & 2 \end{bmatrix}$.

   Answer: Not diagonalizable.

   (c) $A = \begin{bmatrix} 2 & 0 \\ 0 & -7 \end{bmatrix}$.

   Answer: The matrix is already diagonal, $P = I$.

   (d) $A = \begin{bmatrix} 2 & -1 & 1 \\ 0 & 2 & 1 \\ 0 & 0 & 2 \end{bmatrix}$.

   Answer: Not diagonalizable.

   (e) $A = \begin{bmatrix} 1 & 3 & 6 \\ -3 & -5 & -6 \\ 3 & 3 & 4 \end{bmatrix}$. Hint. The eigenvalues and the eigenvectors of this matrix were calculated in the preceding set of exercises.

   Answer: $P = \begin{bmatrix} -2 & -1 & 1 \\ 0 & 1 & -1 \\ 1 & 0 & 1 \end{bmatrix}$, $D = \begin{bmatrix} -2 & 0 & 0 \\ 0 & -2 & 0 \\ 0 & 0 & 4 \end{bmatrix}$.

   (f) $A = \begin{bmatrix} 1 & 1 & 1 \\ 1 & 1 & 1 \\ 1 & 0 & 2 \end{bmatrix}$.

   Answer: $P = \begin{bmatrix} -2 & -1 & 1 \\ 1 & -1 & 1 \\ 1 & 1 & 1 \end{bmatrix}$, $D = \begin{bmatrix} 0 & 0 & 0 \\ 0 & 1 & 0 \\ 0 & 0 & 3 \end{bmatrix}$.

   (g) $A = \begin{bmatrix} 1 & 1 & 1 \\ 1 & 1 & 1 \\ 1 & 1 & 1 \end{bmatrix}$.

   Answer: $P = \begin{bmatrix} -1 & -1 & 1 \\ 0 & 1 & 1 \\ 1 & 0 & 1 \end{bmatrix}$, $D = \begin{bmatrix} 0 & 0 & 0 \\ 0 & 0 & 0 \\ 0 & 0 & 3 \end{bmatrix}$.

   (h) $A = \begin{bmatrix} 1 & 2 & 3 & 4 \\ 0 & 1 & 2 & 3 \\ 0 & 0 & 1 & 2 \\ 0 & 0 & 0 & 1 \end{bmatrix}$.

   Answer: Not diagonalizable.

   (i) $A = \begin{bmatrix} a & b-a \\ 0 & b \end{bmatrix}$, $b \neq a$.

   Answer: $P = \begin{bmatrix} 1 & 1 \\ 0 & 1 \end{bmatrix}$, $D = \begin{bmatrix} a & 0 \\ 0 & b \end{bmatrix}$.

2. Show that $\begin{bmatrix} a & b-a \\ 0 & b \end{bmatrix}^k = \begin{bmatrix} a^k & b^k-a^k \\ 0 & b^k \end{bmatrix}$.

3. Let A be a 2×2 matrix with positive eigenvalues $\lambda_1 \neq \lambda_2$.

   (a) Explain why A is diagonalizable, and how one constructs a nonsingular matrix P such that $A = P \begin{bmatrix} \lambda_1 & 0 \\ 0 & \lambda_2 \end{bmatrix} P^{-1}$.

   (b) Define *the square root of matrix* A as $\sqrt{A} = P \begin{bmatrix} \sqrt{\lambda_1} & 0 \\ 0 & \sqrt{\lambda_2} \end{bmatrix} P^{-1}$. Show that $(\sqrt{A})^2 = A$.

   (c) Let $B = \begin{bmatrix} 14 & -10 \\ 5 & -1 \end{bmatrix}$. Find $\sqrt{B}$.

   Answer: $\sqrt{B} = \begin{bmatrix} 4 & -2 \\ 1 & 1 \end{bmatrix}$.

   (d) Are there any other matrices C with the property $A = C^2$?

   Hint. Try $C = P \begin{bmatrix} \pm\sqrt{\lambda_1} & 0 \\ 0 & \pm\sqrt{\lambda_2} \end{bmatrix} P^{-1}$.

4. Let $A = \begin{bmatrix} 2 & 1 \\ -2 & -1 \end{bmatrix}$. Show that $A^k = A$, where k is any positive integer.

5. Let $A = \begin{bmatrix} 1 & 1 \\ -3/4 & -1 \end{bmatrix}$. Show that $\lim_{k \to \infty} A^k = O$, where *the limit of a sequence of matrices is calculated by taking the limit of each component.*

6. Let A be a 3×3 matrix with the eigenvalues $0, -1, 1$. Show that $A^7 = A$.

7. Let A be a 4×4 matrix with the eigenvalues $-i, i, -1, 1$.
   (a) Show that $A^4 = I$.
   (b) Show that $A^{4n} = I$, and $A^{4n+1} = A$ for any positive integer n.

8. Let A be a diagonalizable 2×2 matrix, so that $A = P \begin{bmatrix} \lambda_1 & 0 \\ 0 & \lambda_2 \end{bmatrix} P^{-1}$. Consider a polynomial $q(x) = 2x^2 - 3x + 5$. Calculate $q(A) = 2A^2 - 3A + 5I$.

   Answer:

   $$q(A) = P \begin{bmatrix} 2\lambda_1^2 - 3\lambda_1 + 5 & 0 \\ 0 & 2\lambda_2^2 - 3\lambda_2 + 5 \end{bmatrix} P^{-1}$$

   $$= P \begin{bmatrix} q(\lambda_1) & 0 \\ 0 & q(\lambda_2) \end{bmatrix} P^{-1}.$$

9. Let A be an $n \times n$ matrix, and let $q(\lambda) = |A - \lambda I|$ be its characteristic polynomial. Write $q(\lambda) = a_0 \lambda^n + a_1 \lambda^{n-1} + \cdots + a_{n-1} \lambda + a_n$, with some coefficients $a_0, a_1, \ldots, a_n$. The *Cayley–Hamilton theorem* asserts that any matrix A is a root of its own characteristic polynomial, so that

   $$q(A) = a_0 A^n + a_1 A^{n-1} + \cdots + a_{n-1} A + a_n I = O,$$

   where O is the zero matrix. Justify this theorem in case A is diagonalizable.

5 Orthogonality and symmetry

5.1 Inner products

Given two vectors in R^n, $a = \begin{bmatrix} a_1 \\ a_2 \\ \vdots \\ a_n \end{bmatrix}$ and $b = \begin{bmatrix} b_1 \\ b_2 \\ \vdots \\ b_n \end{bmatrix}$, define their *inner product* (also known as *scalar product* or *dot product*) as

$$a \cdot b = a_1 b_1 + a_2 b_2 + \cdots + a_n b_n.$$

In three dimensions ($n = 3$), this notion was used in Calculus to calculate the length of a vector $\|a\| = \sqrt{a \cdot a} = \sqrt{a_1^2 + a_2^2 + a_3^2}$, and the angle θ between vectors a and b, given by $\cos\theta = \frac{a \cdot b}{\|a\| \|b\|}$. In particular, a and b are perpendicular if and only if $a \cdot b = 0$. Similarly, the projection of b on a was calculated as follows:

$$\text{Proj}_a b = \|b\| \cos\theta \, \frac{a}{\|a\|} = \frac{\|a\| \|b\| \cos\theta}{\|a\|^2} a = \frac{a \cdot b}{\|a\|^2} a.$$

(Recall that $\|b\| \cos\theta$ is the length of the projection vector, while $\frac{a}{\|a\|}$ gives the unit vector in the direction of a.)

In dimensions $n > 3$, these formulas are taken as the definitions of the corresponding notions. Namely, *the length (or the norm, or the magnitude) of a vector a is defined as*

$$\|a\| = \sqrt{a \cdot a} = \sqrt{a_1^2 + a_2^2 + \cdots + a_n^2}.$$

The angle θ between two vectors in R^n is defined by $\cos\theta = \frac{a \cdot b}{\|a\| \|b\|}$. *Vectors a and b in R^n are called orthogonal if*

$$a \cdot b = 0.$$

Define *the projection of $b \in R^n$ on $a \in R^n$* as

$$\text{Proj}_a b = \frac{a \cdot b}{\|a\|^2} a = \frac{a \cdot b}{a \cdot a} a.$$

Let us verify that subtracting from b its projection on a gives a vector orthogonal to a. In other words, $b - \text{Proj}_a b$ is orthogonal to a. Indeed,

$$a \cdot (b - \text{Proj}_a b) = a \cdot b - \frac{a \cdot b}{\|a\|^2} a \cdot a = a \cdot b - a \cdot b = 0,$$

using the distributive property of inner product (verified in Exercises).

https://doi.org/10.1515/9783111086507-005

For example, if $a = \begin{bmatrix} 1 \\ -2 \\ 0 \\ 2 \end{bmatrix}$ and $b = \begin{bmatrix} 2 \\ 1 \\ -4 \\ 3 \end{bmatrix}$ are two vectors in R^4, then $a \cdot b = 6$, $\|a\| = 3$, and

$$\text{Proj}_a b = \frac{a \cdot b}{\|a\|^2} a = \frac{6}{3^2} a = \frac{2}{3} a = \frac{2}{3} \begin{bmatrix} 1 \\ -2 \\ 0 \\ 2 \end{bmatrix} = \begin{bmatrix} 2/3 \\ -4/3 \\ 0 \\ 4/3 \end{bmatrix}.$$

Given vectors x, y, z in R^n, and a number c, the following *properties* follow immediately from the definition of inner product:

$$x \cdot y = y \cdot x,$$
$$x \cdot (y + z) = x \cdot y + x \cdot z,$$
$$(x + y) \cdot z = x \cdot z + y \cdot z,$$
$$(cx) \cdot y = c(x \cdot y) = x \cdot (cy),$$
$$\|cx\| = |c| \, \|x\|.$$

These rules are similar to multiplication of numbers.

If vectors x and y in R^n are orthogonal, *the Pythagorean theorem* holds:

$$\|x + y\|^2 = \|x\|^2 + \|y\|^2.$$

Indeed, we are given that $x \cdot y = 0$, and then

$$\|x + y\|^2 = (x + y) \cdot (x + y) = x \cdot x + 2x \cdot y + y \cdot y = \|x\|^2 + \|y\|^2.$$

If a vector u has length one, $\|u\| = 1$, u is called a *unit vector*. Of all the multiples kv of a vector $v \in R^n$, one often wishes to select the unit vector. Choosing $k = \frac{1}{\|v\|}$ produces such a vector, $\frac{1}{\|v\|} v = \frac{v}{\|v\|}$. Indeed,

$$\left\| \frac{1}{\|v\|} v \right\| = \frac{1}{\|v\|} \|v\| = 1.$$

The vector $u = \frac{v}{\|v\|}$ is called *the normalization of v*. When projecting on a unit vector u, the formula simplifies to

$$\text{Proj}_u b = \frac{u \cdot b}{\|u\|^2} u = (b \cdot u) u.$$

Vector $x \in R^n$ is a column vector (or an $n \times 1$ matrix), while x^T is a row vector (or a $1 \times n$ matrix). One can express the inner product of two vectors in R^n in terms of the matrix product

$$x \cdot y = x^T y. \tag{1.1}$$

If A is an $n \times n$ matrix, then

$$Ax \cdot y = x \cdot A^T y,$$

for any $x, y \in R^n$. Indeed, using (1.1) twice yields

$$Ax \cdot y = (Ax)^T y = x^T A^T y = x \cdot A^T y.$$

Given two vectors $x, y \in R^n$, the angle θ between them was defined as

$$\cos \theta = \frac{x \cdot y}{\|x\| \, \|y\|}.$$

To see that $-1 \leq \frac{x \cdot y}{\|x\| \, \|y\|} \leq 1$ (so that θ can be determined), we need the following *Cauchy–Schwarz inequality*

$$|x \cdot y| \leq \|x\| \, \|y\|. \tag{1.2}$$

To justify this inequality, for any scalar λ expand

$$0 \leq \|\lambda x + y\|^2 = (\lambda x + y) \cdot (\lambda x + y) = \lambda^2 \|x\|^2 + 2\lambda x \cdot y + \|y\|^2.$$

On the right we have a quadratic polynomial in λ, which is nonnegative for all λ. It follows that this polynomial cannot have two real roots, so that its coefficients satisfy

$$(2x \cdot y)^2 - 4\|x\|^2 \|y\|^2 \leq 0,$$

which implies (1.2).

Exercises

1. Let $x_1 = \begin{bmatrix} 1 \\ 2 \\ 2 \end{bmatrix}$, $x_2 = \begin{bmatrix} 2 \\ 3 \\ -4 \end{bmatrix}$, $x_3 = \begin{bmatrix} 1 \\ 0 \\ -5 \end{bmatrix}$, $y_1 = \begin{bmatrix} 0 \\ 2 \\ 2 \\ -1 \end{bmatrix}$, $y_2 = \begin{bmatrix} 1 \\ 1 \\ -2 \\ -2 \end{bmatrix}$, and $y_3 = \begin{bmatrix} 1 \\ 1 \\ -1 \\ 1 \end{bmatrix}$.

   (a) Verify that x_1 is orthogonal to x_2, and y_1 is orthogonal to y_2.
   (b) Calculate $(2x_1 - x_2) \cdot 3x_3$.
   (c) Calculate $\|x_1\|$, $\|y_1\|$, $\|y_2\|$, and $\|y_3\|$.
   (d) Normalize x_1, y_1, y_2, y_3.
   (e) Find the acute angle between y_1 and y_3.
   Answer: $\pi - \arccos(-\frac{1}{6})$.
   (f) Calculate the projection $\text{Proj}_{x_3} x_1$.
   (g) Calculate $\text{Proj}_{x_1} x_3$.
   Answer: $-x_1$.

(h) Calculate $\text{Proj}_{y_1} y_3$.

(i) Calculate $\text{Proj}_{y_1} y_2$.

Answer: The zero vector.

2. Show that $(x + y) \cdot (x - y) = \|x\|^2 - \|y\|^2$, for any $x, y \in R^n$.

3. Show that the diagonals of a parallelogram are orthogonal if and only if the parallelogram is a rhombus (all sides equal).

Hint. Vectors $x + y$ and $x - y$ give the diagonals in the parallelogram with sides x and y.

4. If $\|x\| = 4$, $\|y\| = 3$, and $x \cdot y = -1$, find $\|x + y\|$ and $\|x - y\|$.

Hint. Begin with $\|x + y\|^2$.

5. Let $x \in R^n$, and suppose $e_1, e_2, \ldots, e_n$ is the standard basis of R^n. Let θ_i denote the angle between the vectors x and e_i, for all i (θ_i is called *the direction angle*, while $\cos \theta_i$ is the *the direction cosine*).

(a) Show that

$$\cos^2 \theta_1 + \cos^2 \theta_2 + \cdots + \cos^2 \theta_n = 1.$$

Hint. Note that $\cos \theta_i = \frac{x_i}{\|x\|}$ (x_i is the ith the component of x).

(b) What familiar formula one gets in case $n = 2$?

6. Show that for $x, y \in R^n$ the following *triangle inequality* holds:

$$\|x + y\| \le \|x\| + \|y\|,$$

and interpret it geometrically.

Hint. Using the Cauchy–Schwarz inequality,

$$\|x + y\|^2 = \|x\|^2 + 2x \cdot y + \|y\|^2 \le \|x\|^2 + 2\|x\| \, \|y\| + \|y\|^2.$$

7. Let $x = \begin{bmatrix} x_1 \\ x_2 \\ \vdots \\ x_n \end{bmatrix}, y = \begin{bmatrix} y_1 \\ y_2 \\ \vdots \\ y_n \end{bmatrix}$, and $z = \begin{bmatrix} z_1 \\ z_2 \\ \vdots \\ z_n \end{bmatrix}$ be arbitrary vectors. Verify that

$$x \cdot (y + z) = x \cdot y + x \cdot z.$$

8. If A is an $n \times n$ matrix, e_i and e_j any two coordinate vectors, show that $Ae_j \cdot e_i = a_{ij}$.

9. True or False?

(a) $\|\text{Proj}_a b\| \le \|b\|$.

Answer: True.

(b) $\|\text{Proj}_a b\| \le \|a\|$.

Answer: False.

(c) $\text{Proj}_{2a} b = \text{Proj}_a b$.

Answer: True.

10. Suppose that $x \in R^n$, $y \in R^m$, and matrix A is of size $m \times n$. Show that $Ax \cdot y = x \cdot A^T y$.

5.2 Orthogonal bases

Vectors $v_1, v_2, \ldots, v_p$ in R^n are said to form an orthogonal set if each of these vectors is orthogonal to every other vector, so that $v_i \cdot v_j = 0$ for all $i \neq j$. (One also says that these vectors are mutually orthogonal.) If vectors $u_1, u_2, \ldots, u_p$ in R^n form an orthogonal set, and in addition they are unit vectors ($\|u_i\| = 1$ for all i), we say that $u_1, u_2, \ldots, u_p$ form an orthonormal set. An orthogonal set $v_1, v_2, \ldots, v_p$ can be turned into an orthonormal set by normalization, or taking $u_i = \frac{v_i}{\|v_i\|}$ for all i. For example, the vectors $v_1 = \begin{bmatrix} 0 \\ 2 \\ 2 \\ -1 \end{bmatrix}$,

$v_2 = \begin{bmatrix} 4 \\ 0 \\ 1 \\ 2 \end{bmatrix}$, and $v_3 = \begin{bmatrix} -1 \\ 1 \\ 0 \\ 2 \end{bmatrix}$ form an orthogonal set. Indeed, $v_1 \cdot v_2 = v_1 \cdot v_3 = v_2 \cdot v_3 = 0$.

Calculate $\|v_1\| = 3$, $\|v_2\| = \sqrt{21}$, and $\|v_3\| = \sqrt{6}$. Then the vectors $u_1 = \frac{1}{3} v_1 = \frac{1}{3} \begin{bmatrix} 0 \\ 2 \\ 2 \\ -1 \end{bmatrix}$,

$u_2 = \frac{1}{\sqrt{21}} v_2 = \frac{1}{\sqrt{21}} \begin{bmatrix} 4 \\ 0 \\ 1 \\ 2 \end{bmatrix}$, and $u_3 = \frac{1}{\sqrt{6}} v_3 = \frac{1}{\sqrt{6}} \begin{bmatrix} -1 \\ 1 \\ 0 \\ 2 \end{bmatrix}$ form an orthonormal set.

Theorem 5.2.1. *Suppose that vectors $v_1, v_2, \ldots, v_p$ in R^n are all nonzero, and they form an orthogonal set. Then they are linearly independent.*

Proof. We need to show that the relation

$$x_1 v_1 + x_2 v_2 + \cdots + x_p v_p = 0 \tag{2.1}$$

is possible only if all of the coefficients are zero, $x_1 = x_2 = \cdots = x_p = 0$. Take the inner product of both sides of (2.1) with v_1 to obtain

$$x_1 v_1 \cdot v_1 + x_2 v_2 \cdot v_1 + \cdots + x_p v_p \cdot v_1 = 0.$$

By orthogonality, all of the terms starting with the second are zero, yielding

$$x_1 \|v_1\|^2 = 0.$$

Since v_1 is nonzero, $\|v_1\| > 0$, and then $x_1 = 0$. Taking the inner product of both sides of (2.1) with v_2, one shows similarly that $x_2 = 0$, and so on, showing that all $x_i = 0$. □

It follows that nonzero vectors forming an orthogonal set provide a basis for the subspace that they span, called *orthogonal basis*. Orthonormal vectors give rise to an *orthonormal basis*. Such bases are very convenient, as is explained next.

Suppose that vectors $v_1, v_2, \ldots, v_p$ form an orthogonal basis of some subspace W in R^n. Then any vector w in W can be expressed as

$$w = x_1 v_1 + x_2 v_2 + \cdots + x_p v_p,$$

and the coordinates $x_1, x_2, \ldots, x_p$ are easy to express. Indeed, take the inner product of both sides with v_1 and use the orthogonality,

$$w \cdot v_1 = x_1 v_1 \cdot v_1,$$

giving

$$x_1 = \frac{w \cdot v_1}{\|v_1\|^2}.$$

Taking the inner product of both sides with v_2, gives a formula for x_2, and so on. We thus obtain

$$x_1 = \frac{w \cdot v_1}{\|v_1\|^2}, x_2 = \frac{w \cdot v_2}{\|v_2\|^2}, \ldots, x_p = \frac{w \cdot v_p}{\|v_p\|^2}. \tag{2.2}$$

The resulting decomposition with respect to an orthogonal basis is

$$w = \frac{w \cdot v_1}{\|v_1\|^2} v_1 + \frac{w \cdot v_2}{\|v_2\|^2} v_2 + \cdots + \frac{w \cdot v_p}{\|v_p\|^2} v_p. \tag{2.3}$$

So that *any vector w in W is equal to the sum of its projections on the elements of an orthogonal basis.*

If vectors $u_1, u_2, \ldots, u_p$ form an orthonormal basis of W, and $w \in W$, then

$$w = x_1 u_1 + x_2 u_2 + \cdots + x_p u_p,$$

and in view of (2.2) the coefficients are

$$x_1 = w \cdot u_1, \quad x_2 = w \cdot u_2, \quad \ldots, \quad x_p = w \cdot u_p.$$

The resulting decomposition with respect to an orthonormal basis is

$$w = (w \cdot u_1) u_1 + (w \cdot u_2) u_2 + \cdots + (w \cdot u_p) u_p.$$

Suppose W is a subspace of R^n with a basis $\{w_1, w_2, \ldots, w_p\}$, not necessarily orthogonal. We say that *a vector $z \in R^n$ is orthogonal to a subspace W if z is orthogonal to any vector in W*, notation $z \perp W$.

Lemma 5.2.1. *If a vector z is orthogonal to the basis elements $w_1, w_2, \ldots, w_p$ of W, then z is orthogonal to W.*

Proof. Indeed, decompose any element $w \in W$ as $w = x_1 w_1 + x_2 w_2 + \cdots + x_p w_p$. Given that $z \cdot w_i = 0$ for all i, obtain

$$z \cdot w = x_1 z \cdot w_1 + x_2 z \cdot w_2 + \cdots + x_p z \cdot w_p = 0,$$

so that $z \perp W$. □

Given any vector $b \in R^n$ and a subspace W of R^n, we say that *the vector $\mathrm{Proj}_W b$ is the projection of b on W if the vector $z = b - \mathrm{Proj}_W b$ is orthogonal to W*. It is easy to project on W in case W has an orthogonal basis.

Theorem 5.2.2. *Assume that $\{v_1, v_2, \ldots, v_p\}$ form an orthogonal basis of a subspace W. Then*

$$\mathrm{Proj}_W b = \frac{b \cdot v_1}{\|v_1\|^2} v_1 + \frac{b \cdot v_2}{\|v_2\|^2} v_2 + \cdots + \frac{b \cdot v_p}{\|v_p\|^2} v_p. \tag{2.4}$$

(So that $\mathrm{Proj}_W b$ equals to the sum of projections of b on the basis elements.)

Proof. We need to show that $z = b - \mathrm{Proj}_W b$ is orthogonal to all basis elements of W (so that $z \perp W$). Using the orthogonality of v_i's, calculate

$$z \cdot v_1 = b \cdot v_1 - (\mathrm{Proj}_W b) \cdot v_1 = b \cdot v_1 - \frac{b \cdot v_1}{\|v_1\|^2} v_1 \cdot v_1 = b \cdot v_1 - b \cdot v_1 = 0,$$

and similarly $z \cdot v_i = 0$ for all i. □

In case $b \in W$, $\mathrm{Proj}_W b = b$, as follows by comparing formulas (2.3) and (2.4). If $\mathrm{Proj}_W b \neq b$, then $b \notin W$.

Example 1. Let $v_1 = \begin{bmatrix} 1 \\ -1 \\ 2 \end{bmatrix}$, $v_2 = \begin{bmatrix} 1 \\ 1 \\ 0 \end{bmatrix}$, $b = \begin{bmatrix} 1 \\ 1 \\ 1 \end{bmatrix}$, and $W = \mathrm{Span}\{v_1, v_2\}$. Let us calculate $\mathrm{Proj}_W b$. Since $v_1 \cdot v_2 = 0$, these vectors are orthogonal, and then by (2.4),

$$\mathrm{Proj}_W b = \frac{b \cdot v_1}{\|v_1\|^2} v_1 + \frac{b \cdot v_2}{\|v_2\|^2} v_2 = \frac{2}{6} v_1 + \frac{2}{2} v_2 = \begin{bmatrix} 4/3 \\ 2/3 \\ 2/3 \end{bmatrix}.$$

The set of all vectors in R^n that are orthogonal to a subspace W of R^n is called the *orthogonal complement of W*, and is denoted by $W^\perp$ (pronounced "W perp"). It is straightforward to verify that $W^\perp$ is a subspace of R^n. By Lemma 5.2.1, $W^\perp$ consists of all vectors in R^n that are orthogonal to any basis of W. In 3D, vectors going along the z-axis give the orthogonal complement to vectors in the xy-plane, and vice versa.

Example 2. Consider a subspace W of R^4, $W = \mathrm{Span}\{w_1, w_2\}$, where $w_1 = \begin{bmatrix} 1 \\ 0 \\ 1 \\ -2 \end{bmatrix}$ and $w_2 = \begin{bmatrix} 0 \\ -1 \\ 0 \\ 1 \end{bmatrix}$. The subspace $W^\perp$ consists of vectors $x = \begin{bmatrix} x_1 \\ x_2 \\ x_3 \\ x_4 \end{bmatrix}$ that are orthogonal to the basis of W, so that $x \cdot w_1 = 0$ and $x \cdot w_2 = 0$, or in components

$$x_1 + x_3 - 2x_4 = 0,$$
$$-x_2 + x_4 = 0.$$

One sees that $W^\perp$ is just the null space $N(A)$ of the matrix

$$A = \begin{bmatrix} 1 & 0 & 1 & -2 \\ 0 & -1 & 0 & 1 \end{bmatrix}$$

of this system, and a short calculation shows that

$$W^{\perp} = \mathrm{Span} \left\{ \begin{bmatrix} 2 \\ 1 \\ 0 \\ 1 \end{bmatrix}, \begin{bmatrix} -1 \\ 0 \\ 1 \\ 0 \end{bmatrix} \right\}.$$

Recall that the vector $z = b - \mathrm{Proj}_W\, b$ is orthogonal to the subspace W. In other words, $z \in W^{\perp}$. We conclude that any vector $b \in R^n$ can be decomposed as

$$b = \mathrm{Proj}_W\, b + z,$$

with $\mathrm{Proj}_W\, b \in W$ and $z \in W^{\perp}$. If b belongs to W, then $b = \mathrm{Proj}_W\, b$ and $z = 0$. In case $b \notin W$, then the vector $\mathrm{Proj}_W\, b$ gives the vector (or the point) in W that is closest to b (which is justified in Exercises), and $\|b - \mathrm{Proj}_W\, b\| = \|z\|$ is defined to be *the distance from b to W*.

Fredholm alternative

We now revisit linear systems

$$Ax = b, \tag{2.5}$$

with a given $m \times n$ matrix A, $x \in R^n$, and a given vector $b \in R^m$. We shall use *the corresponding homogeneous system, with $y \in R^n$,*

$$Ay = 0, \tag{2.6}$$

and *the adjoint homogeneous system, with $z \in R^m$,*

$$A^T z = 0. \tag{2.7}$$

Recall that the system (2.5) has a solution if and only if $b \in C(A)$, the column space of A (or the range of the function Ax, for $x \in R^n$). The column space $C(A)$ is a subspace of R^m. All solutions of the system (2.7) constitute the null space of A^T, $N(A^T)$, which is a subspace of R^m.

Theorem 5.2.3. $C(A)^{\perp} = N(A^T)$.

Proof. To prove that two sets are identical, one shows that each element of either one of the sets belongs to the other set.

(i) Assume that the vector $z \in R^m$ belongs to $C(A)^\perp$. Then

$$z \cdot Ax = z^T Ax = (z^T A)x = 0,$$

for all $x \in R^n$. It follows that

$$z^T A = 0,$$

the zero row vector. Taking the adjoint gives (2.7), so that $z \in N(A^T)$.
(ii) Conversely, assume that the vector $z \in R^m$ belongs to $N(A^T)$, so that $A^T z = 0$. Taking the adjoint gives $z^T A = 0$. Then

$$z^T Ax = z \cdot Ax = 0,$$

for all $x \in R^n$. Hence $z \in C(A)^\perp$. □

For square matrices A we have the following important consequence.

Theorem 5.2.4 (Fredholm alternative). *Let A be an $n \times n$ matrix, $b \in R^n$. Then either*
(i) *The homogeneous system (2.6) has only the trivial solution, and the system (2.5) has a unique solution for any vector b.*

Or else
(ii) *Both homogeneous systems (2.6) and (2.7) have nontrivial solutions, and the system (2.5) has solutions if and only if b is orthogonal to any solution of (2.7).*

Proof. If the determinant $|A| \neq 0$, then A^{-1} exists, $v = A^{-1}0 = 0$ is the only solution of (2.6), and $u = A^{-1}b$ is the unique solution of (2.5). In case $|A| = 0$, one has $|A^T| = |A| = 0$, so that both systems (2.6) and (2.7) have nontrivial solutions. In order for (2.5) to be solvable, b must belong to $C(A)$. By Theorem 5.2.3, $C(A)$ is the orthogonal complement of $N(A^T)$, so that b must be orthogonal to all solutions of (2.7). (In this case the system (2.5) has infinitely many solutions of the form $x + cy$, where y is any solution of (2.6), and c is an arbitrary number.) □

So that if A is invertible, the system $Ax = b$ has a (unique) solution for any vector b. In case A is not invertible, solutions exist only for "lucky" b, those orthogonal to any solution of the adjoint system (2.7).

Least squares
Consider a system

$$Ax = b, \tag{2.8}$$

with an $m \times n$ matrix A, $x \in R^n$, and a vector $b \in R^m$. If $C_1, C_2, \ldots, C_n$ are the columns of A and $x_1, x_2, \ldots, x_n$ are the components of x, then one can write (2.8) as

$$x_1 C_1 + x_2 C_2 + \cdots + x_n C_n = b.$$

The system (2.8) is consistent if and only if b belongs to the span of $C_1, C_2, \ldots, C_n$; in other words, $b \in C(A)$, the column space of A. If b is not in $C(A)$, the system (2.8) is inconsistent (there is no solution). What would be a good substitute for the solution? One answer to this question is presented next.

Assume for simplicity that the columns of A are linearly independent. Let p denote the projection of the vector b on $C(A)$, let $\bar{x}$ be the unique solution of

$$A\bar{x} = p. \tag{2.9}$$

(The solution is unique because the columns of A are linearly independent.) The vector $\bar{x}$ is called *the least squares solution of* (2.8). The vector $A\bar{x} = p$ is the closest vector to b in $C(A)$, so that the value of $\|A\bar{x} - b\|$ is the smallest possible. The formula for $\bar{x}$ is derived next.

By the definition of projection, the vector $b - p$ is orthogonal to $C(A)$, implying that $b - p$ is orthogonal to all columns of A, or $b - p$ is orthogonal to all rows of A^T, so that

$$A^T(b - p) = 0.$$

Write this as $A^T p = A^T b$, and use (2.9) to obtain

$$A^T A \bar{x} = A^T b, \tag{2.10}$$

giving

$$\bar{x} = \left(A^T A\right)^{-1} A^T b,$$

since the matrix $A^T A$ is invertible, as is shown in Exercises.

The vector $\bar{x}$ is the unique solution of the system (2.10), known as *the normal equations*. The projection of b on $C(A)$ is

$$p = A\bar{x} = A\left(A^T A\right)^{-1} A^T b,$$

and the matrix $P = A(A^T A)^{-1} A^T$ *projects* any $b \in R^m$ on $C(A)$.

Example 3. The 3×2 system

$$2x_1 + x_2 = 3,$$
$$x_1 - 2x_2 = 4,$$
$$0x_1 + 0x_2 = 1$$

is clearly inconsistent. Intuitively, the best we can do is to solve the first two equations to obtain $x_1 = 2$, $x_2 = -1$. Let us now apply the least squares method. Here $A = \begin{bmatrix} 2 & 1 \\ 1 & -2 \\ 0 & 0 \end{bmatrix}$, $b = \begin{bmatrix} 3 \\ 4 \\ 1 \end{bmatrix}$, and a calculation gives the least squares solution

$$\bar{x} = (A^T A)^{-1} A^T b = \begin{bmatrix} \frac{1}{5} & 0 \\ 0 & \frac{1}{5} \end{bmatrix} \begin{bmatrix} 2 & 1 & 0 \\ 1 & -2 & 0 \end{bmatrix} \begin{bmatrix} 3 \\ 4 \\ 1 \end{bmatrix} = \begin{bmatrix} 2 \\ -1 \end{bmatrix}.$$

The column space of A consists of vectors in R^3 with the third component zero, and the projection of b on $C(A)$ is

$$p = A\bar{x} = \begin{bmatrix} 2 & 1 \\ 1 & -2 \\ 0 & 0 \end{bmatrix} \begin{bmatrix} 2 \\ -1 \end{bmatrix} = \begin{bmatrix} 3 \\ 4 \\ 0 \end{bmatrix},$$

as expected.

Exercises

1. Verify that the vectors $u_1 = \frac{1}{\sqrt{2}} \begin{bmatrix} 1 \\ -1 \end{bmatrix}$ and $u_2 = \begin{bmatrix} 1/\sqrt{2} \\ 1/\sqrt{2} \end{bmatrix}$ form an orthonormal basis of R^2. Then find the coordinates of the vectors $e_1 = \begin{bmatrix} 1 \\ 0 \end{bmatrix}$ and $e_2 = \begin{bmatrix} 0 \\ 1 \end{bmatrix}$ with respect to this basis $B = \{u_1, u_2\}$.
   Answer: $[e_1]_B = \begin{bmatrix} 1/\sqrt{2} \\ 1/\sqrt{2} \end{bmatrix}$, $[e_2]_B = \begin{bmatrix} -1/\sqrt{2} \\ 1/\sqrt{2} \end{bmatrix}$.

2. Verify that the vectors $u_1 = \frac{1}{\sqrt{3}} \begin{bmatrix} 1 \\ 1 \\ 1 \end{bmatrix}$, $u_2 = \frac{1}{\sqrt{6}} \begin{bmatrix} 1 \\ -2 \\ 1 \end{bmatrix}$, and $u_3 = \frac{1}{\sqrt{2}} \begin{bmatrix} 1 \\ 0 \\ -1 \end{bmatrix}$ form an orthonormal basis of R^3. Then find coordinates of the vectors $w_1 = \begin{bmatrix} 1 \\ 1 \\ 1 \end{bmatrix}$, $w_2 = \begin{bmatrix} -3 \\ 0 \\ 3 \end{bmatrix}$, and of the coordinate vector e_2, with respect to this basis $B = \{u_1, u_2, u_3\}$.
   Answer: $[w_1]_B = \begin{bmatrix} \sqrt{3} \\ 0 \\ 0 \end{bmatrix}$, $[w_2]_B = \begin{bmatrix} 0 \\ 0 \\ -\frac{6}{\sqrt{2}} \end{bmatrix}$, $[e_2]_B = \begin{bmatrix} \frac{1}{\sqrt{3}} \\ -\frac{2}{\sqrt{6}} \\ 0 \end{bmatrix}$.

3. Let $v_1 = \begin{bmatrix} 2 \\ -1 \\ 2 \end{bmatrix}$, $v_2 = \begin{bmatrix} 1 \\ 0 \\ -1 \end{bmatrix}$, $b = \begin{bmatrix} 1 \\ 1 \\ 1 \end{bmatrix}$, and $W = \text{Span}\{v_1, v_2\}$.
   (a) Verify that the vectors v_1 and v_2 are orthogonal, and explain why these vectors form an orthogonal basis of W.
   (b) Calculate $\text{Proj}_W b$. Does b belong to W?
   (c) Calculate the coordinates of $w = \begin{bmatrix} 1 \\ 1 \\ -5 \end{bmatrix}$ with respect to the basis $B = \{v_1, v_2\}$.
   Answer: $[w]_B = \begin{bmatrix} -1 \\ 3 \end{bmatrix}$.
   (d) Calculate $\text{Proj}_W u$. Does u belong to W?
   (e) Describe geometrically the subspace W.
   (f) Find $W^\perp$, the orthogonal complement of W, and describe it geometrically.

4. Let $u_1 = \frac{1}{2} \begin{bmatrix} 1 \\ 1 \\ 1 \\ 1 \end{bmatrix}$, $u_2 = \frac{1}{2} \begin{bmatrix} 1 \\ -1 \\ 1 \\ -1 \end{bmatrix}$, $u_3 = \frac{1}{2} \begin{bmatrix} 1 \\ -1 \\ -1 \\ 1 \end{bmatrix}$, $b = \begin{bmatrix} 2 \\ -1 \\ 0 \\ -2 \end{bmatrix}$, and $W = \text{Span}\{u_1, u_2, u_3\}$.

(a) Verify that the vectors u_1, u_2, u_3 are orthonormal, and explain why these vectors form an orthonormal basis of W.

(b) Calculate $\text{Proj}_W b$.

(c) Does b belong to W? If not, what is the point in W that is closest to b?

(d) What is the distance from b to W?

5. Let W be a subspace of R^n of dimension k. Show that $\dim W^\perp = n - k$.

6. Let W be a subspace of R^n. Show that $(W^\perp)^\perp = W$.

7. Let $q_1, q_2, \ldots, q_k$ be orthonormal vectors, and $a = a_1 q_1 + a_2 q_2 + \cdots + a_k q_k$ their linear combination. Justify *the Pythagorean theorem*

$$\|a\|^2 = a_1^2 + a_2^2 + \cdots + a_k^2.$$

Hint. Write $\|a\|^2 = a \cdot a = a_1^2 \, q_1 \cdot q_1 + a_1 a_2 \, q_1 \cdot q_2 + \cdots$.

8. Let W be a subspace of R^n, and $b \notin W$. Show that $\text{Proj}_W b$ gives the vector in W that is closest to b.

Hint. Let z be any vector in W. Then, by the Pythagorean theorem,

$$\|b - z\|^2 = \|(b - \text{Proj}_W b) + (\text{Proj}_W b - z)\|^2$$
$$= \|b - \text{Proj}_W b\|^2 + \|\text{Proj}_W b - z\|^2.$$

(Observe that the vectors $b - \text{Proj}_W b \in W^\perp$ and $\text{Proj}_W b - z \in W$ are orthogonal.) Then $\|b - z\|^2 \geq \|b - \text{Proj}_W b\|^2$.

9. Let A be an $m \times n$ matrix with linearly independent columns. Show that the matrix $A^T A$ is square, invertible, and symmetric.

Hint. Assume that $A^T A x = 0$ for some $x \in R^n$. Then $0 = x^T A^T A x = (Ax)^T Ax = \|Ax\|^2$, so that $Ax = 0$. This implies that $x = 0$, since the columns of A are linearly independent. It follows that $A^T A$ is invertible.

10. Let $w_1, w_2, \ldots, w_n$ be vectors in R^m. The following $n \times n$ determinant:

$$G = \begin{vmatrix} w_1 \cdot w_1 & w_1 \cdot w_2 & \cdots & w_1 \cdot w_n \\ w_2 \cdot w_1 & w_2 \cdot w_2 & \cdots & w_2 \cdot w_n \\ \vdots & \vdots & \ddots & \vdots \\ w_n \cdot w_1 & w_n \cdot w_2 & \cdots & w_n \cdot w_n \end{vmatrix}$$

is called *the Gram determinant or the Gramian.*

(a) Show that $w_1, w_2, \ldots, w_n$ are linearly dependent if and only if the Gramian $G = 0$.

(b) Let A be an $m \times n$ matrix with linearly independent columns. Show again that the square matrix $A^T A$ is invertible and symmetric.

Hint. The determinant $|A^T A|$ is the Gramian of the columns of A.

11. Consider the system

$$2x_1 + x_2 = 3,$$
$$x_1 - 2x_2 = 4,$$
$$2x_1 - x_2 = -5.$$

(a) Verify that this system is inconsistent.

(b) Calculate the least squares solution.

Answer: $\bar{x}_1 = 0$, $\bar{x}_2 = 0$.

(c) Calculate the projection p of the vector $b = \begin{bmatrix} 3 \\ 4 \\ -5 \end{bmatrix}$ on the column space $C(A)$ of the matrix of this system, and conclude that $b \in C(A)^{\perp}$.

Answer: $p = \begin{bmatrix} 0 \\ 0 \\ 0 \end{bmatrix}$.

5.3 Gram–Schmidt orthogonalization

A given set of linearly independent vectors $w_1, w_2, \ldots, w_p$ in R^n forms a basis for the subspace W that they span. It is desirable to have an orthogonal basis of the subspace $W = \mathrm{Span}\{w_1, w_2, \ldots, w_p\}$. With an orthogonal basis, it is easy to calculate the coordinates of any vector $w \in W$, and if a vector b is not in W, it is easy to calculate the projection of b on W. Given an arbitrary basis of a subspace W, our goal is to produce *an orthonormal basis spanning the same subspace W.*

The *Gram–Schmidt orthogonalization process* produces an orthogonal basis $v_1, v_2, \ldots, v_p$ of the subspace $W = \mathrm{Span}\{w_1, w_2, \ldots, w_p\}$ as follows:

$$v_1 = w_1,$$
$$v_2 = w_2 - \frac{w_2 \cdot v_1}{\|v_1\|^2} v_1,$$
$$v_3 = w_3 - \frac{w_3 \cdot v_1}{\|v_1\|^2} v_1 - \frac{w_3 \cdot v_2}{\|v_2\|^2} v_2,$$
$$\vdots$$
$$v_p = w_p - \frac{w_p \cdot v_1}{\|v_1\|^2} v_1 - \frac{w_p \cdot v_2}{\|v_2\|^2} v_2 - \cdots - \frac{w_p \cdot v_{p-1}}{\|v_{p-1}\|^2} v_{p-1}.$$

The first vector w_1 is included in the new basis as v_1. To obtain v_2, we subtract from w_2 its projection on v_1. It follows that v_2 is orthogonal to v_1. To obtain v_3, we subtract from w_3 its projection on the previously constructed vectors v_1 and v_2, in other words, we subtract from w_3 its projection on the subspace spanned by v_1 and v_2. By the definition of projection on a subspace and Theorem 5.2.2, v_3 is orthogonal to that subspace, and in particular, v_3 is orthogonal to v_1 and v_2. In general, to obtain v_p, we subtract from w_p

its projection on the previously constructed vectors $v_1, v_2, \ldots, v_{p-1}$. By the definition of projection on a subspace and Theorem 5.2.2, v_p is orthogonal to $v_1, v_2, \ldots, v_{p-1}$.

The new vectors v_i belong to the subspace W because they are linear combinations of the old vectors w_i. The vectors $v_1, v_2, \ldots, v_p$ are linearly independent, because they form an orthogonal set, and since their number is p, they form a basis of W, an orthogonal basis of W.

Once the orthogonal basis $v_1, v_2, \ldots, v_p$ is constructed, one can obtain *an orthonormal basis* $u_1, u_2, \ldots, u_p$ by normalization, taking $u_i = \frac{v_i}{\|v_i\|}$.

Example 1. Let $w_1 = \begin{bmatrix} 1 \\ -1 \\ -1 \\ 1 \end{bmatrix}$, $w_2 = \begin{bmatrix} 1 \\ -2 \\ 2 \\ 3 \end{bmatrix}$, and $w_3 = \begin{bmatrix} 0 \\ 1 \\ 1 \\ 2 \end{bmatrix}$. It is easy to check that these vectors are linearly independent, and hence they form a basis of $W = \text{Span}\{w_1, w_2, w_3\}$. We now use the Gram–Schmidt process to obtain an orthonormal basis of W.

Starting with $v_1 = w_1 = \begin{bmatrix} 1 \\ -1 \\ -1 \\ 1 \end{bmatrix}$, we calculate $\|v_1\|^2 = \|w_1\|^2 = 4$, $w_2 \cdot v_1 = w_2 \cdot w_1 = 4$ to obtain

$$v_2 = w_2 - \frac{w_2 \cdot v_1}{\|v_1\|^2} v_1 = w_2 - \frac{4}{4} v_1 = \begin{bmatrix} 1 \\ -2 \\ 2 \\ 3 \end{bmatrix} - \begin{bmatrix} 1 \\ -1 \\ -1 \\ 1 \end{bmatrix} = \begin{bmatrix} 0 \\ -1 \\ 3 \\ 2 \end{bmatrix}.$$

Next, $w_3 \cdot v_1 = 0$, $w_3 \cdot v_2 = 6$, $\|v_2\|^2 = 14$, and then

$$v_3 = w_3 - \frac{w_3 \cdot v_1}{\|v_1\|^2} v_1 - \frac{w_3 \cdot v_2}{\|v_2\|^2} v_2$$

$$= w_3 - 0 \cdot v_1 - \frac{6}{14} v_2 = \begin{bmatrix} 0 \\ 1 \\ 1 \\ 2 \end{bmatrix} - \frac{3}{7} \begin{bmatrix} 0 \\ -1 \\ 3 \\ 2 \end{bmatrix} = \begin{bmatrix} 0 \\ 10/7 \\ -2/7 \\ 8/7 \end{bmatrix}.$$

The orthogonal basis of W is

$$v_1 = \begin{bmatrix} 1 \\ -1 \\ -1 \\ 1 \end{bmatrix}, \quad v_2 = \begin{bmatrix} 0 \\ -1 \\ 3 \\ 2 \end{bmatrix}, \quad v_3 = \frac{1}{7} \begin{bmatrix} 0 \\ 10 \\ -2 \\ 8 \end{bmatrix}.$$

Calculate $\|v_1\| = 2$, $\|v_2\| = \sqrt{14}$, $\|v_3\| = \frac{1}{7}\sqrt{168}$. The orthonormal basis of W is obtained by normalization:

$$u_1 = \frac{1}{2} \begin{bmatrix} 1 \\ -1 \\ -1 \\ 1 \end{bmatrix}, \quad u_2 = \frac{1}{\sqrt{14}} \begin{bmatrix} 0 \\ -1 \\ 3 \\ 2 \end{bmatrix}, \quad u_3 = \frac{1}{\sqrt{168}} \begin{bmatrix} 0 \\ 10 \\ -2 \\ 8 \end{bmatrix}.$$

5.3.1 QR factorization

Let $A = [w_1 \, w_2 \, \ldots \, w_n]$ be an $m \times n$ matrix, and assume that its columns $w_1, w_2, \ldots, w_n$ are linearly independent. Then they form a basis of the column space $C(A)$. Applying Gram–Schmidt process to the columns of A produces an orthonormal basis $\{u_1, u_2, \ldots, u_n\}$ of $C(A)$. Form an $m \times n$ matrix

$$Q = [u_1 \, u_2 \, \ldots \, u_n],$$

using these orthonormal columns.

Turning to matrix R, from the first line of Gram–Schmidt process, express the vector w_1 as a multiple of u_1,

$$w_1 = r_{11} u_1, \tag{3.1}$$

with the coefficient denoted by r_{11} ($r_{11} = w_1 \cdot u_1 = \|w_1\|$). From the second line of Gram–Schmidt process, express w_2 as a linear combination of v_1 and v_2, and then of u_1 and u_2,

$$w_2 = r_{12} u_1 + r_{22} u_2, \tag{3.2}$$

with some coefficients r_{12} and r_{22} ($r_{12} = w_2 \cdot u_1$, $r_{22} = w_2 \cdot u_2$). From the third line of Gram–Schmidt process, express

$$w_3 = r_{13} u_1 + r_{23} u_2 + r_{33} u_3,$$

with the appropriate coefficients ($r_{13} = w_3 \cdot u_1$, $r_{23} = w_3 \cdot u_2$, $r_{33} = w_3 \cdot u_3$). The final line of Gram–Schmidt process gives

$$w_n = r_{1n} u_1 + r_{2n} u_2 + \cdots + r_{nn} u_n.$$

Form the $n \times n$ upper triangular matrix R

$$R = \begin{bmatrix} r_{11} & r_{12} & r_{13} & \cdots & r_{1n} \\ 0 & r_{22} & r_{23} & \cdots & r_{2n} \\ 0 & 0 & r_{33} & \cdots & r_{3n} \\ \vdots & \vdots & \vdots & \ddots & \vdots \\ 0 & 0 & 0 & \cdots & r_{nn} \end{bmatrix}.$$

Then the definition of matrix multiplication implies that

$$A = QR, \tag{3.3}$$

what is known as *the QR decomposition of the matrix A*.

We now justify formula (3.3) by comparing the corresponding columns of the matrices A and QR. The first column of A is w_1, while the first column of QR is the product

of Q and the vector $\begin{bmatrix} r_{11} \\ 0 \\ \vdots \\ 0 \end{bmatrix}$ (the first column of R), which gives $r_{11}u_1$, and by (3.1) the first

columns are equal. The second column of A is w_2, while the second column of QR is the

product of Q and the vector $\begin{bmatrix} r_{12} \\ r_{22} \\ \vdots \\ 0 \end{bmatrix}$ (the second column of R), which is $r_{12}u_1 + r_{22}u_2$, and

by (3.2) the second columns are equal. Similarly, all other columns are equal.

Example 2. Let us find the QR decomposition of

$$A = \begin{bmatrix} 1 & 1 & 0 \\ -1 & -2 & 1 \\ -1 & 2 & 1 \\ 1 & 3 & 2 \end{bmatrix}.$$

The columns of A are the vectors w_1, w_2, w_3 from Example 1 above. Therefore the matrix $Q = [u_1\ u_2\ u_3]$ has the orthonormal columns u_1, u_2, u_3 produced in Example 1. To obtain the entries of the matrix R, we "reverse" our calculations in Example 1, expressing w_1, w_2, w_3 first through v_1, v_2, v_3, and then through u_1, u_2, u_3. Recall that

$$w_1 = v_1 = \|v_1\|u_1 = 2u_1,$$

so that $r_{11} = 2$. Similarly,

$$w_2 = v_1 + v_2 = \|v_1\|u_1 + \|v_2\|u_2 = 2u_1 + \sqrt{14}u_2,$$

giving $r_{12} = 2$ and $r_{22} = \sqrt{14}$. Finally,

$$w_3 = 0v_1 + \frac{3}{7}v_2 + v_3 = 0u_1 + \frac{3}{7}\|v_2\|u_2 + \|v_3\|u_3 = 0u_1 + \frac{3}{7}\sqrt{14}\,u_2 + \frac{1}{7}\sqrt{168}\,u_3,$$

so that $r_{13} = 0, r_{23} = \frac{3}{7}\sqrt{14}, r_{33} = \frac{1}{7}\sqrt{168}$. Then $R = \begin{bmatrix} 2 & 2 & 0 \\ 0 & \sqrt{14} & \frac{3}{7}\sqrt{14} \\ 0 & 0 & \frac{1}{7}\sqrt{168} \end{bmatrix}$, and the QR factorization

is

$$\begin{bmatrix} 1 & 1 & 0 \\ -1 & -2 & 1 \\ -1 & 2 & 1 \\ 1 & 3 & 2 \end{bmatrix} = \begin{bmatrix} \frac{1}{2} & 0 & 0 \\ -\frac{1}{2} & \frac{1}{\sqrt{14}} & \frac{10}{\sqrt{168}} \\ -\frac{1}{2} & \frac{3}{\sqrt{14}} & -\frac{2}{\sqrt{168}} \\ \frac{1}{2} & \frac{2}{\sqrt{14}} & \frac{8}{\sqrt{168}} \end{bmatrix} \begin{bmatrix} 2 & 2 & 0 \\ 0 & \sqrt{14} & \frac{3}{7}\sqrt{14} \\ 0 & 0 & \frac{1}{7}\sqrt{168} \end{bmatrix}.$$

Since the vectors u_1, u_2, u_3 are orthonormal, one has (as mentioned above)

$$w_1 = (w_1 \cdot u_1)u_1,$$
$$w_2 = (w_2 \cdot u_1)u_1 + (w_2 \cdot u_2)u_2,$$
$$w_3 = (w_3 \cdot u_1)u_1 + (w_3 \cdot u_2)u_2 + (w_3 \cdot u_3)u_3.$$

Then

$$R = \begin{bmatrix} w_1 \cdot u_1 & w_2 \cdot u_1 & w_3 \cdot u_1 \\ 0 & w_2 \cdot u_2 & w_3 \cdot u_2 \\ 0 & 0 & w_3 \cdot u_3 \end{bmatrix}$$

gives an alternative way to calculate R.

5.3.2 Orthogonal matrices

The matrix $Q = [u_1\, u_2\, \ldots\, u_n]$ in the QR decomposition has orthonormal columns. If Q is of size $m \times n$, its transpose Q^T is an $n \times m$ matrix with the rows $u_1^T, u_2^T, \ldots, u_n^T$, so that $Q^T = \begin{bmatrix} u_1^T \\ u_2^T \\ \vdots \\ u_n^T \end{bmatrix}$. The product $Q^T Q$ is an $n \times n$ matrix, and we claim that (I is the $n \times n$ identity matrix)

$$Q^T Q = I. \tag{3.4}$$

Indeed, the diagonal entries of the product

$$Q^T Q = \begin{bmatrix} u_1^T \\ u_2^T \\ \vdots \\ u_n^T \end{bmatrix} [u_1\, u_2\, \ldots\, u_n]$$

are $u_i^T u_i = u_i \cdot u_i = \|u_i\|^2 = 1$, while the off-diagonal entries are $u_i^T u_j = u_i \cdot u_j = 0$ for $i \neq j$.

A square $n \times n$ matrix with orthonormal columns is called an orthogonal matrix. For orthogonal matrices, formula (3.4) implies that

$$Q^T = Q^{-1}. \tag{3.5}$$

Conversely, if formula (3.5) holds, then $Q^T Q = I$ so that Q has orthonormal columns. We conclude that matrix Q is orthogonal if and only if (3.5) holds. Formula (3.5) provides an alternative definition of orthogonal matrices.

We claim that

$$\|Qx\| = \|x\|,$$

for any orthogonal matrix Q, and all $x \in R^n$. Indeed,

$$\|Qx\|^2 = Qx \cdot Qx = x \cdot Q^T Qx = x \cdot Q^{-1}Qx = x \cdot Ix = \|x\|^2.$$

One shows similarly that

$$Qx \cdot Qy = x \cdot y$$

for any $x, y \in R^n$. It follows that *the orthogonal transformation* Qx *preserves the length of vectors, and the angles between vectors* (since $\cos \theta = \frac{x \cdot y}{\|x\| \|y\|} = \frac{Qx \cdot Qy}{\|Qx\| \|Qy\|}$).

Equating the determinants of both sides of (3.5), we obtain $|Q^T| = |Q^{-1}|$, giving $|Q| = \frac{1}{|Q|}$ or $|Q|^2 = 1$, which implies that

$$|Q| = \pm 1,$$

for any orthogonal matrix Q.

A product of two orthogonal matrices P and Q is also an orthogonal matrix. Indeed, since $P^T = P^{-1}$ and $Q^T = Q^{-1}$, we obtain

$$(PQ)^T = Q^T P^T = Q^{-1}P^{-1} = (PQ)^{-1}.$$

proving that PQ is orthogonal.

If P is a 2×2 orthogonal matrix, it turns out that

$$\text{either } P = \begin{bmatrix} \cos \theta & -\sin \theta \\ \sin \theta & \cos \theta \end{bmatrix} \quad \text{or} \quad P = \begin{bmatrix} \cos \theta & \sin \theta \\ \sin \theta & -\cos \theta \end{bmatrix},$$

for some number θ. Indeed, let $P = \begin{bmatrix} \alpha & \beta \\ \gamma & \delta \end{bmatrix}$ be any orthogonal matrix. We know that the determinant $|P| = \alpha\delta - \beta\gamma = \pm 1$. Let us assume first that $|P| = \alpha\delta - \beta\gamma = 1$. Then

$$P^{-1} = \frac{1}{\alpha\delta - \beta\gamma} \begin{bmatrix} \delta & -\beta \\ -\gamma & \alpha \end{bmatrix} = \begin{bmatrix} \delta & -\beta \\ -\gamma & \alpha \end{bmatrix},$$

and also

$$P^T = \begin{bmatrix} \alpha & \gamma \\ \beta & \delta \end{bmatrix}.$$

Since $P^{-1} = P^T$, it follows that $\delta = \alpha$ and $\beta = -\gamma$, so that $P = \begin{bmatrix} \alpha & -\gamma \\ \gamma & \alpha \end{bmatrix}$. The columns of the orthogonal matrix P are of unit length, so that $\alpha^2 + \gamma^2 = 1$. We can then find a number θ so that $\alpha = \cos \theta$ and $\gamma = \sin \theta$, and conclude that $P = \begin{bmatrix} \cos \theta & -\sin \theta \\ \sin \theta & \cos \theta \end{bmatrix}$.

In the other case, when $|P| = -1$, observe that the product of two orthogonal matrices $\begin{bmatrix} 1 & 0 \\ 0 & -1 \end{bmatrix} P$ is an orthogonal matrix with determinant equal to 1. By the above, $\begin{bmatrix} 1 & 0 \\ 0 & -1 \end{bmatrix} P = \begin{bmatrix} \cos\theta & -\sin\theta \\ \sin\theta & \cos\theta \end{bmatrix}$ for some θ. Then, with $\theta = -\varphi$,

$$P = \begin{bmatrix} 1 & 0 \\ 0 & -1 \end{bmatrix}^{-1} \begin{bmatrix} \cos\theta & -\sin\theta \\ \sin\theta & \cos\theta \end{bmatrix} = \begin{bmatrix} 1 & 0 \\ 0 & -1 \end{bmatrix} \begin{bmatrix} \cos\theta & -\sin\theta \\ \sin\theta & \cos\theta \end{bmatrix}$$

$$= \begin{bmatrix} \cos\theta & -\sin\theta \\ -\sin\theta & -\cos\theta \end{bmatrix} = \begin{bmatrix} \cos\varphi & \sin\varphi \\ \sin\varphi & -\cos\varphi \end{bmatrix}.$$

Exercises

1. Use the Gram–Schmidt process to find an orthonormal basis for the subspace spanned by the given vectors.

   (a) $w_1 = \begin{bmatrix} 1 \\ 0 \\ 1 \end{bmatrix}, w_2 = \begin{bmatrix} 1 \\ 1 \\ 1 \end{bmatrix}$.

   Answer: $u_1 = \begin{bmatrix} \frac{1}{\sqrt{2}} \\ 0 \\ \frac{1}{\sqrt{2}} \end{bmatrix}, u_2 = \begin{bmatrix} 0 \\ 1 \\ 0 \end{bmatrix}$.

   (b) $w_1 = \begin{bmatrix} 1 \\ -2 \\ 2 \end{bmatrix}, w_2 = \begin{bmatrix} -1 \\ 2 \\ 1 \end{bmatrix}$.

   Answer: $u_1 = \frac{1}{3} \begin{bmatrix} 1 \\ -2 \\ 2 \end{bmatrix}, u_2 = \frac{1}{3\sqrt{5}} \begin{bmatrix} -2 \\ 4 \\ 5 \end{bmatrix}$.

   (c) $w_1 = \begin{bmatrix} 2 \\ 1 \\ -1 \\ 0 \end{bmatrix}, w_2 = \begin{bmatrix} 3 \\ 2 \\ -4 \\ 1 \end{bmatrix}, w_3 = \begin{bmatrix} 1 \\ 1 \\ 0 \\ -2 \end{bmatrix}$.

   Answer: $u_1 = \frac{1}{\sqrt{6}} \begin{bmatrix} 2 \\ 1 \\ -1 \\ 0 \end{bmatrix}, u_2 = \frac{1}{\sqrt{6}} \begin{bmatrix} -1 \\ 0 \\ -2 \\ 1 \end{bmatrix}, u_3 = \frac{1}{2\sqrt{3}} \begin{bmatrix} -1 \\ 1 \\ -1 \\ -3 \end{bmatrix}$.

   (d) $w_1 = \begin{bmatrix} 1 \\ 1 \\ -1 \\ -1 \end{bmatrix}, w_2 = \begin{bmatrix} 1 \\ 0 \\ 0 \\ 1 \end{bmatrix}, w_3 = \begin{bmatrix} 1 \\ 1 \\ 0 \\ 0 \end{bmatrix}$.

   Answer: $u_1 = \frac{1}{2} \begin{bmatrix} 1 \\ 1 \\ -1 \\ -1 \end{bmatrix}, u_2 = \frac{1}{\sqrt{2}} \begin{bmatrix} 1 \\ 0 \\ 0 \\ 1 \end{bmatrix}, u_3 = \frac{1}{\sqrt{2}} \begin{bmatrix} 0 \\ 1 \\ 1 \\ 0 \end{bmatrix}$.

   (e) $w_1 = \begin{bmatrix} 3 \\ -2 \\ 1 \\ 1 \\ -1 \end{bmatrix}, w_2 = \begin{bmatrix} -1 \\ 0 \\ 0 \\ 0 \\ 1 \end{bmatrix}$.

   Answer: $u_1 = \frac{1}{4} \begin{bmatrix} 3 \\ -2 \\ 1 \\ 1 \\ -1 \end{bmatrix}, u_2 = \frac{1}{4} \begin{bmatrix} -1 \\ -2 \\ 1 \\ 1 \\ 3 \end{bmatrix}$.

   (f) Let $W = \text{Span}\{w_1, w_2\}$, where $w_1, w_2 \in R^5$ are the vectors from the preceding exercise (e), and $b = \begin{bmatrix} 1 \\ 0 \\ 1 \\ -1 \\ -1 \end{bmatrix}$.

   Find the projection $\text{Proj}_W b$.

   Answer: $\text{Proj}_W b = u_1 - u_2$.

2. Find an orthogonal basis for the null-space $N(A)$ of the following matrices.
   Hint. Find a basis of $N(A)$, then apply the Gram–Schmidt process.

   (a) $A = \begin{bmatrix} 0 & 2 & -1 & 0 \\ -2 & 1 & 2 & 1 \\ -2 & 3 & 1 & 1 \end{bmatrix}$.

Answer: $u_1 = \frac{1}{\sqrt{5}}\begin{bmatrix} 1 \\ 0 \\ 0 \\ 2 \end{bmatrix}$, $u_2 = \frac{1}{\sqrt{10}}\begin{bmatrix} 2 \\ 1 \\ 2 \\ -1 \end{bmatrix}$.

(b) $A = \begin{bmatrix} 1 & -1 & 0 \\ 1 & 2 & -3 \end{bmatrix}$.

Answer: $u_1 = \frac{1}{\sqrt{3}}\begin{bmatrix} 1 \\ 1 \\ 1 \end{bmatrix}$.

(c) $A = \begin{bmatrix} 1 & -1 & 0 & 1 \end{bmatrix}$.

Answer: $u_1 = \frac{1}{\sqrt{2}}\begin{bmatrix} -1 \\ 0 \\ 0 \\ 1 \end{bmatrix}$, $u_2 = \begin{bmatrix} 0 \\ 0 \\ 1 \\ 0 \end{bmatrix}$, $u_3 = \frac{1}{\sqrt{6}}\begin{bmatrix} 1 \\ 2 \\ 0 \\ 1 \end{bmatrix}$.

3. Let $A = QR$ be the QR decomposition of A.

   (a) Assume that A is a nonsingular square matrix. Show that R is also nonsingular, and all of its diagonal entries are positive.

   (b) Show that $R = Q^T A$ (which gives an alternative to calculate R).

4. Find the QR decomposition of the following matrices:

   (a) $A = \begin{bmatrix} 3 & -1 \\ 4 & 0 \end{bmatrix}$.

   Answer: $Q = \begin{bmatrix} \frac{3}{5} & -\frac{4}{5} \\ \frac{4}{5} & \frac{3}{5} \end{bmatrix}$, $R = \begin{bmatrix} w_1 \cdot u_1 & w_2 \cdot u_1 \\ 0 & w_2 \cdot u_2 \end{bmatrix} = \begin{bmatrix} 5 & -\frac{3}{5} \\ 0 & \frac{4}{5} \end{bmatrix}$.

   (b) $A = \begin{bmatrix} 2 & -1 \\ -1 & 1 \\ 2 & 0 \end{bmatrix}$.

   Answer: $Q = \begin{bmatrix} \frac{2}{3} & -\frac{1}{3} \\ -\frac{1}{3} & \frac{2}{3} \\ \frac{2}{3} & \frac{2}{3} \end{bmatrix}$, $R = \begin{bmatrix} 3 & -1 \\ 0 & 1 \end{bmatrix}$.

   (c) $A = \begin{bmatrix} 1 & -1 \\ -1 & 1 \\ 1 & 0 \\ 1 & 2 \end{bmatrix}$. Hint. The columns of A are orthogonal.

   Answer: $Q = \begin{bmatrix} \frac{1}{2} & -\frac{1}{\sqrt{6}} \\ -\frac{1}{2} & \frac{1}{\sqrt{6}} \\ \frac{1}{2} & 0 \\ \frac{1}{2} & \frac{2}{\sqrt{6}} \end{bmatrix}$, $R = \begin{bmatrix} 2 & 0 \\ 0 & \sqrt{6} \end{bmatrix}$.

   (d) $A = \begin{bmatrix} 1 & 0 & 0 \\ -2 & 1 & 0 \\ 2 & 0 & 1 \end{bmatrix}$.

   Answer: $Q = \begin{bmatrix} \frac{1}{3} & \frac{2}{3\sqrt{5}} & -\frac{2}{\sqrt{5}} \\ -\frac{2}{3} & \frac{\sqrt{5}}{3} & 0 \\ \frac{2}{3} & \frac{4}{3\sqrt{5}} & \frac{1}{\sqrt{5}} \end{bmatrix}$, $R = \begin{bmatrix} 3 & -\frac{2}{3} & \frac{2}{3} \\ 0 & \frac{\sqrt{5}}{3} & \frac{4}{3\sqrt{5}} \\ 0 & 0 & \frac{1}{\sqrt{5}} \end{bmatrix}$.

   (e) $A = \begin{bmatrix} 1 & 1 & -1 \\ -1 & 0 & -1 \\ -1 & -1 & 1 \\ 1 & 2 & -1 \end{bmatrix}$.

   Answer: $Q = \begin{bmatrix} \frac{1}{2} & 0 & -\frac{1}{2} \\ -\frac{1}{2} & \frac{1}{\sqrt{2}} & -\frac{1}{2} \\ -\frac{1}{2} & 0 & \frac{1}{2} \\ \frac{1}{2} & \frac{1}{\sqrt{2}} & \frac{1}{2} \end{bmatrix}$, $R = \begin{bmatrix} 2 & 2 & -1 \\ 0 & \sqrt{2} & -\sqrt{2} \\ 0 & 0 & 1 \end{bmatrix}$.

5. Let Q be an orthogonal matrix.

   (a) Show that Q^T is orthogonal.

   (b) Show that an orthogonal matrix has orthonormal rows.

   (c) Show that Q^{-1} is orthogonal.

6. Fill in the missing entries of the following 3×3 orthogonal matrix:

$$Q = \begin{bmatrix} \cos\theta & -\sin\theta & * \\ \sin\theta & \cos\theta & * \\ * & * & * \end{bmatrix}.$$

7. (a) If an orthogonal matrix Q has a real eigenvalue λ show that $\lambda = \pm 1$.
   Hint. If $Qx = \lambda x$, then $\lambda^2 x \cdot x = Qx \cdot Qx = x \cdot Q^T Qx$.
   (b) Give an example of an orthogonal matrix without real eigenvalues.
   (c) Describe all orthogonal matrices that are upper triangular.

8. The matrix $\begin{bmatrix} -1 & 1 & 1 \\ 1 & -1 & 1 \\ 1 & 1 & -1 \end{bmatrix}$ has eigenvalues $\lambda_1 = \lambda_2 = -2, \lambda_3 = 1$. Find an orthonormal basis of the eigenspace corresponding to $\lambda_1 = \lambda_2 = -2$.

9. For the factorization $A = QR$ assume that $w_1, w_2, \ldots, w_n$ in R^m are the columns of A, and $u_1, u_2, \ldots, u_n$ are the columns of Q. Show that

$$R = \begin{bmatrix} w_1 \cdot u_1 & w_2 \cdot u_1 & w_3 \cdot u_1 & \cdots & w_n \cdot u_1 \\ 0 & w_2 \cdot u_2 & w_3 \cdot u_2 & \cdots & w_n \cdot u_2 \\ 0 & 0 & w_3 \cdot u_3 & \cdots & w_n \cdot u_3 \\ \vdots & \vdots & \vdots & \ddots & \vdots \\ 0 & 0 & 0 & \cdots & w_n \cdot u_n \end{bmatrix}.$$

10. Let A be an $n \times n$ matrix, with mutually orthogonal columns $v_1, v_2, \ldots, v_n$. Show that

$$\det A = \pm \|v_1\| \|v_2\| \cdots \|v_n\|.$$

Hint. Consider the $A = QR$ decomposition, where Q is an orthogonal matrix with $\det Q = \pm 1$. Observe that R is a diagonal matrix with the diagonal entries $\|v_1\|, \|v_2\|, \ldots, \|v_n\|$.

11. (a) Let A be an $n \times n$ matrix, with linearly independent columns $a_1, a_2, \ldots, a_n$. Justify Hadamard's inequality

$$|\det A| \leq \|a_1\| \|a_2\| \cdots \|a_n\|.$$

Hint. Consider the $A = QR$ decomposition, where Q is an orthogonal matrix with the orthonormal columns $q_1, q_2, \ldots, q_n$, and r_{ij} are the entries of R. Then $a_j = r_{1j}q_1 + r_{2j}q_2 + \cdots + r_{jj}q_j$. By the Pythagorean theorem, $\|a_j\|^2 = r_{1j}^2 + r_{2j}^2 + \cdots + r_{jj}^2 \geq r_{jj}^2$, so that $|r_{jj}| \leq \|a_j\|$. It follows that

$$|\det A| = |\det Q| \, |\det R|$$
$$= 1 \cdot (|r_{11}| \, |r_{22}| \cdots |r_{nn}|) \leq \|a_1\| \|a_2\| \cdots \|a_n\|.$$

(b) Give geometrical interpretation of Hadamard's inequality in case of three vectors a_1, a_2, a_3 in R^3.

Hint. In that case the matrix A is of size 3×3, and $|\det A|$ gives the volume of the parallelepiped spanned by the vectors a_1, a_2, a_3 (by a property of triple products from Calculus), while the right-hand side of Hadamard's inequality gives the volume of the rectangular parallelepiped (a box) with edges of the same length.

5.4 Linear transformations

Suppose A is an $m \times n$ matrix, $x \in R^n$. Then the product Ax defines a *transformation of vectors $x \in R^n$ into the vectors $Ax \in R^m$*. Transformations often have geometrical significance as the following examples show.

Let $x = \left[\begin{smallmatrix} x_1 \\ x_2 \end{smallmatrix}\right]$ be any vector in R^2. If $A = \left[\begin{smallmatrix} 1 & 0 \\ 0 & 0 \end{smallmatrix}\right]$, then $Ax = \left[\begin{smallmatrix} x_1 \\ 0 \end{smallmatrix}\right]$, gives *the projection of x on the x_1-axis*. For $B = \left[\begin{smallmatrix} 1 & 0 \\ 0 & -1 \end{smallmatrix}\right]$, $Bx = \left[\begin{smallmatrix} x_1 \\ -x_2 \end{smallmatrix}\right]$ provides *the reflection of x across the x_1-axis*. If $C = \left[\begin{smallmatrix} -2 & 0 \\ 0 & -2 \end{smallmatrix}\right]$, then $Cx = \left[\begin{smallmatrix} -2x_1 \\ -2x_2 \end{smallmatrix}\right]$, so that x is transformed into a vector of the opposite direction, which is also stretched in length by a factor of 2.

Suppose that we have *a transformation* (a function) taking each vector x in R^n into a unique vector $T(x)$ in R^m, with common notation $T(x) : R^n \to R^m$. *We say that $T(x)$ is a linear transformation if for any vectors u and v in R^n and any scalar c,*
(i) $T(cu) = cT(u)$ (*T is homogeneous*);
(ii) $T(u + v) = T(u) + T(v)$ (*T is additive*).

The property (ii) holds true for arbitrary number of vectors, as follows by applying it to two vectors at a time. Taking $c = 0$ in (i), we see that $T(0) = 0$ for any linear transformation. ($T(x)$ takes the zero vector in R^n into the zero vector in R^m.) It follows that in case $T(0) \neq 0$ the transformation $T(x)$ is not linear. For example, the transformation $T(x) : R^3 \to R^2$ given by $T\left(\left[\begin{smallmatrix} x_1 \\ x_2 \\ x_3 \end{smallmatrix}\right]\right) = \left[\begin{smallmatrix} 2x_1-x_2+5x_3 \\ x_1+x_2+1 \end{smallmatrix}\right]$ is not linear, because $T\left(\left[\begin{smallmatrix} 0 \\ 0 \\ 0 \end{smallmatrix}\right]\right) = \left[\begin{smallmatrix} 0 \\ 1 \end{smallmatrix}\right]$, which is not equal to the zero vector $\left[\begin{smallmatrix} 0 \\ 0 \end{smallmatrix}\right]$.

If A is any $m \times n$ matrix, and $x \in R^n$, then $T(x) = Ax$ is a linear transformation from R^n to R^m, since the properties (i) and (ii) clearly hold. The 2×2 matrices A, B, and C above provided examples of linear transformations from R^2 to R^2.

It turns out that any linear transformation $T(x) : R^n \to R^m$ can be represented by a matrix. Indeed, let

$$ e_1 = \begin{bmatrix} 1 \\ 0 \\ \vdots \\ 0 \end{bmatrix}, \quad e_2 = \begin{bmatrix} 0 \\ 1 \\ \vdots \\ 0 \end{bmatrix}, \quad \dots, \quad e_n = \begin{bmatrix} 0 \\ 0 \\ \vdots \\ 1 \end{bmatrix} $$

be the standard basis of R^n. Any $x = \begin{bmatrix} x_1 \\ x_2 \\ \vdots \\ x_n \end{bmatrix}$ in R^n can be written as

$$x = x_1 e_1 + x_2 e_2 + \cdots + x_n e_n.$$

We assume that the vectors $T(x) \in R^m$ are also represented through their coordinates with respect to the standard basis in R^m. By linearity of the transformation $T(x)$,

$$T(x) = x_1 T(e_1) + x_2 T(e_2) + \cdots + x_n T(e_n). \tag{4.1}$$

Form the $m \times n$ matrix $A = [T(e_1) \, T(e_2) \, \ldots \, T(e_n)]$, by using the vectors $T(e_i)$'s as its columns. Then (4.1) implies that

$$T(x) = Ax,$$

by the definition of matrix product. One says that A is *the matrix of linear transformation* $T(x)$.

Example 1. Let $T(x) : R^2 \to R^2$ be the rotation of any vector $x \in R^2$ by the angle θ, counterclockwise. Clearly, this transformation is linear (it does not matter if you stretch a vector by a factor of c and then rotate the result, or if the same vector is rotated first, and then is stretched). The standard basis in R^2 is $e_1 = \begin{bmatrix} 1 \\ 0 \end{bmatrix}$, $e_2 = \begin{bmatrix} 0 \\ 1 \end{bmatrix}$. Then $T(e_1)$ is the rotation of e_1, which is a unit vector at the angle θ with the x_1-axis, so that $T(e_1) = \begin{bmatrix} \cos\theta \\ \sin\theta \end{bmatrix}$. Similarly, $T(e_2)$ is a vector in the second quarter at the angle θ with the x_2-axis, so that $T(e_2) = \begin{bmatrix} -\sin\theta \\ \cos\theta \end{bmatrix}$. Then

$$A = [T(e_1) \, T(e_2)] = \begin{bmatrix} \cos\theta & -\sin\theta \\ \sin\theta & \cos\theta \end{bmatrix},$$

the rotation matrix. Observe that this matrix is orthogonal. Conclusion: $T(x) = Ax$, so that rotation can be performed through matrix multiplication. If $x = \begin{bmatrix} x_1 \\ x_2 \end{bmatrix}$, then the vector

$$\begin{bmatrix} \cos\theta & -\sin\theta \\ \sin\theta & \cos\theta \end{bmatrix} \begin{bmatrix} x_1 \\ x_2 \end{bmatrix}$$

is the rotation of x by the angle θ, counterclockwise. If we take $\theta = \frac{\pi}{2}$, then $A = \begin{bmatrix} 0 & -1 \\ 1 & 0 \end{bmatrix}$, and

$$\begin{bmatrix} 0 & -1 \\ 1 & 0 \end{bmatrix} \begin{bmatrix} x_1 \\ x_2 \end{bmatrix} = \begin{bmatrix} -x_2 \\ x_1 \end{bmatrix}$$

is the rotation of $x = \begin{bmatrix} x_1 \\ x_2 \end{bmatrix}$ by the angle $\frac{\pi}{2}$ counterclockwise.

Matrix representation of a linear transformation depends on the basis used. For example, consider a new basis of R^2, $\{e_2, e_1\}$, obtained by changing the order of elements in the standard basis. Then the matrix of rotation in the new basis is

$$B = [T(e_2) \; T(e_1)] = \begin{bmatrix} -\sin\theta & \cos\theta \\ \cos\theta & \sin\theta \end{bmatrix}.$$

Example 2. Let $T(x) : R^3 \to R^3$ be rotation of any vector $x = \begin{bmatrix} x_1 \\ x_2 \\ x_3 \end{bmatrix}$ around the x_3-axis by an angle θ, counterclockwise.

It is straightforward to verify that $T(x)$ is a linear transformation. Let e_1, e_2, e_3 be the standard basis in R^3. Similarly as in Example 1, $T(e_1) = \begin{bmatrix} \cos\theta \\ \sin\theta \\ 0 \end{bmatrix}$, $T(e_2) = \begin{bmatrix} -\sin\theta \\ \cos\theta \\ 0 \end{bmatrix}$, because for vectors lying in the x_1x_2-plane $T(x)$ is just a rotation in that plane. Clearly, $T(e_3) = e_3 = \begin{bmatrix} 0 \\ 0 \\ 1 \end{bmatrix}$. Then the matrix of this transformation is

$$A = \begin{bmatrix} \cos\theta & -\sin\theta & 0 \\ \sin\theta & \cos\theta & 0 \\ 0 & 0 & 1 \end{bmatrix}.$$

Again, we obtained an orthogonal matrix.

Sometimes one can find the matrix of a linear transformation $T(x)$ without evaluating $T(x)$ on the elements of a basis. For example, fix a vector $a \in R^n$ and define $T(x) = \mathrm{Proj}_a\, x$, the projection of any vector $x \in R^n$ on a. It is straightforward to verify that $T(x)$ is a linear transformation. Recall that $\mathrm{Proj}_a\, x = \frac{x \cdot a}{\|a\|^2}\, a$, which we can rewrite as

$$\mathrm{Proj}_a\, x = a\,\frac{a \cdot x}{\|a\|^2} = \frac{a\,a^T x}{\|a\|^2} = \frac{a\,a^T}{\|a\|^2}\, x. \tag{4.2}$$

Define an $n \times n$ matrix $P = \frac{a\,a^T}{\|a\|^2}$, the *projection matrix*. Then $\mathrm{Proj}_a\, x = Px$.

Example 3. Let $a = \begin{bmatrix} 1 \\ 1 \\ 1 \end{bmatrix} \in R^3$. Then the matrix that projects on the line through a is

$$P = \frac{1}{3}\begin{bmatrix} 1 \\ 1 \\ 1 \end{bmatrix}[1 \;\; 1 \;\; 1] = \frac{1}{3}\begin{bmatrix} 1 & 1 & 1 \\ 1 & 1 & 1 \\ 1 & 1 & 1 \end{bmatrix}.$$

For any $x \in R^3$, $Px = \mathrm{Proj}_a\, x$.

We say that a linear transformation $T(x) : R^n \to R^n$ has an eigenvector x, corresponding to the eigenvalue λ if

$$T(x) = \lambda x, \quad x \neq 0.$$

Theorem 5.4.1. *Vector x is an eigenvector of $T(x)$ if and only if it is an eigenvector of the corresponding matrix representation A (with respect to any basis). The corresponding eigenvalues are the same.*

Proof. Follows immediately from the relation $T(x) = Ax$. □

In Example 2, the vector e_3 is an eigenvector for both the rotation $T(x)$ and its 3×3 matrix A, corresponding to $\lambda = 1$. For Example 3, the vector a is an eigenvector for both the projection on a and its matrix P, corresponding to $\lambda = 1$.

Suppose that we have a linear transformation $T_1(x) : R^n \rightarrow R^m$ with the corresponding $m \times n$ matrix A, and a linear transformation $T_2(x) : R^m \rightarrow R^k$ with the corresponding $k \times m$ matrix B, so that $T_1(x) = Ax$ and $T_2(x) = Bx$. It is straightforward to show that *the composition* $T_2(T_1(x)) : R^n \rightarrow R^k$ is a linear transformation. We have

$$T_2(T_1(x)) = BT_1(x) = BAx,$$

so that $k \times n$ product matrix BA is the matrix of composition $T_2(T_1(x))$.

Exercises

1. Is the following map $T(x) : R^2 \rightarrow R^3$ a linear transformation? In case it is a linear transformation, write down its matrix A.

(a) $T\left(\begin{bmatrix} x_1 \\ x_2 \end{bmatrix}\right) = \begin{bmatrix} 2x_1 - x_2 \\ x_1 + x_2 + 1 \\ 3x_1 \end{bmatrix}$.

Answer: No, $T(0) \neq 0$.

(b) $T\left(\begin{bmatrix} x_1 \\ x_2 \end{bmatrix}\right) = \begin{bmatrix} 2x_1 - x_2 \\ x_1 + x_2 \\ 0 \end{bmatrix}$.

Answer: Yes, $T(x)$ is both homogeneous and additive;

$A = \begin{bmatrix} 2 & -1 \\ 1 & 1 \\ 0 & 0 \end{bmatrix}$.

(c) $T\left(\begin{bmatrix} x_1 \\ x_2 \end{bmatrix}\right) = \begin{bmatrix} -5x_2 \\ 2x_1 + x_2 \\ 3x_1 - 3x_2 \end{bmatrix}$.

Answer: Yes. $A = \begin{bmatrix} 0 & -5 \\ 2 & 1 \\ 3 & -3 \end{bmatrix}$.

(d) $T\left(\begin{bmatrix} x_1 \\ x_2 \end{bmatrix}\right) = \begin{bmatrix} 2x_1 - x_2 \\ x_1 \\ 3 \end{bmatrix}$.

Answer: No.

(e) $T\left(\begin{bmatrix} x_1 \\ x_2 \end{bmatrix}\right) = \begin{bmatrix} ax_1 + bx_2 \\ cx_1 + dx_2 \\ ex_1 + fx_2 \end{bmatrix}$. Here a, b, c, d, e, f are arbitrary scalars.

Answer: Yes. $A = \begin{bmatrix} a & b \\ c & d \\ e & f \end{bmatrix}$.

(f) $T\left(\begin{bmatrix} x_1 \\ x_2 \end{bmatrix}\right) = \begin{bmatrix} x_1 x_2 \\ 0 \\ 0 \end{bmatrix}$.

Answer: No.

2. Determine the matrices of the following linear transformations:

(a) $T\left(\begin{bmatrix} x_1 \\ x_2 \\ x_3 \\ x_4 \end{bmatrix}\right) = \begin{bmatrix} x_4 \\ x_3 \\ x_2 \\ x_1 \end{bmatrix}$.

Answer: $A = \begin{bmatrix} 0 & 0 & 0 & 1 \\ 0 & 0 & 1 & 0 \\ 0 & 1 & 0 & 0 \\ 1 & 0 & 0 & 0 \end{bmatrix}$.

(b) $T\left(\begin{bmatrix} x_1 \\ x_2 \\ x_3 \\ x_4 \end{bmatrix}\right) = \begin{bmatrix} x_1 - 2x_3 - x_4 \\ -x_1 + 5x_2 + x_3 - 2x_4 \\ 5x_2 + 2x_3 - 4x_4 \end{bmatrix}$.

Answer: $A = \begin{bmatrix} 1 & 0 & -2 & -1 \\ -1 & 5 & 1 & -2 \\ 0 & 5 & 2 & -4 \end{bmatrix}$.

(c) $T\left(\begin{bmatrix} x_1 \\ x_2 \\ x_3 \end{bmatrix}\right) = \begin{bmatrix} x_1+x_2-2x_3 \\ -2x_1+3x_2+x_3 \\ 0 \\ 2x_1+6x_2-2x_3 \end{bmatrix}$.

Answer: $A = \begin{bmatrix} 1 & 1 & -2 \\ -2 & 3 & 1 \\ 0 & 0 & 0 \\ 2 & 6 & -2 \end{bmatrix}$.

(d) $T\left(\begin{bmatrix} x_1 \\ x_2 \\ x_3 \end{bmatrix}\right) = 7x_1 + 3x_2 - 2x_3$.

Answer: $A = \begin{bmatrix} 7 & 3 & -2 \end{bmatrix}$.

(e) $T(x)$ projects $x \in R^3$ on the x_1x_2-plane, then reflects the result with respect to the origin, and finally doubles the length.

Answer: $A = \begin{bmatrix} -2 & 0 & 0 \\ 0 & -2 & 0 \\ 0 & 0 & 0 \end{bmatrix}$.

(f) $T(x)$ rotates the projection of $x \in R^3$ on the x_1x_2-plane by the angle θ counter-clockwise, while it triples the projection of x on the x_3-axis.

Answer: $A = \begin{bmatrix} \cos\theta & -\sin\theta & 0 \\ \sin\theta & \cos\theta & 0 \\ 0 & 0 & 3 \end{bmatrix}$.

(g) $T(x)$ reflects $x \in R^3$ with respect to the x_1x_3 plane, and then doubles the length.

Answer: $A = \begin{bmatrix} 2 & 0 & 0 \\ 0 & -2 & 0 \\ 0 & 0 & 2 \end{bmatrix}$.

(h) $T(x)$ projects $x \in R^4$ on the subspace spanned by $a = \begin{bmatrix} 1 \\ -1 \\ 1 \\ -1 \end{bmatrix}$.

Hint. Use (4.2).

3. Show that the composition of two linear transformations is a linear transformation. Hint. Observe that

$$T_2(T_1(x_1 + x_2)) = T_2(T_1(x_1) + T_1(x_2)) = T_2(T_1(x_1)) + T_2(T_1(x_2)).$$

4. A linear transformation $T(u) : R^n \to R^m$ is said to be *one-to-one* if $T(u_1) = T(u_2)$ implies that $u_1 = u_2$.
    (a) Show that $T(u)$ is one-to-one if and only if $T(u) = 0$ implies that $u = 0$.
    (b) Assume that $n > m$. Show that $T(u)$ cannot be one-to-one.
         Hint. Represent $T(u) = Au$ with an $m \times n$ matrix A. The system $Au = 0$ has nontrivial solutions.

5. A linear transformation $T(x) : R^n \to R^m$ is said to be *onto* if for every $y \in R^m$ there is $x \in R^n$ such that $y = T(x)$. (So that R^m is the *range* of $T(x)$.)
    (a) Let A be matrix of $T(x)$. Show that $T(x)$ is onto if and only if rank $A = m$.
    (b) Assume that $m > n$. Show that $T(x)$ cannot be onto.

6. Assume that a linear transformation $T(x) : R^n \to R^n$ has an invertible matrix A.
    (a) Show that $T(x)$ is both one-to-one and onto.
    (b) Show that for any $y \in R^n$ the equation $T(x) = y$ has a unique solution $x \in R^n$. The map $y \to x$ is called *the inverse transformation*, and is denoted by $x = T^{-1}(y)$.
    (c) Show that $T^{-1}(y)$ is a linear transformation.

7.   A linear transformation $T(x) : R^3 \to R^3$ projects vector x on $\begin{bmatrix} 1 \\ 2 \\ -1 \end{bmatrix}$.

   (a) Is $T(x)$ one-to-one? (Or is it "many-to-one"?)

   (b) Is $T(x)$ onto?

   (c) Determine the matrix A of this transformation. Hint. Use (4.2).

   (d) Calculate $N(A)$ and $C(A)$, and relate them to parts (a) and (b).

8.   Consider an orthogonal matrix $P = \begin{bmatrix} \cos\theta & -\sin\theta \\ -\sin\theta & -\cos\theta \end{bmatrix}$.

   (a) Show that $P^{-1} = P$ for any θ.

   (b) Show that P is the matrix of the following linear transformation: rotate $x \in R^2$ by an angle θ counterclockwise, then reflect the result with respect to x_1 axis.

   (c) Explain geometrically why $PP = I$.

   (d) Show that $P = \begin{bmatrix} 1 & 0 \\ 0 & -1 \end{bmatrix}\begin{bmatrix} \cos\theta & -\sin\theta \\ \sin\theta & \cos\theta \end{bmatrix}$, the product of the rotation matrix and the matrix representing reflection with respect to x_1 axis.

   (e) Let Q be the matrix of the following linear transformation: reflect $x \in R^2$ with respect to x_1 axis, then rotate the result by an angle θ counterclockwise. Show that

$$Q = \begin{bmatrix} \cos\theta & -\sin\theta \\ \sin\theta & \cos\theta \end{bmatrix}\begin{bmatrix} 1 & 0 \\ 0 & -1 \end{bmatrix} = \begin{bmatrix} \cos\theta & \sin\theta \\ \sin\theta & -\cos\theta \end{bmatrix}.$$

   (f) Explain geometrically why $QQ = I$.

5.5 Symmetric transformations

A square matrix A is called symmetric if $A^T = A$. If a_{ij} denote the entries of A, then symmetric matrices satisfy

$$a_{ij} = a_{ji}, \quad \text{for all } i \text{ and } j.$$

(Symmetric off-diagonal elements are equal, while the diagonal elements are not restricted.) For example, the matrix $A = \begin{bmatrix} 1 & 3 & -4 \\ 3 & -1 & 0 \\ -4 & 0 & 0 \end{bmatrix}$ is symmetric.

Symmetric matrices have a number of nice properties. For example,

$$Ax \cdot y = x \cdot Ay. \tag{5.1}$$

Indeed, by a property of inner product

$$Ax \cdot y = x \cdot A^T y = x \cdot Ay.$$

Theorem 5.5.1. *All eigenvalues of a symmetric matrix A are real, and eigenvectors corresponding to different eigenvalues are orthogonal.*

Proof. Let us prove the orthogonality part first. Let $x \neq 0$ and λ be an eigenvector–eigenvalue pair, so that

$$Ax = \lambda x. \tag{5.2}$$

Let $y \neq 0$ and μ be another such pair:

$$Ay = \mu y, \tag{5.3}$$

and assume that $\lambda \neq \mu$. Take the inner product of both sides of (5.2) with y:

$$Ax \cdot y = \lambda x \cdot y. \tag{5.4}$$

Similarly, take the inner product of x with both sides of (5.3):

$$x \cdot Ay = \mu x \cdot y. \tag{5.5}$$

From (5.4) subtract (5.5), and use (5.1) to obtain

$$0 = (\lambda - \mu)x \cdot y.$$

Since $\lambda - \mu \neq 0$, it follows that $x \cdot y = 0$, proving that x and y are orthogonal.

Turning to all eigenvalues being real, assume that on the contrary $\lambda = a + ib$, with $b \neq 0$, is a complex eigenvalue and $z = \begin{bmatrix} z_1 \\ z_2 \\ \vdots \\ z_n \end{bmatrix}$ is a corresponding eigenvector with complex valued entries. By Theorem 4.2.2, $\bar{\lambda} = a - ib$ is also an eigenvalue, which is different from $\lambda = a + ib$, and $\bar{z} = \begin{bmatrix} \bar{z}_1 \\ \bar{z}_2 \\ \vdots \\ \bar{z}_n \end{bmatrix}$ is a corresponding eigenvector. We just proved that $z \cdot \bar{z} = 0$. In components,

$$z \cdot \bar{z} = z_1 \bar{z}_1 + z_2 \bar{z}_2 + \cdots + z_n \bar{z}_n = |z_1|^2 + |z_2|^2 + \cdots + |z_n|^2 = 0.$$

But then $z_1 = z_2 = \cdots = z_n = 0$, so that z is the zero vector, a contradiction, because an eigenvector cannot be the zero vector. It follows that all eigenvalues are real.   □

For the rest of this section, W will denote a subspace of R^n, of dimension p. Let $T(x) : R^n \to R^n$ be a linear transformation. We say that W is an *invariant subspace of* $T(x)$ if $T(x) \in W$, *for any* $x \in W$. In other words, $T(x)$ maps W into itself, $T(x) : W \to W$.

Observe that for an $n \times n$ matrix A, and any two coordinate vectors e_i and e_j in R^n, one has $Ae_j \cdot e_i = (A)_{ij}$, the ijth entry of A.

A linear transformation $T(x) : W \to W$ is called *self-adjoint* if

$$T(x) \cdot y = x \cdot T(y), \quad \text{for all } x, y \in W.$$

Using matrix representation $T(x) = Ax$, relative to some basis $w_1, w_2, \ldots, w_p$ of W, we can write this definition as

$$Ax \cdot y = x \cdot Ay = A^T x \cdot y, \quad \text{for all } x, y \in R^p. \tag{5.6}$$

If A is symmetric, so that $A = A^T$, then (5.6) holds and $T(x)$ is self-adjoint. Conversely, if $T(x)$ is self-adjoint, then (5.6) holds. Taking $x = e_j \in R^p$ and $y = e_i \in R^p$ in (5.6) gives $(A)_{ij} = (A^T)_{ij}$, so that $A = A^T$, and A is symmetric. We conclude that *a linear transformation $T(x)$ is self-adjoint if and only if its matrix (in any basis) A is symmetric.*

Theorem 5.5.2. *A self-adjoint transformation $T(x) : W \to W$ has at least one eigenvector $x \in W$.*

Proof. Let symmetric matrix A be a matrix representation of $T(x)$ on W. The eigenvalues of A are the roots of its characteristic equation, and by the fundamental theorem of algebra there is at least one root. Since A is symmetric that root is real, and the corresponding eigenvector has real entries. By Theorem 5.4.1, $T(x)$ has the same eigenvector. □

The following theorem describes one of the central facts of Linear Algebra.

Theorem 5.5.3. *Any symmetric $n \times n$ matrix A has a complete set of n mutually orthogonal eigenvectors.*

Proof. Consider the self-adjoint transformation $T(x) = Ax : R^n \to R^n$. By the preceding theorem, $T(x)$ has an eigenvector, denoted by f_1, and let λ_1 be the corresponding eigenvalue. By Theorem 5.4.1, $Af_1 = \lambda_1 f_1$. Consider the $(n-1)$-dimensional subspace $W = f_1^\perp$, consisting of $x \in R^n$ such that $x \cdot f_1 = 0$ (W is the orthogonal complement of f_1). We claim that for any $x \in W$, one has $T(x) \cdot f_1 = 0$, so that $T(x) \in W$, and W is an invariant subspace of $T(x)$. Indeed,

$$T(x) \cdot f_1 = Ax \cdot f_1 = x \cdot Af_1 = \lambda_1 x \cdot f_1 = 0.$$

We now restrict $T(x)$ to the subspace W, $T(x) : W \to W$. Clearly, $T(x)$ is self-adjoint on W. By the preceding theorem, $T(x)$ has an eigenvector f_2 on W, and by its construction f_2 is orthogonal to f_1. Then we restrict $T(x)$ to the $(n-2)$-dimensional subspace $W_1 = f_2^\perp$, the orthogonal complement of f_2 in W. Similarly to the above, one shows that W_1 is an invariant subspace of $T(x)$, so that $T(x)$ has an eigenvector $f_3 \in W_1$, which by its construction is orthogonal to both f_1 and f_2. Continuing this process, we obtain an orthogonal set of eigenvectors $f_1, f_2, \ldots, f_n$ of $T(x)$, which by Theorem 5.4.1 are eigenvectors of A, too. □

Was it necessary to replace the matrix A by its "abstract" version $T(x)$? Yes. Any matrix representation of $T(x)$ on W is of size $(n-1) \times (n-1)$, and definitely is not equal to A. The above process does not work for A.

Since symmetric matrices have a complete set of eigenvectors, they are diagonalizable.

Theorem 5.5.4. *Let A be a symmetric matrix. There is an orthogonal matrix P so that*

$$P^{-1}AP = D. \tag{5.7}$$

The entries of the diagonal matrix D are the eigenvalues of A, while the columns of P are the corresponding normalized eigenvectors.

Proof. By the preceding theorem, A has a complete orthogonal set of eigenvectors. Normalize these eigenvectors of A, and use them as columns of the diagonalizing matrix P. The columns of P are orthonormal, so that P is an orthogonal matrix. □

Recall that one can rewrite (5.7) as $A = PDP^{-1}$. Since P is orthogonal, $P^{-1} = P^T$, and both of these relations can be further rewritten as $P^T AP = D$, and

$$A = PDP^T. \tag{5.8}$$

Example. The matrix $A = \begin{bmatrix} 0 & -2 \\ -2 & 3 \end{bmatrix}$ is symmetric. It has an eigenvalue $\lambda_1 = 4$ with the corresponding normalized eigenvector $\frac{1}{\sqrt{5}} \begin{bmatrix} -1 \\ 2 \end{bmatrix}$, and an eigenvalue $\lambda_2 = -1$ with the corresponding normalized eigenvector $\frac{1}{\sqrt{5}} \begin{bmatrix} 2 \\ 1 \end{bmatrix}$, Then $P = \frac{1}{\sqrt{5}} \begin{bmatrix} -1 & 2 \\ 2 & 1 \end{bmatrix}$ is the orthogonal diagonalizing matrix. A calculation shows that $P^{-1} = \frac{1}{\sqrt{5}} \begin{bmatrix} -1 & 2 \\ 2 & 1 \end{bmatrix}$ (this is a very rare example of a matrix equal to its inverse). The formula (5.7) becomes

$$\frac{1}{\sqrt{5}} \begin{bmatrix} -1 & 2 \\ 2 & 1 \end{bmatrix} \begin{bmatrix} 0 & -2 \\ -2 & 3 \end{bmatrix} \frac{1}{\sqrt{5}} \begin{bmatrix} -1 & 2 \\ 2 & 1 \end{bmatrix} = \begin{bmatrix} 4 & 0 \\ 0 & -1 \end{bmatrix}.$$

A symmetric matrix A is called *positive definite* if all of its eigenvalues are positive. A symmetric matrix A is called *positive semidefinite* if all of its eigenvalues are nonnegative.

Theorem 5.5.5. *A symmetric matrix A is positive definite if and only if*

$$Ax \cdot x > 0, \quad \text{for all } x \neq 0 \ (x \in R^n). \tag{5.9}$$

Proof. If A is positive definite, then $A = PDP^T$ by (5.8), where the matrix P is orthogonal, and the diagonal matrix $D = \begin{bmatrix} \lambda_1 & 0 & \cdots & 0 \\ 0 & \lambda_2 & \cdots & 0 \\ \vdots & \vdots & \ddots & \vdots \\ 0 & 0 & \cdots & \lambda_n \end{bmatrix}$ has positive diagonal entries. For any $x \neq 0$, consider the vector $y = P^T x$, $y = \begin{bmatrix} y_1 \\ y_2 \\ \vdots \\ y_n \end{bmatrix}$. Observe that $y \neq 0$, for otherwise $P^T x = 0$, or $P^{-1}x = 0$, so that $x = P0 = 0$, a contradiction. Then for any $x \neq 0$,

$$Ax \cdot x = PDP^T x \cdot x = DP^T x \cdot P^T x = Dy \cdot y = \lambda_1 y_1^2 + \lambda_2 y_2^2 + \cdots + \lambda_n y_n^2 > 0.$$

Conversely, assume that (5.9) holds, while λ and $x \neq 0$ is an eigenvalue–eigenvector pair,

$$Ax = \lambda x.$$

Taking the inner product of both sides with x gives $Ax \cdot x = \lambda \|x\|^2$, so that

$$\lambda = \frac{Ax \cdot x}{\|x\|^2} > 0,$$

proving that all eigenvalues are positive, so that A is positive definite.   □

The formula (5.9) provides an alternative definition of positive definite matrices, which is often more convenient to use. Similarly, *a symmetric matrix is positive semidefinite if and only if $Ax \cdot x \geq 0$, for all $x \in R^n$.*

Write a positive definite matrix A in the form

$$A = PDP^T = P \begin{bmatrix} \lambda_1 & 0 & \cdots & 0 \\ 0 & \lambda_2 & \cdots & 0 \\ \vdots & \vdots & \ddots & \vdots \\ 0 & 0 & \cdots & \lambda_n \end{bmatrix} P^T.$$

One can define a *square root of A* as follows:

$$\sqrt{A} = P \begin{bmatrix} \sqrt{\lambda_1} & 0 & \cdots & 0 \\ 0 & \sqrt{\lambda_2} & \cdots & 0 \\ \vdots & \vdots & \ddots & \vdots \\ 0 & 0 & \cdots & \sqrt{\lambda_n} \end{bmatrix} P^T,$$

using that all eigenvalues are positive. It follows that $(\sqrt{A})^2 = A$, by squaring the diagonal entries. (Other choices for $\sqrt{A}$ can be obtained replacing $\sqrt{\lambda_i}$ by $\pm\sqrt{\lambda_i}$.)

If A is any nonsingular $n \times n$ matrix (not necessarily symmetric), then the matrix $A^T A$ is positive definite. Indeed, $(A^T A)^T = A^T (A^T)^T = A^T A$, so that this matrix is symmetric, and for any vector $x \neq 0$ ($x \in R^n$),

$$A^T Ax \cdot x = Ax \cdot (A^T)^T x = Ax \cdot Ax = \|Ax\|^2 > 0,$$

because $Ax \neq 0$ (if $Ax = 0$, then $x = A^{-1}0 = 0$, contrary to $x \neq 0$). By Theorem 5.5.5, the matrix $A^T A$ is positive definite. Let now A be an $m \times n$ matrix. Then $A^T A$ is a square $n \times n$ matrix, and a similar argument shows that $A^T A$ is symmetric and positive semidefinite.

Singular value decomposition

We wish to extend the useful concept of diagonalization to nonsquare matrices. For a matrix A of size $m \times n$, the crucial role will be played by two square matrices $A^T A$ of size $n \times n$, and $A^T A$ of size $m \times m$. Both matrices are positive semidefinite (symmetric), and hence both matrices are diagonalizable, with nonnegative eigenvalues.

An $m \times n$ matrix A maps vectors from R^n to R^m (if $x \in R^n$, then $Ax \in R^m$). We shall use orthonormal bases in both R^n and R^m that are connected to A.

Lemma 5.5.1. *If x is an eigenvector of $A^T A$ corresponding to the eigenvalue λ, then Ax is an eigenvector of AA^T corresponding to the same eigenvalue λ. Moreover, if x is unit vector, then the length $\|Ax\| = \sqrt{\lambda}$.*

If x_1 and x_2 are two orthogonal eigenvectors of $A^T A$, then the vectors Ax_1 and Ax_2 are orthogonal.

Proof. We are given that

$$A^T Ax = \lambda x \tag{5.10}$$

for some nonzero $x \in R^n$. Multiplication by A from the left,

$$AA^T (Ax) = \lambda(Ax),$$

shows that $Ax \in R^m$ is an eigenvector of AA^T corresponding to the eigenvalue λ. If x is a unit eigenvector of $A^T A$, multiply (5.10) by x^T to obtain

$$x^T A^T Ax = \lambda x^T x = \lambda \|x\|^2 = \lambda,$$

or

$$(Ax)^T (Ax) = \lambda,$$

implying

$$\|Ax\|^2 = \lambda \tag{5.11}$$

and justifying the second claim. For the final claim, we are given that $A^T A x_2 = \lambda_2 x_2$ for some number λ_2 and nonzero vector $x_2 \in R^n$, and, moreover, that $x_1 \cdot x_2 = 0$. Then

$$Ax_1 \cdot Ax_2 = x_1 \cdot A^T Ax_2 = \lambda_2 x_1 \cdot x_2 = 0,$$

proving the orthogonality of Ax_1 and Ax_2. □

If λ_i are the eigenvalues of $A^T A$ with corresponding unit eigenvectors x_i, then the numbers $\sigma_i = \sqrt{\lambda_i} \geq 0$ are called *the singular values* of A. Observe that $\sigma_i = \|Ax_i\|$ by (5.11).

For a nonsquare matrix A, the elements a_{ii} are still considered to be diagonal entries. For example, if A is of size 2×7, then *its diagonal consists of a_{11} and a_{22}. An $m \times n$ matrix is called diagonal if all off-diagonal entries are zero.*

Singular value decomposition

Any $m \times n$ matrix A can be factored into

$$A = Q_1 \Sigma Q_2^T,$$

where Q_1 and Q_2 are orthogonal matrices of sizes $m \times m$ and $n \times n$, respectively, and Σ is an $m \times n$ diagonal matrix with singular values of A on the diagonal.

To explain the process, let us assume first that A is of size 3×2, mapping R^2 to R^3. Let x_1 and x_2 be the orthonormal eigenvectors of $A^T A$, which is a 2×2 symmetric matrix. We use them as columns of a 2×2 orthogonal matrix $Q_2 = [x_1 \ x_2]$. Let us begin by assuming that the singular values $\sigma_1 = \|Ax_1\|$ and $\sigma_2 = \|Ax_2\|$ are both nonzero (positive). The vectors $q_1 = \frac{Ax_1}{\sigma_1}$ and $q_2 = \frac{Ax_2}{\sigma_2}$ are orthonormal, in view of Lemma 5.5.1. Let $q_3 \in R^3$ be unit vector perpendicular to both q_1 and q_2 ($q_3 = \pm q_1 \times q_2$). Form a 3×3 orthogonal matrix $Q_1 = [q_1 \ q_2 \ q_3]$. We claim that

$$A = Q_1 \begin{bmatrix} \sigma_1 & 0 \\ 0 & \sigma_2 \\ 0 & 0 \end{bmatrix} Q_2^T. \tag{5.12}$$

Indeed, since $Q^T = Q^{-1}$ for orthogonal matrices, it suffices to justify an equivalent formula

$$Q_1^T A Q_2 = \begin{bmatrix} \sigma_1 & 0 \\ 0 & \sigma_2 \\ 0 & 0 \end{bmatrix}. \tag{5.13}$$

The *ij*th entry on the left is (here $1 \le i \le 3, 1 \le j \le 2$)

$$q_i^T A x_j = \sigma_j q_i^T q_j,$$

which is equal to σ_1 if $i = j = 1$, it is equal to σ_2 if $i = j = 2$, and to zero for all other i, j. The matrix on the right in (5.13) has the same entries. Thus (5.12) is justified.

Let us now consider the case when $\sigma_1 = Ax_1 \ne 0$, but $Ax_2 = 0$. Define $q_1 = \frac{Ax_1}{\sigma_1}$, as above. Form a 3×3 orthogonal matrix $Q_1 = [q_1 \ q_2 \ q_3]$, where q_2 and q_2 are chosen to be orthonormal vectors that are both perpendicular to q_1. With $Q_2 = [x_1 \ x_2]$ as above, we claim that

$$A = Q_1 \begin{bmatrix} \sigma_1 & 0 \\ 0 & 0 \\ 0 & 0 \end{bmatrix} Q_2^T.$$

Indeed, in the equivalent formula (5.13) the $i2$ element is now

$$q_i^T A x_2 = 0,$$

so that all elements of the second column are zero.

We now consider general $m \times n$ matrices that map $R^n \to R^m$. If $x_1, x_2, \ldots, x_n$ are orthonormal eigenvectors of $A^T A$, define an $n \times n$ orthogonal matrix $Q_2 = [x_1 \, x_2 \, \cdots \, x_n]$. Assume that there are exactly $r \le n$ positive singular values $\sigma_1 = A x_1, \sigma_2 = A x_2, \ldots, \sigma_r = A x_r$ (which means that in case $r < n$ one has $A x_i = 0$ for $i > r$). Define $q_1 = \frac{A x_1}{\sigma_1}, \ldots, q_r = \frac{A x_r}{\sigma_r}$. These vectors are mutually orthogonal by Lemma 5.5.1. If $r = m$, these vectors form a basis of R^m. If $r < m$, we augment these vectors with $m - r$ orthonormal vectors to obtain an orthonormal basis $q_1, q_2, \ldots, q_m$ in R^m. (The case $r > m$ is not possible, since the r vectors $q_i \in R^m$ are linearly independent.) Define an $m \times m$ orthogonal matrix $Q_1 = [q_1 \, q_2 \, \cdots \, q_m]$. As above,

$$A = Q_1 \Sigma Q_2^T,$$

where Σ is an $m \times n$ diagonal matrix with r positive diagonal entries $\sigma_1, \sigma_2, \ldots, \sigma_r$, and the rest of the diagonal entries of Σ are zero. It is customary to arrange singular values in decreasing order $\sigma_1 \ge \sigma_2 \ge \cdots \ge \sigma_r > 0$.

Singular value decomposition is useful in *image processing*. Suppose that a spaceship is taking a picture on the planet Jupiter, and encodes it, pixel by pixel, in a large $m \times n$ matrix A. Assume that A has r positive singular values (r may be smaller than m and n). Observe that

$$A = Q_1 \Sigma Q_2^T = \sigma_1 q_1 x_1^T + \sigma_2 q_2 x_2^T + \cdots + \sigma_r q_r x_r^T,$$

which is similar to the spectral decomposition of square matrices considered in Exercises. Then it is sufficient to send to the Earth $2r$ vectors, x_i's and q_i's, and r positive singular values σ_i.

Exercises

1. Given an arbitrary square matrix A, show that the matrices $A + A^T$ and AA^T are symmetric. If A is nonsingular, show that AA^T is positive definite.
2. (a) Given an arbitrary square matrix A and a symmetric B, show that $A^T B A$ is symmetric.
   (b) Suppose that both A and B are symmetric. Show that AB is symmetric if and only if A and B commute.
3. Explain why both determinant and trace of a positive definite matrix are positive.
4. Write the matrix A in the form $A = PDP^T$ with orthogonal P and diagonal D. Determine if A is positive definite (p. d.).

(a) $A = \begin{bmatrix} 0 & 1 \\ 1 & 0 \end{bmatrix}$.

    Answer: $P = \frac{1}{\sqrt{2}} \begin{bmatrix} -1 & 1 \\ 1 & 1 \end{bmatrix}$, $D = \begin{bmatrix} -1 & 0 \\ 0 & 1 \end{bmatrix}$, not p. d.

(b) $A = \begin{bmatrix} -1 & 2 \\ 2 & 2 \end{bmatrix}$.

    Answer: $P = \frac{1}{\sqrt{5}} \begin{bmatrix} -2 & 1 \\ 1 & 2 \end{bmatrix}$, $D = \begin{bmatrix} -2 & 0 \\ 0 & 3 \end{bmatrix}$.

(c) $A = \begin{bmatrix} 0 & 2 & 0 \\ 2 & 0 & 0 \\ 0 & 0 & 5 \end{bmatrix}$.

    Answer: $P = \begin{bmatrix} -\frac{1}{\sqrt{2}} & \frac{1}{\sqrt{2}} & 0 \\ \frac{1}{\sqrt{2}} & \frac{1}{\sqrt{2}} & 0 \\ 0 & 0 & 1 \end{bmatrix}$, $D = \begin{bmatrix} -2 & 0 & 0 \\ 0 & 2 & 0 \\ 0 & 0 & 5 \end{bmatrix}$, not p. d.

(d) $A = \begin{bmatrix} 2 & -1 & 1 \\ -1 & 2 & -1 \\ 1 & -1 & 2 \end{bmatrix}$.

    Answer: $P = \begin{bmatrix} \frac{1}{\sqrt{3}} & -\frac{1}{\sqrt{2}} & \frac{1}{\sqrt{2}} \\ -\frac{1}{\sqrt{3}} & 0 & \frac{1}{\sqrt{2}} \\ \frac{1}{\sqrt{3}} & \frac{1}{\sqrt{2}} & 0 \end{bmatrix}$, $D = \begin{bmatrix} 4 & 0 & 0 \\ 0 & 1 & 0 \\ 0 & 0 & 1 \end{bmatrix}$, p. d.

5. Let an $n \times n$ matrix A be *skew-symmetric*, so that $A^T = -A$.

    (a) Show that each eigenvalue is either zero or a purely imaginary number.
        Hint. If $Ax = \lambda x$ and λ is real, then $x \cdot x > 0$ and $\lambda x \cdot x = Ax \cdot x = x \cdot A^T x = -x \cdot Ax = -\lambda x \cdot x$, so that $\lambda = 0$. If $Az = \lambda z$ and λ is complex, then $A\bar{z} = \bar{\lambda}\bar{z}$ and $z \cdot \bar{z} > 0$. Obtain $\lambda z \cdot \bar{z} = Az \cdot \bar{z} = z \cdot A^T \bar{z} = -z \cdot A\bar{z} = -\bar{\lambda} z \cdot \bar{z}$, so that $\lambda = -\bar{\lambda}$.

    (b) If n is odd show that one of the eigenvalues is zero.
        Hint. What is $|A|$?

    (c) Show that the matrix $I + A$ is nonsingular.
        Hint. What are the eigenvalues of this matrix?

    (d) Show that the matrix $(I - A)(I + A)^{-1}$ is orthogonal.

6. Given an arbitrary square matrix A, show that the matrix $A^T A + I$ is positive definite.

7. Assume that a matrix A is symmetric and invertible. Show that A^{-1} is symmetric.

8. Let

$$A = \lambda_1 u_1 u_1^T + \lambda_2 u_2 u_2^T + \cdots + \lambda_n u_n u_n^T, \tag{5.14}$$

where the vectors $u_1, u_2, \ldots, u_n \in R^n$ form an orthonormal set, and $\lambda_1, \lambda_2, \ldots, \lambda_n$ are real numbers, not necessarily different.

    (a) Show that A is an $n \times n$ symmetric matrix.

    (b) Show that $u_1, u_2, \ldots, u_n \in R^n$ are the eigenvectors of A, and $\lambda_1, \lambda_2, \ldots, \lambda_n$ are the corresponding eigenvalues of A.

    (c) For any $x \in R^n$ show that

$$Ax = \lambda_1 \operatorname{Proj}_{u_1} x + \lambda_2 \operatorname{Proj}_{u_2} x + \cdots + \lambda_n \operatorname{Proj}_{u_n} x.$$

    (Formula (5.14) is known as *the spectral decomposition* of A, and the eigenvalues $\lambda_1, \lambda_2, \ldots, \lambda_n$ are often called *the spectrum* of A.)

9. (a) Determine if $A = \begin{bmatrix} -5 & -1 & 1 & 1 \\ -1 & 2 & -1 & 0 \\ 1 & -1 & 2 & 7 \\ 1 & 0 & 7 & 8 \end{bmatrix}$ is positive definite.

Hint. Let $x = e_1$, then $Ax \cdot x = -5$.

(b) Show that all diagonal entries of a positive definite matrix are positive.

Hint. Note that $0 < Ae_k \cdot e_k = a_{kk}$.

10. Assume that a matrix A is positive definite, and S is a nonsingular matrix of the same size. Show that the matrix $S^T AS$ is positive definite.

11. Let $A = [a_{ij}]$ and $U = [u_{ij}]$ be positive definite $n \times n$ matrices. Show that $\sum_{i,j=1}^{n} a_{ij}u_{ij} > 0$.

Hint. Diagonalize $A = PDP^{-1}$, where the entries of the diagonal matrix D are the positive eigenvalues $\lambda_1, \lambda_2, \ldots, \lambda_n$ of A. Let $V = PUP^{-1}$. The matrix $V = [v_{ij}]$ is positive definite, and hence its diagonal entries are positive, $v_{ii} > 0$. Since similar matrices have the same trace, obtain $\sum_{i,j=1}^{n} a_{ij}u_{ij} = \text{tr}(AU) = \text{tr}(PAUP^{-1}) = \text{tr}(PAP^{-1}PUP^{-1}) = \text{tr}(DV) = \lambda_1 v_{11} + \lambda_2 v_{22} + \cdots + \lambda_n v_{nn} > 0$.

12. Calculate the singular value decomposition of $A = \begin{bmatrix} 2 & -4 \\ -2 & -8 \\ 1 & -8 \end{bmatrix}$.

Answer: $A = \begin{bmatrix} 1/3 & 2/3 & 2/3 \\ 2/3 & -2/3 & 1/3 \\ 2/3 & 1/3 & -2/3 \end{bmatrix} \begin{bmatrix} 12 & 0 \\ 0 & 3 \\ 0 & 0 \end{bmatrix} \begin{bmatrix} 0 & 1 \\ -1 & 0 \end{bmatrix}^T$.

5.6 Quadratic forms

All terms of the function $f(x_1, x_2) = x_1^2 - 3x_1 x_2 + 5x_2^2$ are quadratic in its variables x_1 and x_2, giving an example of *a quadratic form*. If $x = \begin{bmatrix} x_1 \\ x_2 \end{bmatrix}$ and $A = \begin{bmatrix} 1 & -\frac{3}{2} \\ -\frac{3}{2} & 5 \end{bmatrix}$, it is easy to verify that

$$f(x_1, x_2) = Ax \cdot x.$$

This *symmetric matrix A* is called *the matrix of the quadratic form $f(x_1, x_2)$*. The quadratic form $g(x_1, x_2) = x_1^2 + 5x_2^2$ involves only *a sum of squares*. Its matrix is diagonal $\begin{bmatrix} 1 & 0 \\ 0 & 5 \end{bmatrix}$. Such quadratic forms are easier to analyze. For example, the equation

$$x_1^2 + 5x_2^2 = 1$$

defines an ellipse in the $x_1 x_2$-plane, with *the principal axes* going along the x_1 and x_2 axes. We shall see in this section that the graph of

$$x_1^2 - 3x_1 x_2 + 5x_2^2 = 1$$

is also an ellipse, with rotated principal axes.

In general, given a symmetric $n \times n$ matrix A and $x = \begin{bmatrix} x_1 \\ x_2 \\ \vdots \\ x_n \end{bmatrix} \in R^n$, one considers a quadratic form $Ax \cdot x$, with the matrix A. The sum $\sum_{j=1}^{n} a_{ij}x_j$ gives component i of Ax, and then

$$Ax \cdot x = \sum_{i=1}^{n} x_i \sum_{j=1}^{n} a_{ij} x_j = \sum_{i=1}^{n} \sum_{j=1}^{n} a_{ij} x_i x_j.$$

This sum is equal to $\sum_{j=1}^{n} \sum_{i=1}^{n} a_{ij} x_i x_j$, and one often writes $Ax \cdot x = \sum_{i,j=1}^{n} a_{ij} x_i x_j$, meaning double summation in any order. If a quadratic form includes a term $k \, x_i x_j$, with the coefficient k, then its matrix A has the entries $a_{ij} = a_{ji} = \frac{k}{2}$, so that A is symmetric.

A quadratic form is called *positive definite if its matrix A is positive definite*, which implies that $Ax \cdot x > 0$ for all $x \neq 0$ by Theorem 5.5.5.

Example 1. Consider the quadratic form

$$Ax \cdot x = x_1^2 + 2x_2^2 + 3x_3^2 - 2x_1 x_2 + 2x_2 x_3,$$

where $\begin{bmatrix} x_1 \\ x_2 \\ x_3 \end{bmatrix} \in R^3$. The matrix of this form is $A = \begin{bmatrix} 1 & -1 & 0 \\ -1 & 2 & 1 \\ 0 & 1 & 3 \end{bmatrix}$. To see if A is positive definite, let us calculate its eigenvalues. Expanding the characteristic polynomial $|A - \lambda I|$ in the first row gives the characteristic equation

$$\lambda^3 - 6\lambda^2 + 9\lambda - 2 = 0.$$

Guessing a root, $\lambda_1 = 2$, allows one to factor the characteristic equation:

$$(\lambda - 2)(\lambda^2 - 4\lambda + 1) = 0,$$

so that $\lambda_2 = 2 - \sqrt{3}$ and $\lambda_3 = 2 + \sqrt{3}$. All eigenvalues are positive, therefore A is positive definite. By Theorem 5.5.5, $Ax \cdot x > 0$ for all $x \neq 0$, which is the same as saying that

$$x_1^2 + 2x_2^2 + 3x_3^2 - 2x_1 x_2 + 2x_2 x_3 > 0,$$

for all x_1, x_2, x_3, except when $x_1 = x_2 = x_3 = 0$.

For a diagonal matrix

$$D = \begin{bmatrix} \lambda_1 & 0 & \cdots & 0 \\ 0 & \lambda_2 & \cdots & 0 \\ \vdots & \vdots & \ddots & \vdots \\ 0 & 0 & \cdots & \lambda_n \end{bmatrix}, \tag{6.1}$$

the corresponding quadratic form,

$$Dx \cdot x = \lambda_1 x_1^2 + \lambda_2 x_2^2 + \cdots + \lambda_n x_n^2,$$

is a sum of squares. In fact, a quadratic form is a sum of squares if and only if its matrix is diagonal.

It is often advantageous to make a change of variables $x = Sy$ in a quadratic form $Ax \cdot x$, using an invertible $n \times n$ matrix S. The old variables $x_1, x_2, \ldots, x_n$ are replaced by the new variables $y_1, y_2, \ldots, y_n$. (One can express the new variables through the old ones by the transformation $y = S^{-1}x$.) The quadratic form changes as follows:

$$Ax \cdot x = ASy \cdot Sy = S^T ASy \cdot y. \tag{6.2}$$

The matrices $S^T AS$ and A are called congruent. They represent the same quadratic form in different variables.

Recall that for any symmetric matrix A one can find an orthogonal matrix P, so that $P^T AP = D$, where D is the diagonal matrix in (6.1). The entries of D are the eigenvalues of A, and the columns of P are the normalized eigenvectors of A (see (5.8)). Let now $x = Py$. Using (6.2),

$$Ax \cdot x = P^T APy \cdot y = Dy \cdot y = \lambda_1 y_1^2 + \lambda_2 y_2^2 + \cdots + \lambda_n y_n^2.$$

It follows that *any quadratic form can be reduced to a sum of squares by an orthogonal change of variables.* In other words, *any quadratic form can be diagonalized.*

Example 2. Let us return to the quadratic form $x_1^2 - 3x_1 x_2 + 5x_2^2$, with its matrix $A = \begin{bmatrix} 1 & -\frac{3}{2} \\ -\frac{3}{2} & 5 \end{bmatrix}$. One calculates that A has an eigenvalue $\lambda_1 = \frac{11}{2}$ with the corresponding normalized eigenvector $\frac{1}{\sqrt{10}} \begin{bmatrix} -1 \\ 3 \end{bmatrix}$, and an eigenvalue $\lambda_2 = \frac{1}{2}$ with the corresponding normalized eigenvector $\frac{1}{\sqrt{10}} \begin{bmatrix} 3 \\ 1 \end{bmatrix}$. Then $P = \frac{1}{\sqrt{10}} \begin{bmatrix} -1 & 3 \\ 3 & 1 \end{bmatrix}$ is the orthogonal diagonalizing matrix. Write the change of variables $x = Py$, which is $\begin{bmatrix} x_1 \\ x_2 \end{bmatrix} = \frac{1}{\sqrt{10}} \begin{bmatrix} -1 & 3 \\ 3 & 1 \end{bmatrix} \begin{bmatrix} y_1 \\ y_2 \end{bmatrix}$, in components as

$$x_1 = \frac{1}{\sqrt{10}} (-y_1 + 3y_2),$$
$$x_2 = \frac{1}{\sqrt{10}} (3y_1 + y_2). \tag{6.3}$$

Substituting these expressions into the quadratic form $x_1^2 - 3x_1 x_2 + 5x_2^2$, and simplifying, we obtain

$$x_1^2 - 3x_1 x_2 + 5x_2^2 = \frac{11}{2} y_1^2 + \frac{1}{2} y_2^2,$$

so that the quadratic form is a sum of squares in the new coordinates.

We can now identify the curve

$$x_1^2 - 3x_1 x_2 + 5x_2^2 = 1 \tag{6.4}$$

as an ellipse, because in the y_1, y_2 coordinates

$$\frac{11}{2}y_1^2 + \frac{1}{2}y_2^2 = 1 \tag{6.5}$$

is clearly an ellipse. The principal axes of the ellipse (6.5) are $y_1 = 0$ and $y_2 = 0$. Corresponding to $y_2 = 0$ (or the y_1 axis), obtain from (6.3)

$$x_1 = -\frac{1}{\sqrt{10}}y_1,$$

$$x_2 = 3\frac{1}{\sqrt{10}}y_1, \tag{6.6}$$

a principal axis for (6.4), which is a line through the origin in the x_1x_2-plane parallel to the vector $\frac{1}{\sqrt{10}}\begin{bmatrix} -1 \\ 3 \end{bmatrix}$ (one of the eigenvectors of A), with y_1 serving as a parameter on this line. This principal axis can also be written in the form $x_2 = -3x_1$, making it easy to plot in the x_1x_2-plane. Similarly, the line $x_2 = \frac{1}{3}x_1$ through the other eigenvector of A gives the second principal axis (it is obtained by setting $y_1 = 0$ in (6.3)). Observe that the principal axes are perpendicular (orthogonal) to each other, as the eigenvectors of a symmetric matrix. (Here P is an orthogonal 2×2 matrix with determinant $|P| = -1$. Hence, P is of the form $\begin{bmatrix} \cos\theta & \sin\theta \\ \sin\theta & -\cos\theta \end{bmatrix}$, which corresponds to reflection with respect to x_1 axis followed by a rotation. The change of variables $x = Py$ produces the principal axes in the x_1x_2-coordinates from the principal axes in the y_1y_2-coordinates through reflection followed by a rotation.)

Example 3. Let us diagonalize the quadratic form $-x_1^2 - 3x_1x_2 + 3x_2^2$, with the matrix $B = \begin{bmatrix} -1 & -\frac{3}{2} \\ -\frac{3}{2} & 3 \end{bmatrix}$. The matrix B has the same eigenvectors as the matrix A in Example 2 (observe that $B = A - 2I$). Hence the diagonalizing matrix P is the same, and we use the same change of variable (6.3) to obtain

$$-x_1^2 - 3x_1x_2 + 3x_2^2 = \frac{7}{2}y_1^2 - \frac{3}{2}y_2^2.$$

The equation

$$\frac{7}{2}y_1^2 - \frac{3}{2}y_2^2 = 1$$

gives a hyperbola in the y_1y_2-plane ($y_2 = \pm\sqrt{\frac{7}{3}y_1^2 - \frac{2}{3}}$), extending along the y_2-axis. It follows that the curve

$$-x_1^2 - 3x_1x_2 + 3x_2^2 = 1$$

is also a hyperbola, with the principal axes $x_2 = -3x_1$ and $x_2 = \frac{1}{3}x_1$. (This hyperbola extends along the $x_2 = \frac{1}{3}x_1$ axis.)

Simultaneous diagonalization

Suppose that we have two quadratic forms $Ax \cdot x$ and $Bx \cdot x$, with $x \in R^n$. Each form can be diagonalized, or reduced to a sum of squares. Is it possible to *diagonalize both forms simultaneously*, by using the same nonsingular change of variables?

Theorem 5.6.1. *Two quadratic forms can be simultaneously diagonalized, provided that one of them is positive definite.*

Proof. Assume that A is a positive definite matrix. By a change of variables $x = S_1 y$ (where S_1 is an orthogonal matrix), we can diagonalize the corresponding quadratic form:

$$Ax \cdot x = \lambda_1 y_1^2 + \lambda_2 y_2^2 + \cdots + \lambda_n y_n^2.$$

Since A is positive definite, its eigenvalues $\lambda_1, \lambda_2, \ldots, \lambda_n$ are positive. We now make a further change of variables $y_1 = \frac{1}{\sqrt{\lambda_1}} z_1, y_2 = \frac{1}{\sqrt{\lambda_2}} z_2, \ldots, y_n = \frac{1}{\sqrt{\lambda_n}} z_n$, or in matrix form $y = S_2 z$, where

$$S_2 = \begin{bmatrix} \frac{1}{\sqrt{\lambda_1}} & 0 & \cdots & 0 \\ 0 & \frac{1}{\sqrt{\lambda_2}} & \cdots & 0 \\ \vdots & \vdots & \ddots & \vdots \\ 0 & 0 & \cdots & \frac{1}{\sqrt{\lambda_n}} \end{bmatrix},$$

a diagonal matrix. Then

$$Ax \cdot x = z_1^2 + z_2^2 + \cdots + z_n^2 = z \cdot z. \tag{6.7}$$

Denote $S = S_1 S_2$. The change of variables we used to achieve (6.7) is $x = S_1 y = S_1 S_2 z = Sz$.

By the same change of variables $x = Sz$, the second quadratic form $Bx \cdot x$ is transformed to a new quadratic form $S^T B Sz \cdot z$. Let us now diagonalize this new quadratic form by a change of variables $z = Pu$, where P is an orthogonal matrix. With the second quadratic form now diagonalized, let us see what happens to the first quadratic form after the last change of variables. Since $P^T = P^{-1}$ for orthogonal matrices, we obtain in view of (6.7) that

$$Ax \cdot x = z \cdot z = Pu \cdot Pu = u \cdot P^T Pu = u \cdot u = u_1^2 + u_2^2 + \cdots + u_n^2,$$

so that the first quadratic form is also diagonalized. (The change of variables that diagonalized both quadratic forms is $x = Sz = SPu = S_1 S_2 Pu$.) □

The law of inertia

Recall that diagonalization of a quadratic form $Ax \cdot x$ is a sum of square terms $\sum_{i=1}^{n} \lambda_i y_i^2$, where λ_i's are the eigenvalues of the $n \times n$ matrix A. The number of positive eigenvalues of A determines the number of positive terms in the diagonalization. A nonsingular change of variables $x = Sz$ transforms the quadratic forms $Ax \cdot x$ into $S^T ASz \cdot z$, with a congruent matrix $S^T AS$. The diagonalization of $S^T ASz \cdot z$ will be different from that of $Ax \cdot x$, however, the number of positive and negative terms will remain the same. This fact is known as *the law of inertia*, and it is justified next.

Theorem 5.6.2. *If $|S| \neq 0$, then the congruent matrix $S^T AS$ has the same number of positive eigenvalues, and the same number of negative eigenvalues as A.*

Proof. The idea of the proof is to gradually change the matrix S to an orthogonal matrix Q through a family $S(t)$, while preserving the number of positive, negative, and zero eigenvalues of the matrix $S(t)^T AS(t)$ in the process. Once $S(t) = Q$, this matrix becomes $Q^{-1}AQ$, which is a similar matrix to A, with the same eigenvalues.

Assume first that $|A| \neq 0$, so that A has no zero eigenvalue. Write down $S = QR$ decomposition. Observe that $|R| \neq 0$ (because $|Q||R| = |S| \neq 0$), and hence all diagonal entries of the upper triangular matrix R are positive. Consider two families of matrices $S(t) = Q[(1-t)I + tR]$ and $F(t) = S^T(t)AS(t)$ depending on a parameter t, with $0 \leq t \leq 1$. Observe that $|S(t)| \neq 0$ for all $t \in [0,1]$, because $|Q| = \pm 1$, while the matrix $(1-t)I + tR$ is an upper triangular matrix with positive diagonal entries, and hence its determinant is positive. It follows that $|F(t)| \neq 0$ for all $t \in [0,1]$. As t varies from 0 to 1, the eigenvalues of $F(t)$ change continuously. These eigenvalues cannot be zero, since zero eigenvalue would imply $|F(t)| = 0$, which is not possible. It follows that the number of positive eigenvalues of $F(t)$ remains the same for all t. When $t = 0$, $S(0) = Q$ and then $F(0) = Q^T(t)AQ(t) = Q^{-1}(t)AQ(t)$, which is a matrix similar to A, and hence $F(0)$ has the same eigenvalues as A, and in particular the same number of positive eigenvalues as A. At $t = 1$, $F(1) = S^T AS$, since $S(1) = S$. We conclude that the matrices A and $S^T AS$ have the same number of positive eigenvalues. The same argument shows that the matrices A and $S^T AS$ have the same number of negative eigenvalues.

We now turn to the case $|A| = 0$, so that A has zero eigenvalue(s). If $\epsilon > 0$ is small enough, then the matrix $A - \epsilon I$ has no zero eigenvalue, and it has the same number of positive eigenvalues as A, which by above is the same as the number of positive eigenvalues of $S^T(A - \epsilon I)S$, which in turn is the same as the number of positive eigenvalues of $S^T AS$ (decreasing ϵ, if necessary). Considering $A + \epsilon I$, with small $\epsilon > 0$, one shows similarly that the number of negative eigenvalues of $S^T AS$ and A is the same. ☐

Rayleigh quotient

It is often desirable to find the minimum and maximum values of a quadratic form $Ax \cdot x$ over all unit vectors x in R^n (i. e., over the unit ball $\|x\| = 1$ in R^n). Since all eigenvalues of a symmetric $n \times n$ matrix A are real, let us arrange them in increasing order $\lambda_1 \leq \lambda_2 \leq$

$\cdots \leq \lambda_n$, with some eigenvalues possibly repeated. Even with repeated eigenvalues, a symmetric matrix A has a complete set of n orthonormal eigenvectors $\xi_1, \xi_2, \ldots, \xi_n$, according to Theorem 5.5.3. Here $A\xi_1 = \lambda_1\xi_1, A\xi_2 = \lambda_2\xi_2, \ldots, A\xi_n = \lambda_n\xi_n$, and $\|\xi_i\| = 1$ for all i.

When $x = \xi_1$, the quadratic form $Ax \cdot x$ is equal to

$$A\xi_1 \cdot \xi_1 = \lambda_1\xi_1 \cdot \xi_1 = \lambda_1,$$

which turns out to be the minimum value of $Ax \cdot x$. Similarly, the maximum value of $Ax \cdot x$ will be shown to be λ_n, and it occurs at $x = \xi_n$.

Proposition 5.6.1. *The extreme values of $Ax \cdot x$ over the set of all unit vectors are the smallest and largest eigenvalues of A:*

$$\min_{\|x\|=1} Ax \cdot x = \lambda_1, \quad \text{it occurs at } x = \xi_1,$$

$$\max_{\|x\|=1} Ax \cdot x = \lambda_n, \quad \text{taken on at } x = \xi_n.$$

Proof. Since $A\xi_1 \cdot \xi_1 = \lambda_1$ and $A\xi_n \cdot \xi_n = \lambda_n$, it suffices to show that for all unit vectors x,

$$\lambda_1 \leq Ax \cdot x \leq \lambda_n. \tag{6.8}$$

Since the eigenvectors $\xi_1, \xi_2, \ldots, \xi_n$ form an orthonormal basis of R^n, we may represent

$$x = c_1\xi_1 + c_2\xi_2 + \cdots + c_n\xi_n,$$

and, by the Pythagorean theorem,

$$c_1^2 + c_2^2 + \cdots + c_n^2 = \|x\|^2 = 1. \tag{6.9}$$

Also

$$Ax = c_1A\xi_1 + c_2A\xi_2 + \cdots + c_nA\xi_n = c_1\lambda_1\xi_1 + c_2\lambda_2\xi_2 + \cdots + c_n\lambda_n\xi_n.$$

Then, using that $\xi_i \cdot \xi_j = 0$ for $i \neq j$, and $\xi_i \cdot \xi_i = \|\xi_i\|^2 = 1$, we obtain

$$Ax \cdot x = (c_1\lambda_1\xi_1 + c_2\lambda_2\xi_2 + \cdots + c_n\lambda_n\xi_n) \cdot (c_1\xi_1 + c_2\xi_2 + \cdots + c_n\xi_n)$$
$$= \lambda_1c_1^2 + \lambda_2c_2^2 + \cdots + \lambda_nc_n^2 \leq \lambda_n(c_1^2 + c_2^2 + \cdots + c_n^2) = \lambda_n,$$

using (6.9), and the other inequality is proved similarly. □

The ratio $\frac{Ax \cdot x}{x \cdot x}$ is called *the Rayleigh quotient*, where the vector x is no longer assumed to be unit. Set $a = \|x\|$. The vector $z = \frac{1}{a}x$ is unit, and then (since $x = az$)

$$\frac{Ax \cdot x}{x \cdot x} = \frac{Az \cdot z}{z \cdot z} = Az \cdot z.$$

Suppose that $Ax_1 = \lambda_1x_1, Ax_n = \lambda_nx_n$, and eigenvectors x_1, x_n are not assumed to be unit.

Theorem 5.6.3. *The extreme values of the Rayleigh quotient are*

$$\min_{x \in R^n} \frac{Ax \cdot x}{x \cdot x} = \lambda_1, \quad \text{it occurs at } x = x_1 \text{ (or at } x = a\xi_1, \text{ for any } a \neq 0),$$

$$\max_{x \in R^n} \frac{Ax \cdot x}{x \cdot x} = \lambda_n, \quad \text{it occurs at } x = x_n \text{ (or at } x = a\xi_n, \text{ for any } a \neq 0).$$

Proof. In view of Proposition 5.6.1, with $z = \frac{1}{\|x\|}x$, we obtain

$$\min_{x \in R^n} \frac{Ax \cdot x}{x \cdot x} = \min_{\|z\|=1} Az \cdot z = \lambda_1.$$

The minimum occurs at $z = \xi_1$, or at $x = a\xi_1$ with any a. The second part is justified similarly. □

Exercises

1. Given a matrix A, write down the corresponding quadratic form $Ax \cdot x$.
   (a) $A = \begin{bmatrix} 2 & -1 \\ -1 & -3 \end{bmatrix}$.
      Answer: $2x_1^2 - 2x_1x_2 - 3x_2^2$.
   (b) $A = \begin{bmatrix} -1 & \frac{3}{2} \\ \frac{3}{2} & 0 \end{bmatrix}$.
      Answer: $-x_1^2 + 3x_1x_2$.
   (c) $A = \begin{bmatrix} 0 & -\frac{3}{2} & -3 \\ -\frac{3}{2} & 1 & 2 \\ -3 & 2 & -2 \end{bmatrix}$.
      Answer: $x_2^2 - 3x_1x_2 - 6x_1x_3 + 4x_2x_3 - 2x_3^2$.
2. Write down the matrix A of the following quadratic forms:
   (a) $2x_1^2 - 6x_1x_2 + 5x_2^2$.
      Answer: $A = \begin{bmatrix} 2 & -3 \\ -3 & 5 \end{bmatrix}$.
   (b) $-x_1x_2 - 4x_2^2$.
      Answer: $A = \begin{bmatrix} 0 & -\frac{1}{2} \\ -\frac{1}{2} & -4 \end{bmatrix}$.
   (c) $3x_1^2 - 2x_1x_2 + 8x_2x_3 + x_2^2 - 5x_3^2$.
      Answer: $A = \begin{bmatrix} 3 & -1 & 0 \\ -1 & 1 & 4 \\ 0 & 4 & -5 \end{bmatrix}$.
   (d) $3x_1x_2 - 6x_1x_3 + 4x_2x_3$.
      Answer: $A = \begin{bmatrix} 0 & \frac{3}{2} & -3 \\ \frac{3}{2} & 0 & 2 \\ -3 & 2 & 0 \end{bmatrix}$.
   (e) $-x_1^2 + 4x_2^2 + 2x_3^2 - 5x_1x_2 - 4x_1x_3 + 4x_2x_3 - 8x_3x_4$.
      Answer: $A = \begin{bmatrix} -1 & -\frac{5}{2} & -2 & 0 \\ -\frac{5}{2} & 4 & 2 & 0 \\ -2 & 2 & 2 & -4 \\ 0 & 0 & -4 & 0 \end{bmatrix}$.

3. Let A be a 20×20 matrix with $a_{ij} = i + j$.
   (a) Show that A is symmetric.
   (b) In the quadratic form $Ax \cdot x$ find the coefficient of the $x_3 x_8$ term.
   Answer: 22.
   (c) How many terms can the form $Ax \cdot x$ contain?
   Answer: $\frac{20 \cdot 21}{2} = 210$.

4. Diagonalize the following quadratic forms:
   (a) $3x_1^2 + 2x_1 x_2 + 3x_2^2$.
   Answer: $P = \frac{1}{\sqrt{2}} \begin{bmatrix} -1 & 1 \\ 1 & 1 \end{bmatrix}$, the change of variables $x = Py$ gives $2y_1^2 + 4y_2^2$.
   (b) $-4x_1 x_2 + 3x_2^2$.
   Answer: $P = \frac{1}{\sqrt{5}} \begin{bmatrix} 2 & -1 \\ 1 & 2 \end{bmatrix}$, obtain $-y_1^2 + 4y_2^2$.
   (c) $3x_1^2 + x_2^2 - 2x_3^2 + 4x_2 x_3$.
   Answer: $P = \begin{bmatrix} 1 & 0 & 0 \\ 0 & -\frac{1}{\sqrt{5}} & \frac{2}{\sqrt{5}} \\ 0 & \frac{2}{\sqrt{5}} & \frac{1}{\sqrt{5}} \end{bmatrix}$, the change of variables $x_1 = y_1$, $x_2 = -\frac{1}{\sqrt{5}}y_2 + \frac{2}{\sqrt{5}}y_3$,
   $x_3 = \frac{2}{\sqrt{5}}y_2 + \frac{1}{\sqrt{5}}y_3$ produces $3y_1^2 - 3y_2^2 + 2y_3^2$.
   (d) $-x_1^2 - x_2^2 - x_3^2 + 2x_1 x_2 + 2x_1 x_3 + 2x_2 x_3$.
   Hint. The matrix of the quadratic form has eigenvalues $-2, -2, 1$. The eigenvalue -2 has two linearly independent eigenvectors. One needs to apply Gram–Schmidt process to these eigenvectors to obtain the first two columns of the orthogonal matrix P.
   Answer: The orthogonal $P = \begin{bmatrix} -\frac{1}{\sqrt{2}} & -\frac{1}{\sqrt{6}} & \frac{1}{\sqrt{3}} \\ 0 & \frac{2}{\sqrt{6}} & \frac{1}{\sqrt{3}} \\ \frac{1}{\sqrt{2}} & -\frac{1}{\sqrt{6}} & \frac{1}{\sqrt{3}} \end{bmatrix}$, the change of variables $x_1 = -\frac{1}{\sqrt{2}}y_1 -$
   $\frac{1}{\sqrt{6}}y_2 + \frac{1}{\sqrt{3}}y_3$, $x_2 = \frac{2}{\sqrt{6}}y_2 + \frac{1}{\sqrt{3}}y_3$, $x_3 = \frac{1}{\sqrt{2}}y_1 - \frac{1}{\sqrt{6}}y_2 + \frac{1}{\sqrt{3}}y_3$ produces $-2y_1^2 - 2y_2^2 + y_3^2$.

5. Consider congruent matrices A and $S^T A S$, with $|S| \neq 0$. Assume that A has zero eigenvalue. Show that $S^T A S$ also has zero eigenvalue of the same multiplicity as A.
   Hint. By the law of inertia, the matrices $S^T A S$ and A have the same number of positive eigenvalues, and the same number of negative eigenvalues.

6. (a) Let A be a 3×3 symmetric matrix with the eigenvalues $\lambda_1 > 0$, $\lambda_2 > 0$, and $\lambda_3 = 0$. Show that $Ax \cdot x \geq 0$ for all $x \in R^3$. Show also that there is a vector $x_0 \in R^3$ such that $Ax_0 \cdot x_0 = 0$.
   Hint. If P is the orthogonal diagonalizing matrix for A, and $x = Py$, then $Ax \cdot x = \lambda_1 y_1^2 + \lambda_2 y_2^2 \geq 0$.
   (b) Recall that a symmetric $n \times n$ matrix is called *positive semidefinite* if $Ax \cdot x \geq 0$ for all $x \in R^n$. Using quadratic forms, show that a symmetric matrix A is positive semidefinite if and only if all eigenvalues of A are nonnegative.
   (c) Show that a positive semidefinite matrix with nonzero determinant is positive definite.
   (d) A symmetric $n \times n$ matrix is called *negative semidefinite* if $Ax \cdot x \leq 0$ for all $x \in R^n$. Show that a symmetric matrix A is negative semidefinite if and only if all eigenvalues of A are nonpositive.

7. An $n \times n$ matrix with the entries $a_{ij} = \frac{1}{i+j-1}$ is known as *the Hilbert matrix*

$$A = \begin{bmatrix} 1 & \frac{1}{2} & \frac{1}{3} & \cdots & \frac{1}{n} \\ \frac{1}{2} & \frac{1}{3} & \frac{1}{4} & \cdots & \frac{1}{n+1} \\ \frac{1}{3} & \frac{1}{4} & \frac{1}{5} & \cdots & \frac{1}{n+2} \\ \vdots & \vdots & \vdots & \ddots & \vdots \\ \frac{1}{n} & \frac{1}{n+1} & \frac{1}{n+2} & \cdots & \frac{1}{2n-1} \end{bmatrix}.$$

Show that A is positive definite.

Hint. For any $x \in R^n$, $x \neq 0$,

$$Ax \cdot x = \sum_{i,j=1}^{n} \frac{x_i x_j}{i+j-1} = \sum_{i,j=1}^{n} x_i x_j \int_0^1 t^{i+j-2} \, dt$$

$$= \int_0^1 \left(\sum_{i=1}^{n} x_i t^{i-1} \right)^2 dt > 0.$$

5.7 Vector spaces

Vectors in R^n can be added and multiplied by scalars. There are other mathematical objects that can be added and multiplied by numbers (scalars), for example, matrices or functions. We shall refer to such objects as *vectors, belonging to abstract vector spaces,* provided that the operations of addition and scalar multiplication satisfy the familiar properties of vectors in R^n.

Definition 5.7.1. A vector space V is a collection of objects called vectors, which may be added together and multiplied by numbers, so that for any $x, y \in V$ and any number c, one has $x + y \in V$ and $cx \in V$. Moreover, addition and scalar multiplication are required to satisfy the following natural rules, also called *axioms* (which hold for all vectors $x, y, z \in V$ and any numbers c, c_1, c_2):

$$x + y = y + x,$$
$$x + (y + z) = (x + y) + z,$$

There is a unique "zero vector", denoted $\mathbf{0}$, such that $x + \mathbf{0} = x$,
For each x in V there is a unique vector $-x$ such that $x + (-x) = \mathbf{0}$,

$$1x = x,$$
$$(c_1 c_2)x = c_1(c_2 x),$$
$$c(x + y) = cx + cy,$$
$$(c_1 + c_2)x = c_1 x + c_2 x.$$

The following additional rules can be easily deduced from the above axioms:

$$0\,x = \mathbf{0},$$

$$c\,\mathbf{0} = \mathbf{0},$$

$$(-1)\,x = -x.$$

Any subspace in R^n provides an example of a vector space. In particular, any plane through the origin in R^3 is a vector space. Other examples of vector spaces involve matrices and polynomials.

Example 1. Two-by-two matrices can be added and multiplied by scalars, and the above axioms are clearly satisfied, so that 2×2 matrices form a vector space, denoted by $M_{2\times2}$. Each 2×2 matrix is now regarded as *a vector in* $M_{2\times2}$. The role of the zero vector $\mathbf{0}$ is played by the zero matrix $O = \begin{bmatrix} 0 & 0 \\ 0 & 0 \end{bmatrix}$.

The standard basis for $M_{2\times2}$ is provided by the matrices $E_{11} = \begin{bmatrix} 1 & 0 \\ 0 & 0 \end{bmatrix}$, $E_{12} = \begin{bmatrix} 0 & 1 \\ 0 & 0 \end{bmatrix}$, $E_{21} = \begin{bmatrix} 0 & 0 \\ 1 & 0 \end{bmatrix}$, *and* $E_{22} = \begin{bmatrix} 0 & 0 \\ 0 & 1 \end{bmatrix}$, so that the vector space $M_{2\times2}$ is four-dimensional. Indeed, given an arbitrary $A = \begin{bmatrix} a_{11} & a_{12} \\ a_{21} & a_{22} \end{bmatrix} \in M_{2\times2}$, one can decompose

$$A = a_{11}E_{11} + a_{12}E_{12} + a_{21}E_{21} + a_{22}E_{22},$$

so that $a_{11}, a_{12}, a_{21}, a_{22}$ are the coordinates of A with respect to the standard basis.
   One defines similarly the vector space $M_{m\times n}$ of $m \times n$ matrices. The dimension of $M_{m\times n}$ is mn.

Example 2. One checks that the above axioms apply for polynomials of power n of the form $a_n x^n + a_{n-1} x^{n-1} + \cdots + a_2 x^2 + a_1 x + a_0$, with numerical coefficients $a_0, a_1, a_2, \ldots, a_n$. Hence, these polynomials form a vector space, denoted by P_n. Particular polynomials are regarded *as vectors in P_n*. The vectors $1, x, x^2, \ldots, x^n$ form *the standard basis of P_n*, so that P_n is an $(n + 1)$-dimensional vector space.

Example 3. The vector space $P_n(-1, 1)$ consists of polynomials of power n, which are considered only on the interval $x \in (-1, 1)$. What is the reason for restricting polynomials to an interval? We can now define the notion of an *inner (scalar) product*. Given two vectors $p(x), q(x) \in P_n(-1, 1)$ *define their inner product as*

$$p(x) \cdot q(x) = \int_{-1}^{1} p(x)q(x)\,dx.$$

The norm (or the "magnitude") $\|p(x)\|$ of a vector $p(x) \in P_n(-1, 1)$ is defined by the relation

$$\|p(x)\|^2 = p(x) \cdot p(x) = \int_{-1}^{1} p^2(x)\,dx,$$

so that $\|p(x)\| = \sqrt{p(x) \cdot p(x)}$. If $p(x) \cdot q(x) = 0$, we say that *the polynomials are orthogonal.* For example, the vectors $p(x) = x$ and $q(x) = x^2$ are orthogonal, because

$$x \cdot x^2 = \int_{-1}^{1} x^3 \, dx = 0.$$

Calculate

$$\|1\|^2 = 1 \cdot 1 = \int_{-1}^{1} 1 \, dx = 2,$$

so that the norm of the vector $p(x) = 1$ is $\|1\| = \sqrt{2}$. *The projection* of $q(x)$ on $p(x)$,

$$\text{Proj}_{p(x)} \, q(x) = \frac{p(x) \cdot q(x)}{p(x) \cdot p(x)} p(x),$$

is defined similarly to vectors in R^n. For example, the projection of x^2 on 1 is

$$\text{Proj}_1 \, x^2 = \frac{x^2 \cdot 1}{1 \cdot 1} 1 = \frac{1}{3},$$

since $x^2 \cdot 1 = \int_{-1}^{1} x^2 \, dx = \frac{2}{3}$.

The standard basis $1, x, x^2, \ldots, x^n$ of $P_n(-1, 1)$ is not orthogonal. While the vectors 1 and x are orthogonal, the vectors 1 and x^2 are not. We now apply *the Gram–Schmidt process* to produce an orthogonal basis $p_0(x), p_1(x), p_2(x), \ldots, p_n(x)$, but instead of normalization it is customary to *standardize the polynomials by requiring that $p_i(1) = 1$ for all i.* Set $p_0(x) = 1$. Since the second element x of the standard basis is orthogonal to $p_0(x)$, we take $p_1(x) = x$. (Observe that $p_0(x)$ and $p_1(x)$ are already standardized.) According to the Gram–Schmidt process, calculate (subtracting from x^2 its projections on 1, and on x)

$$x^2 - \frac{x^2 \cdot 1}{1 \cdot 1} 1 - \frac{x^2 \cdot x}{x \cdot x} x = x^2 - \frac{1}{3}.$$

Multiply this polynomial by $\frac{3}{2}$ to obtain $p_2(x) = \frac{1}{2}(3x^2 - 1)$, with $p_2(1) = 1$. The next step of the Gram–Schmidt process involves (subtracting from x^3 its projections on $p_0(x)$, $p_1(x)$, $p_2(x)$)

$$x^3 - \frac{x^3 \cdot 1}{1 \cdot 1} 1 - \frac{x^3 \cdot x}{x \cdot x} x - \frac{x^3 \cdot p_2(x)}{p_2(x) \cdot p_2(x)} p_2(x) = x^3 - \frac{3}{5} x.$$

Multiply this polynomial by $\frac{5}{2}$ to obtain $p_3(x) = \frac{1}{2}(5x^3 - 3x)$, with $p_3(1) = 1$, and so on. The *orthogonal polynomials* $p_0(x), p_1(x), p_2(x), p_3(x), \ldots$ are known as the *Legendre polynomials.* They have many applications.

Next, we discuss linear transformations and their matrices. Let V_1, V_2 be two vector spaces. We say that *a map $T : V_1 \to V_2$ is a linear transformation if for any $x, x_1, x_2 \in V_1$, and any number c,*

$$T(cx) = cT(x),$$
$$T(x_1 + x_2) = T(x_1) + T(x_2).$$

Clearly, the second of these properties applies to any number of terms. Letting $c = 0$, we conclude that any linear transformation satisfies $T(\mathbf{0}) = \mathbf{0}$ ($T(x)$ takes the zero vector in V_1 into the zero vector in V_2). It follows that in case $T(\mathbf{0}) \neq \mathbf{0}$, the map is not a linear transformation. For example, the map $T : M_{2\times2} \to M_{2\times2}$ given by $T(A) = 3A - I$ is not a linear transformation, because $T(O) = -I \neq O$.

Example 4. Let $D : P_4 \to P_3$ be a transformation taking any polynomial $p(x) = a_4 x^4 + a_3 x^3 + a_2 x^2 + a_1 x + a_0$ into

$$D(p(x)) = 4a_4 x^3 + 3a_3 x^2 + 2a_2 x + a_1.$$

Clearly, D is just differentiation, and hence this transformation is linear.

Let $T(x)$ be a linear transformation $T : V_1 \to V_2$. Assume that $B_1 = \{w_1, w_2, \ldots, w_p\}$ is a basis of V_1, and $B_2 = \{z_1, z_2, \ldots, z_s\}$ is a basis of V_2. Any vector $x \in V_1$ can be written as

$$x = x_1 w_1 + x_2 w_2 + \cdots + x_p w_p,$$

with *the coordinates* $[x]_{B_1} = \begin{bmatrix} x_1 \\ x_2 \\ \vdots \\ x_p \end{bmatrix} \in R^p$. Any vector $y \in V_2$ can be written as

$$y = y_1 z_1 + y_2 z_2 + \cdots + y_s z_s,$$

with *the coordinates* $[y]_{B_2} = \begin{bmatrix} y_1 \\ y_2 \\ \vdots \\ y_s \end{bmatrix} \in R^s$. We show next that the coordinate vectors $[x]_{B_1} \in R^p$ and $[T(x)]_{B_2} \in R^s$ are related by a matrix multiplication. By the linearity of transformation $T(x)$,

$$T(x) = x_1 T(e_1) + x_2 T(e_2) + \cdots + x_p T(e_p).$$

In coordinates (here $[T(x)]_{B_2}$ is a vector in R^s),

$$[T(x)]_{B_2} = x_1 [T(e_1)]_{B_2} + x_2 [T(e_2)]_{B_2} + \cdots + x_p [T(e_p)]_{B_2}. \tag{7.1}$$

Form a matrix $A = [[T(e_1)]_{B_2} \; [T(e_2)]_{B_2} \; \ldots \; [T(e_p)]_{B_2}]$, of size $s \times p$, by using the vectors $[T(e_i)]_{B_2}$ as its columns. Then (7.1) implies that

$$[T(\dot{x})]_{B_2} = A[x]_{B_1},$$

by the definition of matrix multiplication. One says that A is *the matrix of linear transformation* $T(x)$.

Example 5. Let us return to the differentiation $D : P_4 \to P_3$, and use the standard bases $B_1 = \{1, x, x^2, x^3, x^4\}$ of P_4, and $B_2 = \{1, x, x^2, x^3\}$ of P_3. Since

$$D(1) = 0 = 0 \times 1 + 0 \times x + 0 \times x^2 + 0 \times x^3,$$

obtain the coordinates $[D(1)]_{B_2} = \begin{bmatrix} 0 \\ 0 \\ 0 \\ 0 \end{bmatrix}$. (Here 0×1 means zero times the vector 1, $0 \times x$ is zero times the vector x, etc.) Similarly,

$$D(x) = 1 = 1 \times 1 + 0 \times x + 0 \times x^2 + 0 \times x^3,$$

giving $[D(x)]_{B_2} = \begin{bmatrix} 1 \\ 0 \\ 0 \\ 0 \end{bmatrix}$. Next, $D(x^2) = 2x$, giving $[D(x^2)]_{B_2} = \begin{bmatrix} 0 \\ 2 \\ 0 \\ 0 \end{bmatrix}$, $D(x^3) = 3x^2$, giving

$[D(x^3)]_{B_2} = \begin{bmatrix} 0 \\ 0 \\ 3 \\ 0 \end{bmatrix}$, $D(x^4) = 4x^3$, giving $[D(x^4)]_{B_2} = \begin{bmatrix} 0 \\ 0 \\ 0 \\ 4 \end{bmatrix}$. The matrix of the transformation D is then

$$A = \begin{bmatrix} 0 & 1 & 0 & 0 & 0 \\ 0 & 0 & 2 & 0 & 0 \\ 0 & 0 & 0 & 3 & 0 \\ 0 & 0 & 0 & 0 & 4 \end{bmatrix}.$$

This matrix A allows one to perform differentiation of polynomials in P_4 through matrix multiplication. For example, let $p(x) = -2x^4 + x^3 + 5x - 6$, with $p'(x) = -8x^3 + 3x^2 + 5$. Then $[p(x)]_{B_1} = \begin{bmatrix} -6 \\ 5 \\ 0 \\ 1 \\ -2 \end{bmatrix}$, $[p'(x)]_{B_2} = \begin{bmatrix} 5 \\ 0 \\ 3 \\ -8 \end{bmatrix}$, and one verifies that

$$\begin{bmatrix} 5 \\ 0 \\ 3 \\ -8 \end{bmatrix} = \begin{bmatrix} 0 & 1 & 0 & 0 & 0 \\ 0 & 0 & 2 & 0 & 0 \\ 0 & 0 & 0 & 3 & 0 \\ 0 & 0 & 0 & 0 & 4 \end{bmatrix} \begin{bmatrix} -6 \\ 5 \\ 0 \\ 1 \\ -2 \end{bmatrix}.$$

The matrix A transforms the coefficients of $p(x)$ into those of $p'(x)$.

Exercises

1. Write down the standard basis S in $M_{2 \times 3}$, and then find the coordinates of $A = \begin{bmatrix} 1 & -3 & 2 \\ -5 & 0 & 4 \end{bmatrix}$ with respect to this basis.

Answer: $E_{11} = \begin{bmatrix} 1 & 0 & 0 \\ 0 & 0 & 0 \end{bmatrix}$, $E_{12} = \begin{bmatrix} 0 & 1 & 0 \\ 0 & 0 & 0 \end{bmatrix}$, $E_{13} = \begin{bmatrix} 0 & 0 & 1 \\ 0 & 0 & 0 \end{bmatrix}$, $E_{21} = \begin{bmatrix} 0 & 0 & 0 \\ 1 & 0 & 0 \end{bmatrix}$, $E_{22} = \begin{bmatrix} 0 & 0 & 0 \\ 0 & 1 & 0 \end{bmatrix}$, $E_{23} =$

$\begin{bmatrix} 0 & 0 & 0 \\ 0 & 0 & 1 \end{bmatrix}$; $[A]_S = \begin{bmatrix} 1 \\ -3 \\ 2 \\ -5 \\ 0 \\ 4 \end{bmatrix}$.

2. (a) Show that the matrices $A_1 = \begin{bmatrix} 1 & 0 \\ 0 & 0 \end{bmatrix}$, $A_2 = \begin{bmatrix} 1 & 2 \\ 0 & 0 \end{bmatrix}$, and $A_3 = \begin{bmatrix} 1 & 2 \\ 3 & 0 \end{bmatrix}$ are linearly inde-
    pendent vectors of $M_{2\times2}$.

    (b) Let $C = \begin{bmatrix} 3 & 4 \\ 3 & 0 \end{bmatrix}$. Show that the matrices A_1, A_2, A_3, C are linearly dependent vectors
    of $M_{2\times2}$.

    Hint. Express C as a linear combination of A_1, A_2, A_3.

    (c) Let $A_4 = \begin{bmatrix} 0 & 0 \\ 0 & 1 \end{bmatrix}$. Show that $B = \{A_1, A_2, A_3, A_4\}$ is a basis of $M_{2\times2}$.

    (d) $F = \begin{bmatrix} 3 & 4 \\ 0 & -7 \end{bmatrix}$. Find the coordinates of F with respect to the basis B.

    Answer: $[F]_B = \begin{bmatrix} 1 \\ 2 \\ 0 \\ -7 \end{bmatrix}$.

3. Calculate the norm of the following vectors in $P_2(-1, 1)$:
    (a) x. Hint. Note that $\|x\|^2 = x \cdot x = \int_{-1}^{1} x^2 \, dx$.
    (b) $p(x) = x^2 - 1$.

    Answer: $\|x^2 - 1\| = \frac{4}{\sqrt{15}}$.

    (c) $q(x) = \sqrt{2}$.

    Answer: $\| \sqrt{2} \| = 2$.

4. Apply the Gram–Schmidt process to the vectors $1, x + 2, x^2 - x$ of $P_2(-1, 1)$ to obtain
    a standardized orthogonal basis of $P_2(-1, 1)$.

5. Let $I : P_3 \rightarrow P_4$ be a map taking any polynomial $p(x) = a_3x^3 + a_2x^2 + a_1x + a_0$ into
    $I(p(x)) = a_3\frac{x^4}{4} + a_2\frac{x^3}{3} + a_1\frac{x^2}{2} + a_0x$.
    (a) Identify I with a calculus operation, and explain why I is a linear transforma-
    tion.

    (b) Find the matrix representation of I (using the standard bases in both P_3 and P_4).

    Answer: $\begin{bmatrix} 0 & 0 & 0 & 0 \\ 1 & 0 & 0 & 0 \\ 0 & 1/2 & 0 & 0 \\ 0 & 0 & 1/3 & 0 \\ 0 & 0 & 0 & 1/4 \end{bmatrix}$.

    (c) Is the map I onto?

6. Let $T : M_{2\times2} \rightarrow M_{2\times2}$ be a map taking matrices $A = \begin{bmatrix} a & b \\ c & d \end{bmatrix}$ into $T(A) = \begin{bmatrix} 2c & 2d \\ a & b \end{bmatrix}$.
    (a) Show that T is a linear transformation.
    (b) Find the matrix representation of T (using the standard bases).

    Answer: $\begin{bmatrix} 0 & 0 & 2 & 0 \\ 0 & 0 & 0 & 2 \\ 1 & 0 & 0 & 0 \\ 0 & 1 & 0 & 0 \end{bmatrix}$.

7. Let $T : M_{2\times2} \rightarrow M_{2\times2}$ be a map taking matrices $A = \begin{bmatrix} a & b \\ c & d \end{bmatrix}$ into $T(A) = \begin{bmatrix} c & a \\ 1 & b \end{bmatrix}$. Show that
    T is not a linear transformation.
    Hint. Consider $T(O)$.

8. Justify *Rodrigues' formula* for Legendre polynomials, namely

$$P_n(x) = \frac{1}{2^n n!} \frac{d^n}{dx^n} [(x^2 - 1)^n].$$

Hint. Differentiations produce a polynomial of degree n, with $P_n(0) = 1$. To see that $\int_{-1}^{1} P_n(x) P_m(x) \, dx = 0$, with $n < m$, perform m integrations by parts, shifting all derivatives on $P_n(x)$.

6 Systems of differential and difference equations

Solving of systems of differential equations provides one of the most useful applications of eigenvectors and eigenvalues. *Generalized eigenvectors* and *eigenvector chains* are introduced in the process. For simplicity, the presentation begins with 3×3 systems, and then the general theory is developed. Functions of matrices are developed, particularly *matrix exponentials*. *Fundamental solution matrices* are applied to systems with periodic coefficients, including Hamiltonian systems and Massera's theorem. The last section covers systems of difference equations, and their applications to Markov matrices and Jacobi's iterations.

6.1 Linear systems with constant coefficients

We wish to find the functions $x_1(t)$, $x_2(t)$, $x_3(t)$ that solve a system of differential equations, with given numerical coefficients a_{ij},

$$
\begin{aligned}
x_1' &= a_{11}x_1 + a_{12}x_2 + a_{13}x_3, \\
x_2' &= a_{21}x_1 + a_{22}x_2 + a_{23}x_3, \\
x_3' &= a_{31}x_1 + a_{32}x_2 + a_{33}x_3,
\end{aligned}
\tag{1.1}
$$

subject to the initial conditions

$$
x_1(t_0) = \alpha, \quad x_2(t_0) = \beta, \quad x_3(t_0) = \gamma,
$$

with given numbers t_0, α, β, and γ. Using matrix notation, we may write this system as

$$
x' = Ax, \quad x(t_0) = x_0,
\tag{1.2}
$$

where $x(t) = \begin{bmatrix} x_1(t) \\ x_2(t) \\ x_3(t) \end{bmatrix}$ is *the unknown vector function*, $A = \begin{bmatrix} a_{11} & a_{12} & a_{13} \\ a_{21} & a_{22} & a_{23} \\ a_{31} & a_{32} & a_{33} \end{bmatrix}$ is a 3×3 *matrix of the coefficients*, and $x_0 = \begin{bmatrix} \alpha \\ \beta \\ \gamma \end{bmatrix}$ is *the vector of initial conditions*. Indeed, the left-hand side in (1.1) contains components of the vector $x'(t) = \begin{bmatrix} x_1'(t) \\ x_2'(t) \\ x_3'(t) \end{bmatrix}$, while components of the vector function Ax are on the right-hand side.

If two vector functions $y(t)$ and $z(t)$ are solutions of the system $x' = Ax$, their *linear combination* $c_1y(t) + c_2z(t)$ is also a solution of the same system, for any numbers c_1 and c_2, which is straightforward to justify. It is known from the theory of differential equations that *the initial value problem* (1.2) has a unique solution $x(t)$, valid for all $t \in (-\infty, \infty)$.

Let us search for solution of (1.2) in the form

$$
x(t) = e^{\lambda t}\xi,
\tag{1.3}
$$

https://doi.org/10.1515/9783111086507-006

where λ is a number, and ξ is a vector with entries independent of t. Substitution of $x(t)$ into (1.2) gives

$$\lambda e^{\lambda t}\xi = A(e^{\lambda t}\xi),$$

simplifying to

$$A\xi = \lambda\xi.$$

So that if λ is an eigenvalue of A, and ξ is a corresponding eigenvector, then (1.3) provides a solution of the system in (1.2). Let $\lambda_1, \lambda_2, \lambda_3$ be the eigenvalues of the matrix A. There are several cases to consider.

Case 1. The eigenvalues of A are real and distinct. Then the corresponding eigenvectors ξ_1, ξ_2, and ξ_3 are linearly independent by Theorem 4.2.1. Since $e^{\lambda_1 t}\xi_1, e^{\lambda_2 t}\xi_2$, and $e^{\lambda_3 t}\xi_3$ are solutions of the system (1.2), their linear combination

$$x(t) = c_1 e^{\lambda_1 t}\xi_1 + c_2 e^{\lambda_2 t}\xi_2 + c_3 e^{\lambda_3 t}\xi_3 \tag{1.4}$$

also solves the system (1.2). We claim that (1.4) gives *the general solution of the system* (1.2), meaning that it is possible to choose the constants c_1, c_2, c_3 to satisfy any initial condition:

$$x(t_0) = c_1 e^{\lambda_1 t_0}\xi_1 + c_2 e^{\lambda_2 t_0}\xi_2 + c_3 e^{\lambda_3 t_0}\xi_3 = x_0. \tag{1.5}$$

We need to solve a system of three linear equations with three unknowns c_1, c_2, c_3. The matrix of this system is nonsingular, because its columns $e^{\lambda_1 t_0}\xi_1, e^{\lambda_2 t_0}\xi_2, e^{\lambda_3 t_0}\xi_3$ are linearly independent (as multiples of linearly independent vectors ξ_1, ξ_2, ξ_3). Therefore, there is a unique solution triple $\bar{c}_1, \bar{c}_2, \bar{c}_3$ of the system (1.5). Then $x(t) = \bar{c}_1 e^{\lambda_1 t}\xi_1 + \bar{c}_2 e^{\lambda_2 t}\xi_2 + \bar{c}_3 e^{\lambda_3 t}\xi_3$ gives the solution of the initial value problem (1.2).

Example 1. Solve the system

$$x' = \begin{bmatrix} 2 & 1 & 1 \\ 1 & 2 & 1 \\ 0 & 0 & 4 \end{bmatrix} x, \quad x(0) = \begin{bmatrix} -2 \\ 2 \\ -1 \end{bmatrix}.$$

Calculate the eigenvalues $\lambda_1 = 1$, with corresponding eigenvector $\xi_1 = \begin{bmatrix} -1 \\ 1 \\ 0 \end{bmatrix}$, $\lambda_2 = 3$, with corresponding eigenvector $\xi_2 = \begin{bmatrix} 1 \\ 1 \\ 0 \end{bmatrix}$, and $\lambda_3 = 4$, with corresponding eigenvector $\xi_3 = \begin{bmatrix} 1 \\ 1 \\ 1 \end{bmatrix}$. The general solution is then

$$x(t) = c_1 e^t \begin{bmatrix} -1 \\ 1 \\ 0 \end{bmatrix} + c_2 e^{3t} \begin{bmatrix} 1 \\ 1 \\ 0 \end{bmatrix} + c_3 e^{4t} \begin{bmatrix} 1 \\ 1 \\ 1 \end{bmatrix},$$

or in components

$$x_1(t) = -c_1 e^t + c_2 e^{3t} + c_3 e^{4t},$$
$$x_2(t) = c_1 e^t + c_2 e^{3t} + c_3 e^{4t},$$
$$x_3(t) = c_3 e^{4t}.$$

Turning to the initial conditions, we obtain a system of equations

$$x_1(0) = -c_1 + c_2 + c_3 = -2,$$
$$x_2(0) = c_1 + c_2 + c_3 = 2,$$
$$x_3(0) = c_3 = -1.$$

Calculate $c_1 = 2$, $c_2 = 1$, and $c_3 = -1$. Answer:

$$x_1(t) = -2e^t + e^{3t} - e^{4t},$$
$$x_2(t) = 2e^t + e^{3t} - e^{4t},$$
$$x_3(t) = -e^{4t}.$$

The answer can also be presented in the vector form as

$$x(t) = 2e^t \begin{bmatrix} -1 \\ 1 \\ 0 \end{bmatrix} + e^{3t} \begin{bmatrix} 1 \\ 1 \\ 0 \end{bmatrix} - e^{4t} \begin{bmatrix} 1 \\ 1 \\ 1 \end{bmatrix}.$$

Case 2. The eigenvalue λ_1 is double, so that $\lambda_2 = \lambda_1$, while $\lambda_3 \neq \lambda_1$, and assume that λ_1 has two linearly independent eigenvectors ξ_1 and ξ_2. Let ξ_3 denote an eigenvector corresponding to λ_3. This vector does not lie in the plane spanned by ξ_1 and ξ_2, and then the vectors ξ_1, ξ_2, ξ_3 are linearly independent. Claim: the general solution of (1.2) is given by the formula (1.4), with λ_2 replaced by λ_1,

$$x(t) = c_1 e^{\lambda_1 t} \xi_1 + c_2 e^{\lambda_1 t} \xi_2 + c_3 e^{\lambda_3 t} \xi_3.$$

Indeed, this vector function solves (1.2) for any c_1, c_2, c_3. To satisfy the initial conditions, obtain a linear system for c_1, c_2, c_3, namely

$$c_1 e^{\lambda_1 t_0} \xi_1 + c_2 e^{\lambda_1 t_0} \xi_2 + c_3 e^{\lambda_3 t_0} \xi_3 = x_0,$$

which has a unique solution for any x_0, because its matrix has linearly independent columns, and hence is nonsingular. The existence of *a complete set of eigenvectors* is the key here!

Example 2. Solve the system

$$x' = \begin{bmatrix} 2 & 1 & 1 \\ 1 & 2 & 1 \\ 1 & 1 & 2 \end{bmatrix} x, \quad x(0) = \begin{bmatrix} 1 \\ 0 \\ -4 \end{bmatrix}.$$

The eigenvalues and eigenvectors of this matrix were calculated in Section 4.2. The eigenvalues are: $\lambda_1 = 1$, $\lambda_2 = 1$, and $\lambda_3 = 4$. The double eigenvalue $\lambda_1 = 1$ has two linearly independent eigenvectors $\xi_1 = \begin{bmatrix} -1 \\ 0 \\ 1 \end{bmatrix}$ and $\xi_2 = \begin{bmatrix} -1 \\ 1 \\ 0 \end{bmatrix}$. The other eigenvalue $\lambda_3 = 4$ comes with corresponding eigenvector $\xi_3 = \begin{bmatrix} 1 \\ 1 \\ 1 \end{bmatrix}$. The general solution is then

$$x(t) = c_1 e^t \begin{bmatrix} -1 \\ 0 \\ 1 \end{bmatrix} + c_2 e^t \begin{bmatrix} -1 \\ 1 \\ 0 \end{bmatrix} + c_3 e^{4t} \begin{bmatrix} 1 \\ 1 \\ 1 \end{bmatrix}.$$

In components,

$$x_1(t) = -c_1 e^t - c_2 e^t + c_3 e^{4t},$$
$$x_2(t) = c_2 e^t + c_3 e^{4t},$$
$$x_3(t) = c_1 e^t + c_3 e^{4t}.$$

Using the initial conditions, we obtain a system of equations

$$x_1(0) = -c_1 - c_2 + c_3 = 1,$$
$$x_2(0) = c_2 + c_3 = 0,$$
$$x_3(0) = c_1 + c_3 = -4.$$

Calculate $c_1 = -3$, $c_2 = 1$, and $c_3 = -1$. Answer:

$$x_1(t) = 2e^t - e^{4t},$$
$$x_2(t) = e^t - e^{4t},$$
$$x_3(t) = -3e^t - e^{4t}.$$

Proceeding similarly, one can solve the initial value problem (1.2) for any $n \times n$ matrix A, provided that all of its eigenvalues are real, and A has *a complete set of n linearly independent eigenvectors*. For example, if matrix A is symmetric, then all of its eigenvalues are real, and *there is always a complete set of n linearly independent eigenvectors* (even though some eigenvalues may be repeated).

Case 3. The eigenvalue λ_1 has multiplicity two (λ_1 is a double root of the characteristic equation, $\lambda_2 = \lambda_1$), $\lambda_3 \neq \lambda_1$, but λ_1 has only one linearly independent eigenvector ξ. The eigenvalue λ_1 brings in only one solution $e^{\lambda_1 t}\xi$. By analogy with the second order equations, one can try $te^{\lambda_1 t}\xi$ for the second solution. However, this vector function is a scalar multiple of the first solution, linearly dependent with it, at any $t = t_0$. Modify the guess,

$$x(t) = te^{\lambda_1 t}\xi + e^{\lambda_1 t}\eta, \tag{1.6}$$

and search for a constant vector η, to obtain a second linearly independent solution of (1.2). Substituting (1.6) into (1.2), and using that $A\xi = \lambda_1 \xi$, we obtain

$$e^{\lambda_1 t}\xi + \lambda_1 te^{\lambda_1 t}\xi + \lambda_1 e^{\lambda_1 t}\eta = \lambda_1 te^{\lambda_1 t}\xi + e^{\lambda_1 t}A\eta.$$

Canceling a pair of terms and dividing by $e^{\lambda_1 t}$ gives

$$(A - \lambda_1 I)\eta = \xi. \tag{1.7}$$

Even though the matrix $A - \lambda_1 I$ is singular (its determinant is zero), it can be shown (using the Jordan normal forms) that the linear system (1.7) has a solution η, called *generalized eigenvector*. It follows from (1.7) that η is not a multiple of ξ. (Indeed, if $\eta = c\xi$, then $(A - \lambda_1 I)\eta = c(A - \lambda_1 I)\xi = c(A\xi - \lambda_1 \xi) = 0$, while $\xi \neq 0$ in (1.7).) Using this vector η in (1.6) provides the second linearly independent solution, corresponding to $\lambda = \lambda_1$.

Example 3. Solve the system

$$x' = \begin{bmatrix} 1 & -1 \\ 1 & 3 \end{bmatrix} x.$$

This matrix has a double eigenvalue $\lambda_1 = \lambda_2 = 2$, and only one linearly independent eigenvector $\xi = \begin{bmatrix} 1 \\ -1 \end{bmatrix}$, giving only one solution, $x_1(t) = e^{2t}\begin{bmatrix} 1 \\ -1 \end{bmatrix}$. The system (1.7) to determine the generalized eigenvector $\eta = \begin{bmatrix} \eta_1 \\ \eta_2 \end{bmatrix}$ takes the form $(A - 2I)\eta = \xi$, or in components

$$-\eta_1 - \eta_2 = 1,$$
$$\eta_1 + \eta_2 = -1.$$

Discard the second equation, because it is a multiple of the first. The first equation has infinitely many solutions, but all we need is just one solution, that is not a multiple of ξ. Set $\eta_2 = 0$, which gives $\eta_1 = -1$. So that $\eta = \begin{bmatrix} -1 \\ 0 \end{bmatrix}$ is a generalized eigenvector. (Observe that infinitely many generalized eigenvectors can be obtained by choosing an arbitrary $\eta_2 \neq 0$.) The second linearly independent solution is (in view of (1.6))

$$x_2(t) = te^{2t}\begin{bmatrix} 1 \\ -1 \end{bmatrix} + e^{2t}\begin{bmatrix} -1 \\ 0 \end{bmatrix}.$$

The general solution is then

$$x(t) = c_1 e^{2t}\begin{bmatrix} 1 \\ -1 \end{bmatrix} + c_2\left(te^{2t}\begin{bmatrix} 1 \\ -1 \end{bmatrix} + e^{2t}\begin{bmatrix} -1 \\ 0 \end{bmatrix}\right).$$

Example 4. Let us solve the system

$$x' = \begin{bmatrix} 1 & 4 & 0 \\ -4 & -7 & 0 \\ 0 & 0 & 5 \end{bmatrix} x.$$

This matrix has a double eigenvalue $\lambda_1 = \lambda_2 = -3$, with only one linearly independent eigenvector $\xi_1 = \begin{bmatrix} -1 \\ 1 \\ 0 \end{bmatrix}$, giving only one solution, $x_1(t) = e^{-3t} \begin{bmatrix} -1 \\ 1 \\ 0 \end{bmatrix}$. The system (1.7) to determine the generalized eigenvector $\eta = \begin{bmatrix} \eta_1 \\ \eta_2 \\ \eta_3 \end{bmatrix}$ takes the form $(A + 3I)\eta = \xi_1$, or in components

$$4\eta_1 + 4\eta_2 = -1,$$
$$-4\eta_1 - 4\eta_2 = 1,$$
$$8\eta_3 = 0.$$

Obtain $\eta_3 = 0$. Discard the second equation. Set $\eta_2 = 0$ in the first equation, so that $\eta_1 = -\frac{1}{4}$. Conclude that $\eta = \begin{bmatrix} -\frac{1}{4} \\ 0 \\ 0 \end{bmatrix}$ is a generalized eigenvector. The second linearly independent solution is

$$x_2(t) = e^{-3t} \left(t \begin{bmatrix} -1 \\ 1 \\ 0 \end{bmatrix} + \begin{bmatrix} -\frac{1}{4} \\ 0 \\ 0 \end{bmatrix} \right).$$

The third eigenvalue $\lambda_3 = 5$ is simple, with corresponding eigenvector $\begin{bmatrix} 0 \\ 0 \\ 1 \end{bmatrix}$, so that $x_3(t) = e^{5t} \begin{bmatrix} 0 \\ 0 \\ 1 \end{bmatrix}$. The general solution is then

$$x(t) = c_1 e^{-3t} \begin{bmatrix} -1 \\ 1 \\ 0 \end{bmatrix} + c_2 e^{-3t} \left(t \begin{bmatrix} -1 \\ 1 \\ 0 \end{bmatrix} + \begin{bmatrix} -\frac{1}{4} \\ 0 \\ 0 \end{bmatrix} \right) + c_3 e^{5t} \begin{bmatrix} 0 \\ 0 \\ 1 \end{bmatrix}.$$

For larger matrices A, it is possible to have repeated eigenvalues of multiplicity greater than 2, missing more than one eigenvector compared to a complete set. We shall cover such a possibility later on.

Exercises

1. Find the general solution for the following systems of differential equations:
    (a) $x'(t) = \begin{bmatrix} 3 & 4 \\ -1 & -2 \end{bmatrix} x(t)$.

    Answer: $x(t) = c_1 e^{-t} \begin{bmatrix} -1 \\ 1 \end{bmatrix} + c_2 e^{2t} \begin{bmatrix} -4 \\ 1 \end{bmatrix}$.

(b) $x'(t) = \begin{bmatrix} 4 & -2 \\ -2 & 1 \end{bmatrix} x(t)$.

Answer: $x(t) = c_1 e^{5t} \begin{bmatrix} -2 \\ 1 \end{bmatrix} + c_2 \begin{bmatrix} 1 \\ 2 \end{bmatrix}$.

(c) $x'(t) = \begin{bmatrix} 1 & 2 \\ -1 & 4 \end{bmatrix} x(t)$.

Answer: $x(t) = c_1 e^{2t} \begin{bmatrix} 2 \\ 1 \end{bmatrix} + c_2 e^{3t} \begin{bmatrix} 1 \\ 1 \end{bmatrix}$.

(d) $x'(t) = \begin{bmatrix} 1 & 1 & 1 \\ 2 & 2 & 1 \\ 4 & -2 & 1 \end{bmatrix} x(t)$.

Answer: $x(t) = c_1 e^{-t} \begin{bmatrix} -1 \\ 0 \\ 2 \end{bmatrix} + c_2 e^{2t} \begin{bmatrix} -1 \\ -3 \\ 2 \end{bmatrix} + c_3 e^{3t} \begin{bmatrix} 1 \\ 2 \\ 0 \end{bmatrix}$.

(e) $x'(t) = \begin{bmatrix} 1 & 1 & -1 \\ 2 & 0 & -1 \\ 0 & -2 & 1 \end{bmatrix} x(t)$.

Answer: $x(t) = c_1 e^{-t} \begin{bmatrix} 0 \\ 1 \\ 1 \end{bmatrix} + c_2 e^{3t} \begin{bmatrix} -1 \\ -1 \\ 1 \end{bmatrix} + c_3 \begin{bmatrix} 1 \\ 1 \\ 2 \end{bmatrix}$.

(f) $x'(t) = \begin{bmatrix} 0 & 1 & 0 & 0 \\ 1 & 0 & 0 & 0 \\ 0 & 0 & 4 & -5 \\ 0 & 0 & 1 & -2 \end{bmatrix} x(t)$.

Answer:

$$x(t) = c_1 e^{-t} \begin{bmatrix} 0 \\ 0 \\ 1 \\ 1 \end{bmatrix} + c_2 e^{-t} \begin{bmatrix} -1 \\ 1 \\ 0 \\ 0 \end{bmatrix} + c_3 e^{t} \begin{bmatrix} 1 \\ 1 \\ 0 \\ 0 \end{bmatrix} + c_4 e^{3t} \begin{bmatrix} 0 \\ 0 \\ 5 \\ 1 \end{bmatrix}.$$

(g) $x'(t) = \begin{bmatrix} 2 & 1 & 1 \\ 1 & 2 & 1 \\ 1 & 1 & 2 \end{bmatrix} x(t)$.

Answer: $x(t) = c_1 e^{4t} \begin{bmatrix} 1 \\ 1 \\ 1 \end{bmatrix} + c_2 e^{t} \begin{bmatrix} -1 \\ 0 \\ 1 \end{bmatrix} + c_3 e^{t} \begin{bmatrix} -1 \\ 1 \\ 0 \end{bmatrix}$.

2. Solve the following initial value problems:

(a) $x'(t) = \begin{bmatrix} 1 & 2 \\ 4 & 3 \end{bmatrix} x(t)$, $x(0) = \begin{bmatrix} 1 \\ -2 \end{bmatrix}$.

Answer: $x_1(t) = \frac{4e^{-t}}{3} - \frac{e^{5t}}{3}$, $x_2(t) = -\frac{4e^{-t}}{3} - \frac{2e^{5t}}{3}$.

(b) $x'(t) = \begin{bmatrix} 1 & -1 & 1 \\ 2 & 1 & 2 \\ 3 & 0 & 3 \end{bmatrix} x(t)$, $x(0) = \begin{bmatrix} 0 \\ -1 \\ 1 \end{bmatrix}$.

Answer:

$$x_1(t) = -1 + e^{2t}, \quad x_2(t) = -4e^{2t} + 3e^{3t}, \quad x_3(t) = 1 - 3e^{2t} + 3e^{3t}.$$

3. Given a vector function $x(t) = \begin{bmatrix} x_1(t) \\ x_2(t) \\ x_3(t) \end{bmatrix}$, define its derivative as $x'(t) = \lim_{h \to 0} \frac{x(t+h) - x(t)}{h}$.

(a) Show that $x'(t) = \begin{bmatrix} x_1'(t) \\ x_2'(t) \\ x_3'(t) \end{bmatrix}$.

(b) If $x(t) = e^{\lambda t} \xi$, where λ is a number, and a vector ξ has constant entries, show that $x'(t) = \lambda e^{\lambda t} \xi$.

4. Let $x(t)$ and $y(t)$ be two vector functions in R^n. Show that the product rule holds for the scalar product, namely

$$\frac{d}{dt} x(t) \cdot y(t) = x'(t) \cdot y(t) + x(t) \cdot y'(t).$$

5. Solve
   (a) $x'(t) = \begin{bmatrix} 1 & -1 \\ 4 & -3 \end{bmatrix} x(t)$.
   (b)

$$x_1' = x_1 - x_2, \qquad x_1(0) = 1,$$
$$x_2' = 4x_1 - 3x_2, \quad x_2(0) = -1.$$

   Answer: $x_1(t) = e^{-t}(3t + 1)$, $x_2(t) = e^{-t}(6t - 1)$.
   (c) $x'(t) = \begin{bmatrix} 0 & -1 & 1 \\ 2 & -3 & 1 \\ 1 & -1 & -1 \end{bmatrix} x(t)$.
   Answer:

$$x(t) = c_1 e^{-t} \begin{bmatrix} 1 \\ 1 \\ 0 \end{bmatrix} + c_2 e^{-t} \left(t \begin{bmatrix} 1 \\ 1 \\ 0 \end{bmatrix} + \begin{bmatrix} 1 \\ 1 \\ 1 \end{bmatrix} \right) + c_3 e^{-2t} \begin{bmatrix} 0 \\ 1 \\ 1 \end{bmatrix}.$$

   Show also that all solutions tend to zero, as $t \to \infty$.

6. Calculate the eigenvalues of $A = \begin{bmatrix} -3 & 5 & -1 & 1 \\ 0 & -1 & 0 & 0 \\ 2 & 4 & -3 & 2 \\ 3 & -5 & 0 & -1 \end{bmatrix}$.

   Without calculating the general solution, explain why all solutions of $x' = Ax$ tend to zero, as $t \to \infty$.
   Hint. The eigenvalues are $-4, -2, -1, -1$.

7. Show that the system $x' = \begin{bmatrix} a & 1 \\ 2 & -a \end{bmatrix} x$ has solutions that tend to zero, and solutions that tend to infinity, as $t \to \infty$. Here a is any number.
   Hint. The eigenvalues are real, of opposite sign.

8. Let η be a generalized eigenvector corresponding to an eigenvector ξ. Show that 2η is not a generalized eigenvector.

9. Show that generalized eigenvector is not unique.
   Hint. Consider $\eta + c\xi$, with an arbitrary number c.

10. Explain why generalized eigenvectors are not possible for symmetric matrices, and why generalized eigenvectors are not needed to solve $x' = Ax$ for symmetric A.
    Hint. If $(A - \lambda I)\eta = \xi$ and $A^T = A$, then $\xi \cdot \xi = (A - \lambda I)\eta \cdot \xi = \eta \cdot (A - \lambda I)\xi = 0$.

6.2 A pair of complex conjugate eigenvalues

Complex- and real-valued solutions

Recall that one differentiates complex-valued functions similarly to the real-valued ones. For example,

$$\frac{d}{dt} e^{it} = i e^{it},$$

where $i = \sqrt{-1}$ is treated the same way as any other number. Any complex-valued function $x(t)$ can be written in the form $x(t) = u(t) + iv(t)$, where $u(t)$ and $v(t)$ are real-valued functions. It follows by the definition of derivative, $x'(t) = \lim_{h \to 0} \frac{x(t+h)-x(t)}{h}$, that $x'(t) = u'(t) + iv'(t)$. For example, using Euler's formula,

$$\frac{d}{dt} e^{it} = \frac{d}{dt}(\cos t + i \sin t) = -\sin t + i \cos t = i(\cos t + i \sin t) = ie^{it}.$$

If a system

$$x' = Ax \tag{2.1}$$

has a complex valued solution $x(t) = u(t) + iv(t)$, then

$$u'(t) + iv'(t) = A(u(t) + iv(t)).$$

Equating the real and imaginary parts, we obtain $u' = Au$ and $v' = Av$, so that both $u(t)$ and $v(t)$ are real-valued solutions of the system (2.1).

The general solution

Assume that matrix A of the system (2.1) has a pair of complex conjugate eigenvalues $p + iq$ and $p - iq$, $q \neq 0$. They need to contribute two linearly independent solutions. The eigenvector corresponding to $p + iq$ is complex-valued, which we may write as $\xi + i\eta$, where ξ and η are real-valued vectors. Then $x(t) = e^{(p+iq)t}(\xi + i\eta)$ is a complex-valued solution of the system (2.1). To get two real-valued solutions, take the real and imaginary parts of this solution. Using Euler's formula,

$$x(t) = e^{pt}(\cos qt + i \sin qt)(\xi + i\eta)$$
$$= e^{pt}(\cos qt \, \xi - \sin qt \, \eta) + ie^{pt}(\sin qt \, \xi + \cos qt \, \eta),$$

so that

$$u(t) = e^{pt}(\cos qt \, \xi - \sin qt \, \eta),$$
$$v(t) = e^{pt}(\sin qt \, \xi + \cos qt \, \eta) \tag{2.2}$$

are two real-valued solutions of (2.1), corresponding to a pair of eigenvalues $p \pm iq$.

In case of a 2×2 matrix A (when there are no other eigenvalues), the general solution is

$$x(t) = c_1 u(t) + c_2 v(t), \tag{2.3}$$

since it is shown in Exercises that the vectors $u(t)$ and $v(t)$ are linearly independent, so that it is possible to choose c_1 and c_2 to satisfy any initial condition $x(t_0) = x_0$. If one applies the same procedure to the other eigenvalue $p-iq$, and corresponding eigenvector $\xi - i\eta$, the answer is the same, as is easy to check.

For larger matrices A, the solutions in (2.2) contribute to the general solution, along with other solutions.

Example 1. Solve the system

$$x' = \begin{bmatrix} 1 & -2 \\ 2 & 1 \end{bmatrix} x, \quad x(0) = \begin{bmatrix} 2 \\ 1 \end{bmatrix}.$$

Calculate the eigenvalues $\lambda_1 = 1+2i$ and $\lambda_2 = 1-2i$. An eigenvector corresponding to λ_1 is $\begin{bmatrix} i \\ 1 \end{bmatrix}$. So that there is a complex valued solution $e^{(1+2i)t} \begin{bmatrix} i \\ 1 \end{bmatrix}$. Using Euler's formula, rewrite this solution as

$$e^t(\cos 2t + i \sin 2t) \begin{bmatrix} i \\ 1 \end{bmatrix} = e^t \begin{bmatrix} -\sin 2t \\ \cos 2t \end{bmatrix} + ie^t \begin{bmatrix} \cos 2t \\ \sin 2t \end{bmatrix}.$$

Taking the real and imaginary parts, we obtain two linearly independent real-valued solutions, so that the general solution is

$$x(t) = c_1 e^t \begin{bmatrix} -\sin 2t \\ \cos 2t \end{bmatrix} + c_2 e^t \begin{bmatrix} \cos 2t \\ \sin 2t \end{bmatrix}.$$

In components,

$$x_1(t) = -c_1 e^t \sin 2t + c_2 e^t \cos 2t,$$
$$x_2(t) = c_1 e^t \cos 2t + c_2 e^t \sin 2t.$$

From the initial conditions,

$$x_1(0) = c_2 = 2,$$
$$x_2(0) = c_1 = 1,$$

so that $c_1 = 1$ and $c_2 = 2$. Answer:

$$x_1(t) = -e^t \sin 2t + 2e^t \cos 2t,$$
$$x_2(t) = e^t \cos 2t + 2e^t \sin 2t.$$

Example 2. Solve the system

$$x' = \begin{bmatrix} 2 & -1 & 2 \\ 1 & 0 & 2 \\ -2 & 1 & -1 \end{bmatrix} x.$$

One of the eigenvalues is $\lambda_1 = 1$, with an eigenvector $\begin{bmatrix} 0 \\ 2 \\ 1 \end{bmatrix}$. Then $e^t \begin{bmatrix} 0 \\ 2 \\ 1 \end{bmatrix}$ gives a solution. The other two eigenvalues are $\lambda_2 = i$ and $\lambda_3 = -i$. An eigenvector corresponding to $\lambda_2 = i$ is $\begin{bmatrix} -1-i \\ -1-i \\ 1 \end{bmatrix}$, giving a complex-valued solution $e^{it} \begin{bmatrix} -1-i \\ -1-i \\ 1 \end{bmatrix}$ that can be rewritten as

$$(\cos t + i \sin t) \begin{bmatrix} -1-i \\ -1-i \\ 1 \end{bmatrix} = \begin{bmatrix} -\cos t + \sin t \\ -\cos t + \sin t \\ \cos t \end{bmatrix} + i \begin{bmatrix} -\cos t - \sin t \\ -\cos t - \sin t \\ \sin t \end{bmatrix}.$$

Taking the real and imaginary parts gives us two more real-valued linearly independent solutions, so that the general solution is

$$x(t) = c_1 e^t \begin{bmatrix} 0 \\ 2 \\ 1 \end{bmatrix} + c_2 \begin{bmatrix} -\cos t + \sin t \\ -\cos t + \sin t \\ \cos t \end{bmatrix} + c_3 \begin{bmatrix} -\cos t - \sin t \\ -\cos t - \sin t \\ \sin t \end{bmatrix}.$$

Exercises

1. Solve the following systems:

(a) $x' = \begin{bmatrix} 1 & -1 \\ 1 & 1 \end{bmatrix} x$.

Answer: $x(t) = c_1 e^t \begin{bmatrix} \cos t \\ \sin t \end{bmatrix} + c_2 e^t \begin{bmatrix} -\sin t \\ \cos t \end{bmatrix}$.

(b) $x' = \begin{bmatrix} 3 & -2 \\ 2 & 3 \end{bmatrix} x$, $x(0) = \begin{bmatrix} 0 \\ 1 \end{bmatrix}$.

Answer: $x_1(t) = -e^{3t} \sin 2t$, $x_2(t) = e^{3t} \cos 2t$.

(c)

$$x_1' = 3x_1 + 5x_2, \quad x_1(0) = -1,$$
$$x_2' = -5x_1 - 3x_2, \quad x_2(0) = 2.$$

Answer: $x_1(t) = \frac{7}{4} \sin 4t - \cos 4t$, $x_2(t) = 2 \cos 4t - \frac{1}{4} \sin 4t$.

(d) $x' = \begin{bmatrix} 1 & 2 & -1 \\ -2 & -1 & 1 \\ -1 & 1 & 0 \end{bmatrix} x$.

Answer:

$$x(t) = c_1 \begin{bmatrix} 1 \\ 1 \\ 3 \end{bmatrix} + c_2 \begin{bmatrix} \cos t + \sin t \\ \cos t - \sin t \\ 2 \cos t \end{bmatrix} + c_3 \begin{bmatrix} -\cos t + \sin t \\ \cos t + \sin t \\ 2 \sin t \end{bmatrix}.$$

(e) $x' = \begin{bmatrix} -1 & -1 & 1 \\ -1 & 0 & 2 \\ -2 & -1 & 2 \end{bmatrix} x$, $x(0) = \begin{bmatrix} 3 \\ -1 \\ 1 \end{bmatrix}$.

Answer: $x_1(t) = e^t - 2 \sin t + 2 \cos t$, $x_2(t) = -3e^t + 2 \sin t + 2 \cos t$, $x_3(t) = -e^t - 2 \sin t + 2 \cos t$.

(f) $x' = \begin{bmatrix} 2 & -1 & -3 \\ 1 & 1 & 4 \\ 2 & -1 & -3 \end{bmatrix} x$, $x(0) = \begin{bmatrix} 0 \\ 0 \\ 4 \end{bmatrix}$.

Answer: $x_1(t) = \cos 2t - 6 \sin 2t - 1$, $x_2(t) = 11 \cos 2t + 8 \sin 2t - 11$, $x_3(t) = \cos 2t - 6 \sin 2t + 3$.

2. Without calculating the eigenvectors show that all solutions of $x' = \begin{bmatrix} -3 & -2 \\ 4 & 1 \end{bmatrix} x$ tend to zero as $t \to \infty$.

   Hint. The eigenvalues $\lambda = -1 \pm 2i$ have negative real parts, so that the vectors $u(t)$ and $v(t)$ in (2.2) tend to zero as $t \to \infty$, for any ξ and η.

3. Solve the system

$$x' = \begin{bmatrix} 0 & -1 \\ 1 & 0 \end{bmatrix} x, \quad x(0) = \begin{bmatrix} \alpha \\ \beta \end{bmatrix},$$

   with given numbers α and β. Show that the solution $x(t)$ represents rotation of the initial vector $\begin{bmatrix} \alpha \\ \beta \end{bmatrix}$ by an angle t counterclockwise.

4. Define the derivative of a complex valued function $x(t) = u(t) + iv(t)$ as $x'(t) = \lim_{h \to 0} \frac{x(t+h)-x(t)}{h}$. Show that $x'(t) = u'(t) + iv'(t)$.

5. Consider the system

$$x_1' = ax_1 + bx_2,$$
$$x_2' = cx_1 + dx_2,$$

   with given numbers a, b, c, and d. Assume that $a + d < 0$ and $ad - bc > 0$. Show that all solutions tend to zero, as $t \to \infty$ (meaning that $x_1(t) \to 0$, and $x_2(t) \to 0$, as $t \to \infty$).

   Hint. Show that the eigenvalues for the matrix of this system are either negative, or have negative real parts.

6. (a) Let A be a 3×3 constant matrix. Suppose that all solutions of $x' = Ax$ are bounded as $t \to +\infty$, and as $t \to -\infty$. Show that every nonconstant solution is periodic, and there is a common period for all nonconstant solutions.

   Hint. One of the eigenvalues of A must be zero, and the other two purely imaginary.

   (b) Assume that a nonzero 3×3 matrix A is *skew-symmetric*, which means that $A^T = -A$. Show that one of the eigenvalues of A is zero, and the other two are purely imaginary.

   Hint. Write A in the form $A = \begin{bmatrix} 0 & p & q \\ -p & 0 & r \\ -q & -r & 0 \end{bmatrix}$, and calculate its characteristic polynomial.

   (c) Show that all nonconstant solutions of $x' = \begin{bmatrix} 0 & p & q \\ -p & 0 & r \\ -q & -r & 0 \end{bmatrix} x$ are periodic, with the period $\frac{2\pi}{\sqrt{p^2+q^2+r^2}}$.

7. Let A be a 2×2 matrix with a negative determinant. Show that the system $x' = Ax$ does not have periodic solutions.

8. Suppose that $p + iq$ is an eigenvalue of A, and $\xi + i\eta$ is a corresponding eigenvector.

(a) Show that ξ and η are linearly independent. (There is no complex number c such that $\eta = c\,\xi$.)

Hint. Linear dependence of ξ and η would imply linear dependence of the distinct eigenvectors $\xi + i\eta$ and $\xi - i\eta$.

(b) Show that the vectors $u(t)$ and $v(t)$ defined in (2.2) are linearly independent for all t.

6.3 The exponential of a matrix

In matrix notation a system of differential equations

$$x' = Ax, \quad x(0) = x_0,$$ (3.1)

looks like a single equation. In case A and $x(t)$ are scalars, the solution of (3.1) is

$$x(t) = e^{At}x_0.$$ (3.2)

In order to extend this formula to systems, we shall define the notion of *the exponential of a matrix*. Recall the powers of square matrices: $A^2 = A \cdot A$, $A^3 = A^2 \cdot A$, and so on. Starting with the Maclauren series

$$e^x = 1 + x + \frac{x^2}{2!} + \frac{x^3}{3!} + \frac{x^4}{4!} + \cdots = \sum_{n=0}^{\infty} \frac{x^n}{n!},$$

define (I is the identity matrix)

$$e^A = I + A + \frac{A^2}{2!} + \frac{A^3}{3!} + \frac{A^4}{4!} + \cdots = \sum_{n=0}^{\infty} \frac{A^n}{n!}.$$ (3.3)

So that e^A is the sum of infinitely many matrices, and each entry of e^A is an infinite series. We shall justify later on that all of these series are convergent for *any matrix A*. If O denotes the zero matrix (with all entries equal to zero), then $e^O = I$.

For a scalar t, the formula (3.3) implies

$$e^{At} = I + At + \frac{A^2 t^2}{2!} + \frac{A^3 t^3}{3!} + \frac{A^4 t^4}{4!} + \cdots,$$

and then differentiating term by term we obtain

$$\frac{d}{dt}e^{At} = A + A^2 t + \frac{A^3 t^2}{2!} + \frac{A^4 t^3}{3!} + \cdots = A\left(I + At + \frac{A^2 t^2}{2!} + \cdots\right) = Ae^{At}.$$

By direct substitution, one verifies that the formula (3.2) gives the solution of the initial-value problem (3.1). (Observe that $x(0) = e^O x_0 = x_0$.)

Example 1. Let $A = \begin{bmatrix} a & 0 \\ 0 & b \end{bmatrix}$, where a and b are given numbers. Then $A^n = \begin{bmatrix} a^n & 0 \\ 0 & b^n \end{bmatrix}$, and addition of diagonal matrices in (3.3) gives

$$e^A = \begin{bmatrix} 1 + a + \frac{a^2}{2!} + \frac{a^3}{3!} + \cdots & 0 \\ 0 & 1 + b + \frac{b^2}{2!} + \frac{b^3}{3!} + \cdots \end{bmatrix} = \begin{bmatrix} e^a & 0 \\ 0 & e^b \end{bmatrix}.$$

Exponentials of larger diagonal matrices are calculated similarly.

The next example connects matrix exponentials to geometrical reasoning.

Example 2. Let $A = \begin{bmatrix} 0 & -1 \\ 1 & 0 \end{bmatrix}$. Calculate: $A^2 = -I, A^3 = -A$, and $A^4 = I$. After that the powers repeat (for example, $A^{61} = A$). Then for any scalar t,

$$e^{At} = \begin{bmatrix} 1 - t^2/2! + t^4/4! + \cdots & -t + t^3/3! - t^5/5! + \cdots \\ t - t^3/3! + t^5/5! + \cdots & 1 - t^2/2! + t^4/4! + \cdots \end{bmatrix}$$

$$= \begin{bmatrix} \cos t & -\sin t \\ \sin t & \cos t \end{bmatrix},$$

the rotation matrix. Then one can express solutions of the system

$$\begin{aligned} x_1' &= -x_2, & x_1(0) &= \alpha, \\ x_2' &= x_1, & x_2(0) &= \beta, \end{aligned} \tag{3.4}$$

with prescribed initial conditions α and β, in the form

$$\begin{bmatrix} x_1(t) \\ x_2(t) \end{bmatrix} = e^{At} \begin{bmatrix} \alpha \\ \beta \end{bmatrix} = \begin{bmatrix} \cos t & -\sin t \\ \sin t & \cos t \end{bmatrix} \begin{bmatrix} \alpha \\ \beta \end{bmatrix},$$

involving the rotation matrix, and representing rotation of the initial position vector $\begin{bmatrix} \alpha \\ \beta \end{bmatrix}$ by an angle t, counterclockwise. We see that solution curves of the system (3.4) are circles in the (x_1, x_2) plane. This is consistent with velocity vector $\begin{bmatrix} x_1' \\ x_2' \end{bmatrix} = \begin{bmatrix} -x_2 \\ x_1 \end{bmatrix}$ being perpendicular to the position vector $\begin{bmatrix} x_1 \\ x_2 \end{bmatrix}$ at all t.

In general, $e^{A+B} \neq e^A e^B$. This is because $AB \neq BA$, for general $n \times n$ matrices. One way to show that $e^{x+y} = e^x e^y$ holds for numbers is to expand all three exponentials in power series, and show that the series on the left is the same as that on the right. In the process, we use that $xy = yx$ for numbers. The same argument shows that $e^{A+B} = e^A e^B$, provided that $BA = AB$, or *the matrices commute.* In particular, $e^{aI+A} = e^{aI} e^A$, because $(aI)A = A(aI)$ (a is any number).

Example 3. Let $A = \begin{bmatrix} 3 & -1 \\ 1 & 3 \end{bmatrix}$, then $A = 3I + \begin{bmatrix} 0 & -1 \\ 1 & 0 \end{bmatrix}$, and therefore

$$e^{At} = e^{3tI} e^{\begin{bmatrix} 0 & -t \\ t & 0 \end{bmatrix}} = \begin{bmatrix} e^{3t} & 0 \\ 0 & e^{3t} \end{bmatrix} \begin{bmatrix} \cos t & -\sin t \\ \sin t & \cos t \end{bmatrix}$$

$$= e^{3t} \begin{bmatrix} 1 & 0 \\ 0 & 1 \end{bmatrix} \begin{bmatrix} \cos t & -\sin t \\ \sin t & \cos t \end{bmatrix} = e^{3t} \begin{bmatrix} \cos t & -\sin t \\ \sin t & \cos t \end{bmatrix}.$$

Since $e^{aI} = e^a I$, it follows that $e^{aI+A} = e^a e^A$ holds for any number a and square matrix A. For nilpotent matrices, the series for e^A has only finitely many terms.

Example 4. Let $K = \begin{bmatrix} 0 & 1 & 0 & 0 \\ 0 & 0 & 1 & 0 \\ 0 & 0 & 0 & 1 \\ 0 & 0 & 0 & 0 \end{bmatrix}$. Calculate $K^2 = \begin{bmatrix} 0 & 0 & 1 & 0 \\ 0 & 0 & 0 & 1 \\ 0 & 0 & 0 & 0 \\ 0 & 0 & 0 & 0 \end{bmatrix}$, $K^3 = \begin{bmatrix} 0 & 0 & 0 & 1 \\ 0 & 0 & 0 & 0 \\ 0 & 0 & 0 & 0 \\ 0 & 0 & 0 & 0 \end{bmatrix}$, $K^4 = O$, the zero matrix, and therefore $K^m = O$ for all powers $m \geq 4$, so that K is nilpotent. The series for e^{Kt} terminates:

$$e^{Kt} = I + Kt + \frac{1}{2!}K^2 t^2 + \frac{1}{3!}K^3 t^3 = \begin{bmatrix} 1 & t & \frac{1}{2!}t^2 & \frac{1}{3!}t^3 \\ 0 & 1 & t & \frac{1}{2!}t^2 \\ 0 & 0 & 1 & t \\ 0 & 0 & 0 & 1 \end{bmatrix}.$$

Example 5. Let $J = \begin{bmatrix} -2 & 1 & 0 & 0 \\ 0 & -2 & 1 & 0 \\ 0 & 0 & -2 & 1 \\ 0 & 0 & 0 & -2 \end{bmatrix}$, a Jordan block. Writing $J = -2I + K$, with K from Example 4, and proceeding as in Example 3, obtain

$$e^{Jt} = e^{-2t} \begin{bmatrix} 1 & t & \frac{1}{2!}t^2 & \frac{1}{3!}t^3 \\ 0 & 1 & t & \frac{1}{2!}t^2 \\ 0 & 0 & 1 & t \\ 0 & 0 & 0 & 1 \end{bmatrix}.$$

Norm of a matrix, and convergence of the series for e^A

Recall the concept of length (or magnitude, or norm) $\|x\|$ of an n-dimensional vector $x \in R^n$, defined by

$$\|x\|^2 = x \cdot x = \sum_{i=1}^{n} x_i^2,$$

and the Cauchy–Schwarz inequality that states

$$|x \cdot y| = \left| \sum_{i=1}^{n} x_i y_i \right| \leq \|x\| \, \|y\|.$$

Let A be an $n \times n$ matrix, given by its columns $A = [C_1 \, C_2 \, \dots \, C_n]$. (Here $C_1 \in R^n$ is the first column of A, etc.) Define *the norm* $\|A\|$ *of* A, as follows:

$$\|A\|^2 = \sum_{i=1}^{n} \|C_i\|^2 = \sum_{i=1}^{n} \sum_{j=1}^{n} a_{ji}^2 = \sum_{i,j=1}^{n} a_{ij}^2. \tag{3.5}$$

Clearly,

$$|a_{ij}| \leq \|A\|, \quad \text{for all } i \text{ and } j, \tag{3.6}$$

since the double sum on the right in (3.5) is greater than or equal to any one of its terms. If $x \in R^n$, we claim that

$$\|Ax\| \le \|A\| \, \|x\|. \tag{3.7}$$

Indeed, using the Cauchy–Schwarz inequality,

$$\|Ax\|^2 = \sum_{i=1}^{n}\left(\sum_{j=1}^{n} a_{ij}x_j\right)^2 \le \sum_{i=1}^{n}\left(\sum_{j=1}^{n} a_{ij}^2 \sum_{j=1}^{n} x_j^2\right)$$

$$= \|x\|^2 \sum_{i=1}^{n}\sum_{j=1}^{n} a_{ij}^2 = \|A\|^2 \, \|x\|^2.$$

Let B be another $n \times n$ matrix, given by its columns $B = [K_1 \, K_2 \ldots K_n]$. Recall that $AB = [AK_1 \, AK_2 \ldots AK_n]$. (Here AK_1 is the first column of the product AB, etc.) Then, using (3.7),

$$\|AB\|^2 = \sum_{i=1}^{n} \|AK_i\|^2 \le \|A\|^2 \sum_{i=1}^{n} \|K_i\|^2 = \|A\|^2 \|B\|^2,$$

which implies that

$$\|AB\| \le \|A\| \, \|B\|.$$

Applying this inequality to two matrices at a time, one shows that

$$\|A_1 A_2 \cdots A_m\| \le \|A_1\| \, \|A_2\| \cdots \|A_m\|, \tag{3.8}$$

for any integer $m \ge 2$, and in particular that $\|A^m\| \le \|A\|^m$. We show in Exercises that *the triangle inequality* holds for matrices:

$$\|A + B\| \le \|A\| + \|B\|,$$

and by applying the triangle inequality to two matrices at a time

$$\|A_1 + A_2 + \cdots + A_m\| \le \|A_1\| + \|A_2\| + \cdots + \|A_m\| \tag{3.9}$$

holds for an arbitrary number of square matrices of the same size.

The above inequalities imply that the exponential of any matrix A,

$$e^A = \sum_{k=0}^{\infty} \frac{1}{k!} A^k,$$

is a convergent series in each component. Indeed, by (3.6), we estimate the absolute value of the (i,j)th component of each term of this series as

$$\frac{1}{k!}\left|(A^k)_{ij}\right| \le \frac{1}{k!}\|A^k\| \le \frac{1}{k!}\|A\|^k.$$

The series $\sum_{k=0}^{\infty} \frac{1}{k!}\|A\|^k$ is convergent (its sum is $e^{\|A\|}$). The series for e^A converges absolutely in each component by the comparison test.

Exercises

1. Find the exponentials e^{At} of the following matrices:

   (a) $A = \begin{bmatrix} 0 & -1 \\ 0 & 0 \end{bmatrix}$.

   Answer: $e^{At} = \begin{bmatrix} 1 & -t \\ 0 & 1 \end{bmatrix}$.

   (b) $D = \begin{bmatrix} 1 & 0 \\ 0 & -4 \end{bmatrix}$.

   Answer: $e^{Dt} = \begin{bmatrix} e^t & 0 \\ 0 & e^{-4t} \end{bmatrix}$.

   (c) $D = \begin{bmatrix} 2 & 0 & 0 \\ 0 & 0 & 0 \\ 0 & 0 & -3 \end{bmatrix}$.

   Answer: $e^{Dt} = \begin{bmatrix} e^{2t} & 0 & 0 \\ 0 & 1 & 0 \\ 0 & 0 & e^{-3t} \end{bmatrix}$.

   (d) $A = \begin{bmatrix} 0 & 1 & 0 \\ 0 & 0 & 1 \\ 0 & 0 & 0 \end{bmatrix}$.

   Answer: $e^{At} = \begin{bmatrix} 1 & t & \frac{1}{2}t^2 \\ 0 & 1 & t \\ 0 & 0 & 1 \end{bmatrix}$.

   (e) $A = \begin{bmatrix} -2 & 1 & 0 \\ 0 & -2 & 1 \\ 0 & 0 & -2 \end{bmatrix}$.

   Answer: $e^{At} = e^{-2t} \begin{bmatrix} 1 & t & \frac{1}{2}t^2 \\ 0 & 1 & t \\ 0 & 0 & 1 \end{bmatrix}$.

   (f) $A = \begin{bmatrix} 0 & -1 & 0 \\ 1 & 0 & 0 \\ 0 & 0 & -2 \end{bmatrix}$.

   Answer: $e^{At} = \begin{bmatrix} \cos t & -\sin t & 0 \\ \sin t & \cos t & 0 \\ 0 & 0 & e^{-2t} \end{bmatrix}$.

2. Show that $(e^A)^{-1} = e^{-A}$.

   Hint. The matrices A and $-A$ commute.

3. Show that $(e^A)^m = e^{mA}$, for any positive integer m.

4. Show that $(e^A)^T = e^{A^T}$.

5. (a) Let λ be an eigenvalue of a square matrix A, corresponding to an eigenvector x. Show that e^A has an eigenvalue e^λ, corresponding to the same eigenvector x. Hint. If $Ax = \lambda x$, then

$$e^A x = \sum_{k=0}^{\infty} \frac{A^k x}{k!} = \sum_{k=0}^{\infty} \frac{\lambda^k}{k!} x = e^\lambda x.$$

   (b) Show that $\det e^A = e^{\operatorname{tr} A}$.

   Hint. Determinant equals to the product of eigenvalues.

   (c) Explain why e^A is nonsingular, for any A.

6. If A is symmetric, show that e^A is positive definite.

    Hint. Observe that $e^A x \cdot x = e^{A/2} e^{A/2} x \cdot x = \|e^{A/2} x\|^2 > 0$, for any $x \neq 0$.

7. Let A be a skew-symmetric matrix (so that $A^T = -A$).

    (a) Show that e^{At} is an orthogonal matrix for any t.

    Hint. If $Q = e^{At}$, then $Q^T = e^{A^T t} = e^{-At} = Q^{-1}$.

    (b) Show that the solution $x(t)$ of

    $$x' = Ax, \quad x(0) = x_0$$

    satisfies $\|x(t)\| = \|x_0\|$ for all t.

8. For any square matrix X, define the sine of X as

    $$\sin X = X - \frac{1}{3!} X^3 + \frac{1}{5!} X^5 - \cdots.$$

    (a) Show that the series converges for any X.

    (b) Let $K = \begin{bmatrix} 0 & 1 & 0 & 0 \\ 0 & 0 & 1 & 0 \\ 0 & 0 & 0 & 1 \\ 0 & 0 & 0 & 0 \end{bmatrix}$. Show that

    $$\sin Kt = \begin{bmatrix} 0 & t & 0 & -\frac{1}{6} t^3 \\ 0 & 0 & t & 0 \\ 0 & 0 & 0 & t \\ 0 & 0 & 0 & 0 \end{bmatrix}.$$

    Hint. Observe that $K^k = O$ for any $k \geq 4$. (The matrix K is *nilpotent*.)

9. Show that $A^2 e^{At} = e^{At} A^2$ for any square matrix A and number t.

10. Show that *the triangle inequality*

    $$\|A + B\| \leq \|A\| + \|B\|$$

    holds for any two $n \times n$ matrices.

    Hint. Using that $(a + b)^2 = a^2 + b^2 + 2ab$, one has

    $$\|A + B\|^2 = \sum_{i,j=1}^{n} (a_{ij} + b_{ij})^2 = \|A\|^2 + \|B\|^2 + 2 \sum_{i,j=1}^{n} a_{ij} b_{ij}.$$

    Using the Cauchy–Schwarz inequality, one obtains

    $$\sum_{i,j=1}^{n} a_{ij} b_{ij} \leq \left(\sum_{i,j=1}^{n} a_{ij}^2 \right)^{\frac{1}{2}} \left(\sum_{i,j=1}^{n} b_{ij}^2 \right)^{\frac{1}{2}} = \|A\| \, \|B\|,$$

    so that $\|A + B\|^2 \leq \|A\|^2 + \|B\|^2 + 2\|A\| \, \|B\| = (\|A\| + \|B\|)^2$.

11. Show that $\|e^A\| \leq e^{\|A\|}$.

12. (a) Assume that a square matrix A is diagonalizable, and that its eigenvalue λ_k is the largest in modulus, so that $|\lambda_k| \geq |\lambda_i|$ for all i. Show that

$$\|A^m\| \leq c|\lambda_k|^m,$$

for all positive integers m, and some number $c > 0$.

Hint. Diagonalize $A = PDP^{-1}$, $A^m = PD^mP^{-1}$, where the diagonal matrix D has λ_i's as its entries. Begin by showing that $\|D^m\| \leq c|\lambda_k|^m$.

(b) Assume that all eigenvalues of a diagonalizable matrix A have modulus $|\lambda_i| < 1$. Show that $\lim_{m\to\infty} A^m = O$ (the zero matrix), the series $\sum_{m=0}^{\infty} A^m$ converges, and

$$I + A + A^2 + \cdots + A^m + \cdots = (I - A)^{-1}.$$

6.4 The general case of repeated eigenvalues

We now return to solving linear systems

$$x' = Ax, \quad x(0) = w, \tag{4.1}$$

with a given $n \times n$ matrix A, and a given initial vector $w \in R^n$. Suppose that $\lambda = r$ is an eigenvalue of A of multiplicity s, meaning that $\lambda = r$ is a root of multiplicity s of the corresponding characteristic equation $|A - \lambda I| = 0$. This root must bring in s linearly independent solutions for the general solution. If there are s linearly independent eigenvectors $\xi_1, \xi_2, \ldots, \xi_s$ corresponding to $\lambda = r$, then $e^{rt}\xi_1, e^{rt}\xi_2, \ldots, e^{rt}\xi_s$ give the desired s linearly independent solutions. However, if there are only $k < s$ linearly independent eigenvectors, one needs the notion of generalized eigenvectors. (Recall that the case $s = 2$, $k = 1$ was considered previously.)

A vector w_m is called a *generalized eigenvector of rank* m, corresponding to an eigenvalue $\lambda = r$, provided that

$$(A - rI)^m w_m = 0, \tag{4.2}$$

but

$$(A - rI)^{m-1} w_m \neq 0. \tag{4.3}$$

Assume that w_m is known. Through matrix multiplications define *a chain of vectors*

$$w_{m-1} = (A - rI)w_m,$$

$$w_{m-2} = (A - rI)w_{m-1} = (A - rI)^2 w_m,$$

$$\vdots$$

$$w_1 = (A - rI)w_2 = (A - rI)^{m-1}w_m,$$

or

$$w_{m-i} = (A - rI)^i w_m, \quad \text{for } i = 1, 2, \ldots, m - 1.$$

Since i steps in the chain bring us down from w_m to w_{m-i}, it follows that $m - i$ steps take us down from w_m to w_i,

$$w_i = (A - rI)^{m-i} w_m. \tag{4.4}$$

Observe that

$$(A - rI)^i w_i = (A - rI)^m w_m = 0, \tag{4.5}$$

using (4.2), and then

$$(A - rI)^j w_i = 0, \quad \text{for } j \geq i. \tag{4.6}$$

Notice also that w_1 is an eigenvector of A corresponding to $\lambda = r$ because

$$(A - rI)w_1 = (A - rI)^m w_m = 0,$$

giving $A w_1 = r w_1$, with $w_1 = (A - rI)^{m-1} w_m \neq 0$ by (4.3). So that a chain begins with a generalized eigenvector w_m and ends with an eigenvector w_1.

Lemma 6.4.1. *The vectors $w_m, w_{m-1}, \ldots, w_1$ of a chain are linearly independent.*

Proof. We need to show that

$$c_m w_m + c_{m-1} w_{m-1} + \cdots + c_1 w_1 = 0 \tag{4.7}$$

is possible only if all of the coefficients $c_i = 0$. Multiply all the terms of equation (4.7) by $(A - rI)^{m-1}$. Using (4.6), obtain

$$c_m (A - rI)^{m-1} w_m = 0.$$

Since $(A - rI)^{m-1} w_m \neq 0$, by the definition of the generalized eigenvector, it follows that $c_m = 0$, so that (4.7) becomes

$$c_{m-1} w_{m-1} + c_{m-2} w_{m-2} + \cdots + c_1 w_1 = 0.$$

Multiplying this equation by $(A - rI)^{m-2}$ gives $c_{m-1}(A - rI)^{m-2} w_{m-1} = 0$, which implies that $c_{m-1} = 0$ because

$$(A - rI)^{m-2} w_{m-1} = (A - rI)^{m-1} w_m \neq 0.$$

Proceed similarly to obtain $c_m = c_{m-1} = \cdots = c_1 = 0$. □

This lemma implies that all $w_i \neq 0$. Since $(A - rI)^i w_i = 0$ by (4.5), while $(A - rI)^{i-1} w_i = w_1 \neq 0$, it follows that all elements of a chain w_i are generalized eigenvectors of rank i.

Solution of the system (4.1) can be written as

$$x(t) = e^{At} w = e^{rtI} e^{(A-rI)t} w = e^{rt} e^{(A-rI)t} w$$

$$= e^{rt} \left[I + (A - rI)t + \frac{1}{2!}(A - rI)^2 t^2 + \frac{1}{3!}(A - rI)^3 t^3 + \cdots \right] w$$

$$= e^{rt} \left[w + (A - rI)w\, t + (A - rI)^2 w \frac{1}{2!} t^2 + (A - rI)^3 w \frac{1}{3!} t^3 + \cdots \right]. \qquad (4.8)$$

Here we used matrix exponentials, and the fact that the matrices $A - rI$ and rI commute. In case w is any vector of the chain, it follows by (4.6) that this series terminates after finitely many terms, and we obtain m linearly independent solutions of the system (4.1), corresponding to the eigenvalue $\lambda = r$, by setting w equal to $w_1, w_2, \ldots, w_m$, and using (4.8) and (4.4):

$$x_1(t) = e^{At} w_1 = e^{rt} w_1,$$

$$x_2(t) = e^{At} w_2 = e^{rt}[w_2 + w_1 t],$$

$$x_3(t) = e^{At} w_3 = e^{rt} \left[w_3 + w_2 t + w_1 \frac{1}{2!} t^2 \right], \qquad (4.9)$$

$$\vdots$$

$$x_m(t) = e^{At} w_m = e^{rt} \left[w_m + w_{m-1} t + w_{m-2} \frac{1}{2!} t^2 + \cdots + w_1 \frac{1}{(m-1)!} t^{m-1} \right].$$

These solutions are linearly independent because each w_k does not belong to the span of $w_1, \ldots, w_{k-1}$, by Lemma 6.4.1.

The formulas (4.9) are sufficient to deal with a repeated eigenvalue of multiplicity $m > 1$ that has only one linearly independent eigenvector. It is then not hard to find a generalized eigenvector w_m, and construct m linearly independent solutions. This fact, and the general case are discussed later on.

Example 1. Let us find the general solution of

$$x'(t) = \begin{bmatrix} 2 & 1 & 2 \\ -5 & -1 & -7 \\ 1 & 0 & 2 \end{bmatrix} x(t).$$

The matrix of this system $A = \begin{bmatrix} 2 & 1 & 2 \\ -5 & -1 & -7 \\ 1 & 0 & 2 \end{bmatrix}$ has an eigenvalue $r = 1$ of multiplicity $m = 3$, and only one linearly independent eigenvector $\begin{bmatrix} -1 \\ -1 \\ 1 \end{bmatrix}$. Calculate

$$A - rI = A - I = \begin{bmatrix} 1 & 1 & 2 \\ -5 & -2 & -7 \\ 1 & 0 & 1 \end{bmatrix},$$

$$(A - rI)^2 = (A - I)^2 = \begin{bmatrix} -2 & -1 & -3 \\ -2 & -1 & -3 \\ 2 & 1 & 3 \end{bmatrix},$$

$$(A - rI)^3 = (A - I)^3 = \begin{bmatrix} 0 & 0 & 0 \\ 0 & 0 & 0 \\ 0 & 0 & 0 \end{bmatrix}.$$

Clearly, any vector $w \in R^3$ satisfies $(A-I)^3 w = 0$, for example, $w_3 = \begin{bmatrix} 1 \\ 0 \\ 0 \end{bmatrix}$. Since $(A-I)^2 w_3 \neq 0$, it follows that w_3 is a generalized eigenvector. We now calculate the chain

$$w_2 = (A - I)w_3 = \begin{bmatrix} 1 & 1 & 2 \\ -5 & -2 & -7 \\ 1 & 0 & 1 \end{bmatrix} \begin{bmatrix} 1 \\ 0 \\ 0 \end{bmatrix} = \begin{bmatrix} 1 \\ -5 \\ 1 \end{bmatrix},$$

$$w_1 = (A - I)w_1 = \begin{bmatrix} 1 & 1 & 2 \\ -5 & -2 & -7 \\ 1 & 0 & 1 \end{bmatrix} \begin{bmatrix} 1 \\ -5 \\ 1 \end{bmatrix} = \begin{bmatrix} -2 \\ -2 \\ 2 \end{bmatrix}.$$

(Observe that w_1 is an eigenvector corresponding to $r = 1$.) The three linearly independent solutions are

$$x_1(t) = \begin{bmatrix} -2 \\ -2 \\ 2 \end{bmatrix} e^t,$$

$$x_2(t) = \begin{bmatrix} 1 \\ -5 \\ 1 \end{bmatrix} e^t + \begin{bmatrix} -2 \\ -2 \\ 2 \end{bmatrix} te^t,$$

$$x_3(t) = \begin{bmatrix} 1 \\ 0 \\ 0 \end{bmatrix} e^t + \begin{bmatrix} 1 \\ -5 \\ 1 \end{bmatrix} te^t + \begin{bmatrix} -2 \\ -2 \\ 2 \end{bmatrix} \frac{1}{2} t^2 e^t.$$

The general solution is then

$$x(t) = c_1 x_1(t) + c_2 x_2(t) + c_3 x_3(t).$$

The constants c_1, c_2, c_3 are determined by the initial conditions.

Chains can be constructed "from the other end," beginning with eigenvectors. Assume that w_1 is an eigenvector of A corresponding to a repeated eigenvalue r. Let w_2 be any solution of

$$(A - rI)w_2 = w_1, \tag{4.10}$$

provided that such solution exists. The matrix of this system is singular (since $|A - rI| = 0$), so that solution w_2 may or may not exist. (If a solution exists, there are infinitely many solutions, in the form $w_2 + cw_1$.) In case w_2 does not exist, we say that the chain ends at w_1, and denote it $\{w_1\}$, *a chain of length one*. If a solution w_2 exists, to get w_3 we find (if possible) any solution of

$$(A - rI)w_3 = w_2.$$

If solution w_3 does not exist, we say that the chain ends at w_2, and denote it $\{w_1, w_2\}$, *a chain of length two*. In such a case, w_2 is a generalized eigenvector. Indeed, by (4.10),

$$(A - rI)^2 w_2 = (A - rI)w_1 = 0,$$

while using (4.10) again

$$(A - rI)w_2 \neq 0.$$

If a solution w_3 exists, solve if possible

$$(A - rI)w_4 = w_3$$

to get w_4. In case w_4 does not exist, obtain the chain $\{w_1, w_2, w_3\}$ of length three. As above, w_1, w_2, w_3 are linearly independent, w_2, w_3 are generalized eigenvectors, and there are infinitely many choices for w_2, w_3. Continue in the same fashion. All chains eventually end, since their elements are linearly independent vectors in R^n.

We now turn to the general case of repeated eigenvalues. Suppose that a matrix A has several linearly independent eigenvectors $\xi_1, \xi_2, \xi_3, \ldots, \xi_p$ corresponding to a repeated eigenvalue $\lambda = r$ of multiplicity m, with $p < m$. One can construct a chain beginning with any eigenvector. We shall employ the following notation for these chains: $(\xi_1, \xi_{12}, \xi_{13}, \ldots), (\xi_2, \xi_{22}, \xi_{23}, \ldots), (\xi_3, \xi_{32}, \xi_{33}, \ldots)$, and so on. By Lemma 6.4.1, the elements of each chain are linearly independent. It turns out that if all elements of each chain are put together, they form a linearly independent set of m vectors, see Proposition 6.4.1 below.

Example 2. Consider a 6×6 matrix

$$J = \begin{bmatrix} 2 & 1 & 0 & 0 & 0 & 0 \\ 0 & 2 & 1 & 0 & 0 & 0 \\ 0 & 0 & 2 & 0 & 0 & 0 \\ 0 & 0 & 0 & 2 & 1 & 0 \\ 0 & 0 & 0 & 0 & 2 & 0 \\ 0 & 0 & 0 & 0 & 0 & 2 \end{bmatrix}.$$

Its only eigenvalue $\lambda = 2$ has multiplicity 6, but there are only 3 linearly independent eigenvectors, which happened to be the coordinate vectors $\xi_1 = e_1$, $\xi_2 = e_4$, and $\xi_3 = e_6$ in R^6, at which three chains will begin. To find the next element of each chain, we need to solve the systems $(J - 2I)\xi_{12} = \xi_1 = e_1$, $(J - 2I)\xi_{22} = \xi_2 = e_4$, and $(J - 2I)\xi_{32} = \xi_3 = e_6$. These systems have the same matrix, and hence they can be solved in parallel, with the augmented matrix $[J - 2I \vdots e_1 \vdots e_4 \vdots e_6]$, which is

$$\begin{bmatrix} 0 & 1 & 0 & 0 & 0 & 0 & 1 & 0 & 0 \\ 0 & 0 & 1 & 0 & 0 & 0 & 0 & 0 & 0 \\ 0 & 0 & 0 & 0 & 0 & 0 & 0 & 0 & 0 \\ 0 & 0 & 0 & 0 & 1 & 0 & 0 & 1 & 0 \\ 0 & 0 & 0 & 0 & 0 & 0 & 0 & 0 & 0 \\ 0 & 0 & 0 & 0 & 0 & 0 & 0 & 0 & 1 \end{bmatrix}.$$

The last of these systems has no solutions. The chain beginning with $\xi_3 = e_6$ terminates immediately, giving $\{e_6\}$ a chain of length one. The systems for ξ_{12} and ξ_{22} have infinitely many solutions, out of which we select the simple ones: $\xi_{12} = e_2$, $\xi_{22} = e_5$. For the next elements of the chains, we solve $(J - 2I)\xi_{13} = \xi_{12} = e_2$ and $(J - 2I)\xi_{23} = \xi_{22} = e_5$. The second of these systems has no solutions, so that the corresponding chain terminates at the second step, giving $\{\xi_2, \xi_{22}\} = \{e_4, e_5\}$, a chain of length two. The first of these systems has a solution $\xi_{13} = e_3$. We obtain a chain $\{\xi_1, \xi_{12}, \xi_{13}\} = \{e_1, e_2, e_3\}$ of length three. This chain cannot be continued any further, because $(J - 2I)\xi_{14} = e_3$ has no solutions. Conclusion: the eigenvalue $\lambda = 2$ of the matrix J has three chains $\{e_1, e_2, e_3\}$, $\{e_4, e_5\}$, and $\{e_6\}$, of total length six. Observe that putting together all elements of the three chains produces a linearly independent set of six vectors, giving a basis in R^6. The matrix J provides an example for the following general facts, see e. g., S. H. Friedberg et al. [8].

Proposition 6.4.1. *If some matrix A has a repeated eigenvalue $\lambda = r$ of multiplicity m, then putting together all elements of all chains, beginning with linearly independent eigenvectors corresponding to $\lambda = r$, produces a set of m linearly independent vectors.*

The matrix J of the above example is very special, with all elements of the chains being the coordinate vectors in R^6. Let now A_0 be a 6×6 matrix that has an eigenvalue $\lambda = 2$ of multiplicity 6 with three-dimensional eigenspace, spanned by the eigenvectors $\xi_1, \xi_2,$ and ξ_3. Form the 3 chains, beginning with $\xi_1, \xi_2,$ and ξ_3, respectively. The length of each chain is between 1 and 4. Indeed, putting together all elements of the three chains produces a linearly independent set of six vectors in R^6, so that there is "no room" for a chain of length 5 or more. So that matrix A_0 has three chains of total length 6. Possible combinations of their length: $6 = 4 + 1 + 1$, $6 = 3 + 2 + 1$, $6 = 2 + 2 + 2$. Let us assume that it is the second possibility, so that the eigenspace of $\lambda = 2$ is spanned by ξ_1, ξ_2, ξ_3, and the chains are: $\{\xi_1, \xi_{12}, \xi_{13}\}, \{\xi_2, \xi_{22}\}, \{\xi_3\}$. The general solution of the system

$$x' = A_0 x$$

is (here $c_1, c_2, \ldots, c_6$ are arbitrary numbers)

$$x(t) = c_1 x_1(t) + c_2 x_2(t) + c_3 x_3(t) + c_4 x_4(t) + c_5 x_5(t) + c_6 x_6(t),$$

where, according to (4.9),

$$x_1(t) = e^{2t}\xi_1,$$
$$x_2(t) = e^{2t}(\xi_{12} + \xi_1 t),$$
$$x_3(t) = e^{2t}\left(\xi_{13} + \xi_{12}t + \xi_1\frac{t^2}{2}\right),$$
$$x_4(t) = e^{2t}\xi_2,$$
$$x_5(t) = e^{2t}(\xi_{22} + \xi_2 t),$$
$$x_6(t) = e^{2t}\xi_3.$$

(4.11)

These solutions are linearly independent, in view of Proposition 6.4.1.

We now use the above chains to reduce matrix A_0 to a simpler form. Since $(A_0 - 2I)\xi_{12} = \xi_1$, it follows that

$$A_0\xi_{12} = 2\xi_{12} + \xi_1.$$

Similarly,

$$A_0\xi_{13} = 2\xi_{13} + \xi_{12},$$
$$A_0\xi_{22} = 2\xi_{22} + \xi_2.$$

Form a 6×6 nonsingular matrix $S = [\xi_1\ \xi_{12}\ \xi_{13}\ \xi_2\ \xi_{22}\ \xi_3]$ using the vectors of the chains as its columns. By the definition of matrix multiplication,

$$A_0 S = [A_0\xi_1\ A_0\xi_{12}\ A_0\xi_{13}\ A_0\xi_2\ A_0\xi_{22}\ A_0\xi_3]$$
$$= [2\xi_1\ 2\xi_{12} + \xi_1\ 2\xi_{13} + \xi_{12}\ 2\xi_2\ 2\xi_{22} + \xi_2\ 2\xi_3]$$
$$= S\begin{bmatrix} 2 & 1 & 0 & 0 & 0 & 0 \\ 0 & 2 & 1 & 0 & 0 & 0 \\ 0 & 0 & 2 & 0 & 0 & 0 \\ 0 & 0 & 0 & 2 & 1 & 0 \\ 0 & 0 & 0 & 0 & 2 & 0 \\ 0 & 0 & 0 & 0 & 0 & 2 \end{bmatrix} = SJ,$$

where J denotes the second matrix on the right, which is the matrix considered in Example 2 above. (For example, the second column of SJ is equal to the product of S and the second column of J, giving $2\xi_{12} + \xi_1$, the second column of $A_0 S$.) We conclude that

$$A_0 = SJS^{-1}.$$

The matrix J is called *the Jordan normal form* of A_0. The matrix A_0 is not diagonalizable, since it has only 3 linearly independent eigenvectors, not a complete set of 6 linearly independent eigenvectors. The Jordan normal form provides a substitute. The matrix A_0 is *similar* to its Jordan normal form J. The columns of S form *the Jordan canonical basis* of R^6.

The matrix J is a block diagonal matrix, with the 3×3, 2×2, and 1×1 blocks:

$$\begin{bmatrix} 2 & 1 & 0 \\ 0 & 2 & 1 \\ 0 & 0 & 2 \end{bmatrix}, \quad \begin{bmatrix} 2 & 1 \\ 0 & 2 \end{bmatrix}, \quad [2], \tag{4.12}$$

called *the Jordan block matrices*.

The Jordan normal form

We now describe the Jordan normal form for a general $n \times n$ matrix with multiple eigenvalues, some possibly repeated. Basically, it is a block-diagonal matrix consisting of Jordan blocks, and A is similar to it. If A has an eigenvalue $\lambda = 2$ of multiplicity six, with a three-dimensional eigenspace, we construct the three chains. If the lengths of the chains happen to be 3, 2, and 1, then the Jordan normal form contains the three blocks listed in (4.12). If an eigenvalue $\lambda = -3$ is simple, it contributes a diagonal entry of -3 in the Jordan normal form. If an eigenvalue $\lambda = -1$ has multiplicity 4 but only one linearly independent eigenvector, it contributes a Jordan block

$$\begin{bmatrix} -1 & 1 & 0 & 0 \\ 0 & -1 & 1 & 0 \\ 0 & 0 & -1 & 1 \\ 0 & 0 & 0 & -1 \end{bmatrix},$$

and so on. For more details, see the following example, and S. H. Friedberg et al. [8].

Example 3. Assume that a 9×9 matrix B has an eigenvalue $\lambda_1 = -2$ of multiplicity 4 with only two linearly independent eigenvectors, each giving rise to a chain of length two; an eigenvalue $\lambda_2 = 3$ of multiplicity two with only one linearly independent eigenvector; an eigenvalue $\lambda_3 = 0$ of multiplicity two with only one linearly independent eigenvector; and finally, a simple eigenvalue $\lambda_4 = 4$. The Jordan normal form will be

$$J_0 = \begin{bmatrix} -2 & 1 & 0 & 0 & 0 & 0 & 0 & 0 & 0 \\ 0 & -2 & 0 & 0 & 0 & 0 & 0 & 0 & 0 \\ 0 & 0 & -2 & 1 & 0 & 0 & 0 & 0 & 0 \\ 0 & 0 & 0 & -2 & 0 & 0 & 0 & 0 & 0 \\ 0 & 0 & 0 & 0 & 3 & 1 & 0 & 0 & 0 \\ 0 & 0 & 0 & 0 & 0 & 3 & 0 & 0 & 0 \\ 0 & 0 & 0 & 0 & 0 & 0 & 0 & 1 & 0 \\ 0 & 0 & 0 & 0 & 0 & 0 & 0 & 0 & 0 \\ 0 & 0 & 0 & 0 & 0 & 0 & 0 & 0 & 4 \end{bmatrix},$$

and

$$B = S J_0 S^{-1}. \tag{4.13}$$

The columns of S consist of all eigenvectors of B, together with all elements of the chains that they generate, as is explained next. Assume that ξ_1 and ξ_2 are two linearly independent eigenvectors corresponding to $\lambda_1 = 2$, giving rise to the chains $\{\xi_1, \xi_{12}\}$ and $\{\xi_2, \xi_{22}\}$; ξ_3 is an eigenvector corresponding to $\lambda_2 = 3$, giving rise to the chain $\{\xi_3, \xi_{32}\}$; ξ_4 is an eigenvector corresponding to $\lambda_3 = 0$, giving rise to the chain $\{\xi_4, \xi_{42}\}$, and finally, ξ_5 is an eigenvector corresponding to the simple eigenvalue $\lambda_4 = 4$. The matrix S in (4.13) is $S = [\xi_1 \; \xi_{12} \; \xi_2 \; \xi_{22} \; \xi_3 \; \xi_{32} \; \xi_4 \; \xi_{42} \; \xi_5]$.

The general solution of the corresponding system of differential equations $x' = Bx$, according to the procedure in (4.9), is

$$x(t) = c_1 e^{-2t}\xi_1 + c_2 e^{-2t}(\xi_{12} + \xi_1 t) + c_3 e^{-2t}\xi_2 + c_4 e^{-2t}(\xi_{22} + \xi_2 t)$$
$$+ c_5 e^{3t}\xi_3 + c_6 e^{3t}(\xi_{32} + \xi_3 t) + c_7\xi_4 + c_8(\xi_{42} + \xi_4 t) + c_9 e^{4t}\xi_5,$$

with arbitrary constants c_i.

The methods we developed for solving systems $x' = Ax$ work for complex eigenvalues as well. For example, consider a complex-valued solution $z = t^k e^{(p+iq)t}\xi$ corresponding to an eigenvalue $\lambda = p + iq$ of A, with a complex-valued eigenvector $\xi = \alpha + i\beta$. Taking the real and imaginary parts of z, we obtain (as previously) two real-valued solutions of the form $t^k e^{pt} \cos qt\, \alpha$ and $t^k e^{pt} \sin qt\, \beta$, with real-valued vectors α and β.

The following important theorem is a consequence.

Theorem 6.4.1. *Assume that all eigenvalues of the matrix A are either negative or have negative real parts. Then all solutions of the system*

$$x' = Ax$$

tend to zero as $t \to \infty$.

Proof. If $\lambda < 0$ is a simple eigenvalue, it contributes the term $e^{\lambda t}\xi$ to the general solution (ξ is the corresponding eigenvector) that tends to zero as $t \to \infty$. If $\lambda < 0$ is a repeated eigenvalue, the vectors it contributes to the general solution are of the form $t^k e^{\lambda t}\xi$, where k is a positive integer. By L'Hospital's rule, $\lim_{t\to\infty} t^k e^{\lambda t}\xi = 0$. In case $\lambda = p + iq$ is simple and $p < 0$, it contributes the terms $e^{pt} \cos qt\, \alpha$ and $e^{pt} \sin qt\, \beta$, both tending to zero as $t \to \infty$. A repeated complex eigenvalue contributes the terms of the form $t^k e^{pt} \cos qt\, \alpha$ and $t^k e^{pt} \sin qt\, \beta$, also tending to zero as $t \to \infty$. □

Fundamental solution matrix

For an arbitrary $n \times n$ system

$$x' = Ax, \tag{4.14}$$

it is always possible to find n linearly independent solutions $x_1(t), x_2(t), \dots, x_n(t)$, as we saw above. Their linear combination

$$x(t) = c_1 x_1(t) + c_2 x_2(t) + \cdots + c_n x_n(t), \tag{4.15}$$

with arbitrary coefficients $c_1, c_2, \ldots, c_n$, is also a solution of (4.14). Form an $n \times n$ *solution matrix* $X(t) = [x_1(t) \, x_2(t) \ldots x_n(t)]$, using these solutions as its columns, and consider the vector $c = \begin{bmatrix} c_1 \\ c_2 \\ \vdots \\ c_n \end{bmatrix}$. Then the solution in (4.15) can be written as a matrix product

$$x(t) = X(t)c. \tag{4.16}$$

The matrix $X(t)$ is invertible for all t because its columns are linearly independent. The formula (4.15) (or (4.16)) gives the *general solution* of the system (4.14), meaning that solution of the initial value problem

$$x' = Ax, \quad x(0) = x_0, \tag{4.17}$$

with any given vector $x_0 \in R^n$, can be found among the solutions in (4.15) (or in (4.16)) for some choice of numbers $c_1, c_2, \ldots, c_n$. Indeed, the solution $x(t)$ needs to satisfy

$$x(0) = X(0)c = x_0,$$

and one can solve this $n \times n$ system for the vector c, because $X(0)$ is an invertible matrix.

If one chooses the solutions satisfying $x_1(0) = e_1, x_2(0) = e_2, \ldots, x_n(0) = e_n$, the coordinate vectors, then the corresponding solution matrix $X(t) = [x_1(t) \, x_2(t) \ldots x_n(t)]$ is called *the fundamental solution matrix of* (4.14), or *the fundamental matrix*, for short. Its advantage is that $x(t) = X(t)x_0$ gives the solution of the initial value problem (4.17). Indeed, this solution satisfies

$$x(0) = X(0)x_0 = [x_1(0) \, x_2(0) \ldots x_n(0)]x_0 = [e_1 \, e_2 \ldots e_n]x_0 = Ix_0 = x_0.$$

Example 4. Find the fundamental solution matrix for

$$x'(t) = \begin{bmatrix} 0 & -1 \\ 1 & 0 \end{bmatrix} x(t),$$

and use it to express the solution with $x_1(0) = -2$, $x_2(0) = 3$.

Solution $x_1(t) = \begin{bmatrix} \cos t \\ \sin t \end{bmatrix}$ satisfies $x_1(0) = e_1$, and solution $x_2(t) = \begin{bmatrix} -\sin t \\ \cos t \end{bmatrix}$ satisfies $x_2(0) = e_2$. It follows that $X(t) = \begin{bmatrix} \cos t & -\sin t \\ \sin t & \cos t \end{bmatrix}$ (the rotation matrix) is the fundamental solution matrix, and the solution with the prescribed initial conditions is

$$x(t) = X(t) \begin{bmatrix} -2 \\ 3 \end{bmatrix} = \begin{bmatrix} \cos t & -\sin t \\ \sin t & \cos t \end{bmatrix} \begin{bmatrix} -2 \\ 3 \end{bmatrix} = \begin{bmatrix} -2\cos t - 3\sin t \\ -2\sin t + 3\cos t \end{bmatrix}.$$

Solution matrices $X(t)$ are constructed the same way for systems with variable coefficients

$$x' = A(t)x, \tag{4.18}$$

where $n \times n$ matrix $A = [a_{ij}(t)]$ has functions $a_{ij}(t)$ as its entries. Namely, the columns of $X(t)$ are linearly independent solutions of (4.18). Again, $X(t)c$ provides the general solution of (4.18). Even though $X(t)$ can be explicitly calculated only rarely, unless $A(t)$ is a constant matrix, it will have some theoretical applications later on in the text.

Finally, observe that any solution matrix $X(t)$ satisfies the following *matrix differential equation*:

$$X'(t) = A(t)X(t). \tag{4.19}$$

Indeed, the first column on the left is $x'_1(t)$, and on the right the first column is $A(t)x_1$, and their equality $x'_1 = A(t)x_1$ reflects the fact that x_1 is a solution of our system $x' = A(t)x$. Similarly one shows that the other columns are identical.

Exercises

1. Solve the following systems:

    (a) $x'(t) = \begin{bmatrix} 2 & 1 & 2 \\ -5 & -1 & -7 \\ 1 & 0 & 2 \end{bmatrix} x(t)$, $x(0) = \begin{bmatrix} 2 \\ -1 \\ 1 \end{bmatrix}$.

    Answer: $x_1(t) = -e^t(3t^2 - 3t - 2)$, $x_2(t) = -e^t(3t^2 + 15t + 1)$, $x_3(t) = e^t(3t^2 + 3t + 1)$.

    (b) $x'(t) = \begin{bmatrix} 2 & 1 & 1 \\ 0 & 2 & -3 \\ 0 & 0 & -1 \end{bmatrix} x(t)$, $x(0) = \begin{bmatrix} 0 \\ -1 \\ 3 \end{bmatrix}$.

    Answer: $x_1(t) = -2e^{-t} + 2e^{2t} - 4te^{2t}$, $x_2(t) = 3e^{-t} - 4e^{2t}$, $x_3(t) = 3e^{-t}$.

    (c) $x'(t) = \begin{bmatrix} 0 & 1 & 2 \\ 1 & -2 & 1 \\ -2 & -1 & -4 \end{bmatrix} x(t)$, $x(0) = \begin{bmatrix} 6 \\ 0 \\ 2 \end{bmatrix}$.

    Answer: $x(t) = \begin{bmatrix} 2e^{-2t}(2t^2 + 8t + 3) \\ 8te^{-2t} \\ -2e^{-2t}(2t^2 + 8t - 1) \end{bmatrix}$.

    (d) $x'(t) = \begin{bmatrix} 0 & 0 & 1 \\ 0 & 1 & 0 \\ 1 & 0 & 0 \end{bmatrix} x(t)$, $x(0) = \begin{bmatrix} 1 \\ 2 \\ 3 \end{bmatrix}$.

    Hint. Here the matrix has a repeated eigenvalue, but a complete set of eigenvectors.

    Answer: $x(t) = \begin{bmatrix} 2e^t - e^{-t} \\ 2e^t + e^{-t} \\ 2e^t \end{bmatrix}$.

2. Construct the fundamental solution matrices of the system $x' = Ax$ for the following matrices A:

    (a) $A = \begin{bmatrix} 0 & -4 \\ 1 & 0 \end{bmatrix}$.

    Answer: $X(t) = \begin{bmatrix} \cos 2t & -2\sin 2t \\ \frac{1}{2}\sin 2t & \cos 2t \end{bmatrix}$.

    (b) $A = \begin{bmatrix} -2 & 1 \\ 4 & 1 \end{bmatrix}$.

    Answer: $X(t) = \begin{bmatrix} \frac{1}{5}(e^{2t} + 4e^{-3t}) & \frac{1}{5}(e^{2t} - e^{-3t}) \\ \frac{4}{5}(e^{2t} - e^{-3t}) & \frac{1}{5}(4e^{2t} + e^{-3t}) \end{bmatrix}$.

    (c) $A = \begin{bmatrix} 3 & -1 \\ 2 & 1 \end{bmatrix}$.

Answer: $X(t) = \begin{bmatrix} e^{2t}(\cos t + \sin t) & -e^{2t}\sin t \\ 2e^{2t}\sin t & e^{2t}(\cos t - \sin t) \end{bmatrix}$.

(d) $A = \begin{bmatrix} -1 & 1 \\ 0 & -1 \end{bmatrix}$.

Answer: $X(t) = \begin{bmatrix} e^{-t} & te^{-t} \\ 0 & e^{-t} \end{bmatrix}$.

(e) $A = \begin{bmatrix} 0 & 0 & -1 \\ 0 & -3 & 0 \\ 1 & 0 & 0 \end{bmatrix}$.

Answer: $X(t) = \begin{bmatrix} \cos t & 0 & -\sin t \\ 0 & e^{-3t} & 0 \\ \sin t & 0 & \cos t \end{bmatrix}$.

3. Consider an $n \times n$ system $x' = A(t)x$, where the matrix $A(t)$ is skew-symmetric ($A^T = -A$).

(a) If $x(t)$ and $y(t)$ are two solutions, show that

$$x(t) \cdot y(t) = x(0) \cdot y(0) \quad \text{for all } t,$$

and in particular $x(t)$ and $y(t)$ are orthogonal, provided that $x(0)$ and $y(0)$ are orthogonal.

Hint. Differentiate $x(t) \cdot y(t)$.

(b) Show that $\|x(t)\| = \|x(0)\|$ for all t.

(c) Show that the fundamental solution matrix $X(t)$ is an orthogonal matrix for all t.

Hint. The columns of $X(t)$ are orthonormal for all t.

4. (a) Verify that the matrix $A = \begin{bmatrix} -1 & 0 & 0 & 1 \\ 1 & -1 & 0 & 0 \\ 0 & 0 & -1 & 0 \\ 0 & 0 & 1 & -1 \end{bmatrix}$ has an eigenvalue $\lambda = -1$ of multiplicity four, with a one-dimensional eigenspace, spanned by e_2, the second coordinate vector in R^4.

(b) Let $w_4 = e_3$, where e_3 is the third coordinate vector in R^4. Verify that w_4 is a generalized eigenvector of rank 4 (so that $(A + I)^4 w_4 = 0$, but $(A + I)^3 w_4 \neq 0$).

(c) Construct the chain $w_3 = (A + I)w_4$, $w_2 = (A + I)w_3$, $w_1 = (A + I)w_2$, and verify that w_1 is an eigenvector of A.

Answer: $w_3 = e_4$, $w_2 = e_1$, $w_1 = e_2$.

(d) Find the general solution of the system $x' = Ax$.

Answer:

$$x(t) = c_1 e^{-t} e_2 + c_2 e^{-t}(e_1 + e_2 t) + c_3 e^{-t}\left(e_4 + e_1 t + e_2 \frac{t^2}{2}\right)$$

$$+ c_4 e^{-t}\left(e_3 + e_4 t + e_1 \frac{t^2}{2} + e_2 \frac{t^3}{3!}\right).$$

5. Suppose that the eigenvalues of a matrix A are $\lambda_1 = 0$ with one linearly independent eigenvector ξ_1, giving rise to the chain $\{\xi_1, \xi_{12}, \xi_{13}, \xi_{14}\}$, and $\lambda_2 = -4$ with two linearly independent eigenvectors ξ_2 and ξ_3; ξ_2 giving rise to the chain $\{\xi_2, \xi_{22}\}$, and ξ_3 giving rise to the chain $\{\xi_3\}$.

(a) What is the size of A?

Answer: 7×7.

(b) Write down the general solution of $x' = Ax$.
Answer:

$$x(t) = c_1\xi_1 + c_2(\xi_{12} + \xi_1 t) + c_3\left(\xi_{13} + \xi_{12}t + \xi_1\frac{t^2}{2}\right)$$

$$+ c_4\left(\xi_{14} + \xi_{13}t + \xi_{12}\frac{t^2}{2} + \xi_1\frac{t^3}{3!}\right)$$

$$+ c_5 e^{-4t}\xi_2 + c_6 e^{-4t}(\xi_{22} + \xi_2 t) + c_7 e^{-4t}\xi_3.$$

6. Show that e^{At} gives the fundamental solution matrix of $x' = Ax$.
   Hint. Note that $e^{At}e_1$ gives the first column of e^{At}, and also a solution of $x' = Ax$, so that the columns of e^{At} are solutions of $x' = Ax$. These solutions are linearly independent, since $\det e^{At} = e^{\mathrm{tr}(At)} > 0$. Hence e^{At} is a solution matrix. Setting $t = 0$, makes $e^{At} = I$.

7. Consider $J_0 = \begin{bmatrix} \lambda & 1 & 0 \\ 0 & \lambda & 1 \\ 0 & 0 & \lambda \end{bmatrix}$, where λ is either real or complex number.
   (a) Show that

$$J_0^n = \begin{bmatrix} \lambda^n & n\lambda^{n-1} & \frac{n(n-1)}{2}\lambda^{n-2} \\ 0 & \lambda^n & n\lambda^{n-1} \\ 0 & 0 & \lambda^n \end{bmatrix}.$$

   Hint. Write $J_0 = \lambda I + N$, with a nilpotent N.
   (b) Assume that the modulus $|\lambda| < 1$. Show that $\lim_{n\to\infty} J_0^n = O$, the zero matrix.
   (c) Assume that all eigenvalues of an $n \times n$ matrix A have modulus less than one, $|\lambda_i| < 1$ for all i. Show that $\lim_{n\to\infty} A^n = O$, the series $\sum_{k=0}^{\infty} A^k$ is convergent, and $\sum_{k=0}^{\infty} A^k = (I - A)^{-1}$.

6.5 Nonhomogeneous systems

We now consider nonhomogeneous systems

$$x' = A(t)x + f(t). \tag{5.1}$$

Here $A(t) = [a_{ij}(t)]$ is an $n \times n$ matrix, with given functions $a_{ij}(t)$, and a vector-function $f(t) \in R^n$ is also prescribed. For *the corresponding homogeneous system*

$$x' = A(t)x \tag{5.2}$$

the general solution, which is denoted by $z(t)$, can be obtained by the methods studied above ($z(t)$ depends on n arbitrary constants, by the representation $z(t) = X(t)c$ using solution matrix). Let vector-function $Y(t)$ be any particular solution of (5.1) so that

$$Y' = A(t)Y + f(t). \tag{5.3}$$

Subtracting (5.3) from (5.1) gives

$$(x - Y)' = A(t)(x - Y),$$

so that $x - Y$ is a solution of the corresponding homogeneous system (5.2), and then $x(t) - Y(t) = z(t) = X(t)c$ for some choice of arbitrary constants c. It follows that

$$x(t) = Y(t) + z(t) = Y(t) + X(t)c.$$

Conclusion: *the general solution of the nonhomogeneous system* (5.1) *is equal to the sum of any particular solution $Y(t)$ of this system and the general solution $X(t)c$ of the corresponding homogeneous system.*

Sometimes one can guess the form of a particular solution $Y(t)$, and then calculate $Y(t)$.

Example 1. Solve the system

$$x_1' = 2x_1 + x_2 - 8e^{-t},$$
$$x_2' = x_1 + 2x_2.$$

Search for a particular solution in the form $Y(t) = \begin{bmatrix} Ae^{-t} \\ Be^{-t} \end{bmatrix}$, or $x_1 = Ae^{-t}$, $x_2 = Be^{-t}$, with numbers A and B to be determined. Substitution produces an algebraic system for A and B:

$$-A = 2A + B - 8,$$
$$-B = A + 2B.$$

Calculate $A = 3$, $B = -1$, so that $Y(t) = \begin{bmatrix} 3e^{-t} \\ -e^{-t} \end{bmatrix}$. The general solution of the corresponding homogeneous system $x' = \begin{bmatrix} 2 & 1 \\ 1 & 2 \end{bmatrix} x$ is

$$z(t) = c_1 e^t \begin{bmatrix} -1 \\ 1 \end{bmatrix} + c_2 e^{3t} \begin{bmatrix} 1 \\ 1 \end{bmatrix}.$$

Answer: $x(t) = \begin{bmatrix} 3e^{-t} \\ -e^{-t} \end{bmatrix} + c_1 e^t \begin{bmatrix} -1 \\ 1 \end{bmatrix} + c_2 e^{3t} \begin{bmatrix} 1 \\ 1 \end{bmatrix}$ is the general solution.

Example 2. Solve the system

$$x' = \begin{bmatrix} 2 & 1 \\ 1 & 2 \end{bmatrix} x + \begin{bmatrix} -3t + 1 \\ -6 \end{bmatrix}.$$

Search for a particular solution in the form $Y(t) = \begin{bmatrix} At+B \\ Ct+D \end{bmatrix}$, and calculate $Y(t) = \begin{bmatrix} 2t-1 \\ -t+3 \end{bmatrix}$. The corresponding homogeneous system is the same as in Example 1.

Answer: $x(t) = \begin{bmatrix} 2t-1 \\ -t+3 \end{bmatrix} + c_1 e^t \begin{bmatrix} -1 \\ 1 \end{bmatrix} + c_2 e^{3t} \begin{bmatrix} 1 \\ 1 \end{bmatrix}$.

Guessing the form of a particular solution $Y(t)$ is not possible in most cases. A more general method for finding a particular solution of nonhomogeneous systems, called *the variation of parameters*, is described next.

If $X(t)$ is a solution matrix of the corresponding homogeneous system (5.2), then $x(t) = X(t)c = c_1 x_1(t) + c_2 x_2(t) + \cdots + c_n x_n(t)$ gives the general solution of (5.2), as we saw in the preceding section. Let us search for a particular solution of (5.1) in the form

$$x(t) = X(t)c(t) = c_1(t)x_1(t) + c_2(t)x_2(t) + \cdots + c_n(t)x_n(t). \tag{5.4}$$

Here the parameters $c_1, c_2, \ldots, c_n$ from general solution are replaced by the unknown functions $c_1(t), c_2(t), \ldots, c_n(t)$. (Functions are "variable quantities," explaining the name of this method.) By the product rule, $x'(t) = X'(t)c(t) + X(t)c'(t)$ so that substitution of $x(t) = X(t)c(t)$ into (5.1) gives

$$X'(t)c(t) + X(t)c'(t) = A(t)X(t)c(t) + f(t).$$

Since $X'(t) = A(t)X(t)$ by the formula (4.19), two terms cancel, giving

$$X(t)c'(t) = f(t). \tag{5.5}$$

Because solution matrix $X(t)$ is nonsingular, one can solve this $n \times n$ system of linear equations for $c_1'(t), c_2'(t), \ldots, c_n'(t)$, and then obtain $c_1(t), c_2(t), \ldots, c_n(t)$ by integration.

Example 3. Find the general solution to the system

$$x_1' = 2x_1 - x_2 + te^{-t},$$
$$x_2' = 3x_1 - 2x_2 - 2.$$

Here $A = \begin{bmatrix} 2 & -1 \\ 3 & -2 \end{bmatrix}$ and $f(t) = \begin{bmatrix} te^{-t} \\ -2 \end{bmatrix}$. The matrix A has an eigenvalue $\lambda_1 = -1$ with corresponding eigenvector $\xi_1 = \begin{bmatrix} 1 \\ 3 \end{bmatrix}$, and an eigenvalue $\lambda_2 = 1$ with corresponding eigenvector $\xi_2 = \begin{bmatrix} 1 \\ 1 \end{bmatrix}$. It follows that the general solution of the corresponding homogeneous system

$$x' = \begin{bmatrix} 2 & -1 \\ 3 & -2 \end{bmatrix} x$$

(denoting $x = \begin{bmatrix} x_1 \\ x_2 \end{bmatrix}$) is $z(t) = c_1 e^{-t} \begin{bmatrix} 1 \\ 3 \end{bmatrix} + c_2 e^t \begin{bmatrix} 1 \\ 1 \end{bmatrix} = c_1 \begin{bmatrix} e^{-t} \\ 3e^{-t} \end{bmatrix} + c_2 \begin{bmatrix} e^t \\ e^t \end{bmatrix}$, and $X(t) = \begin{bmatrix} e^{-t} & e^t \\ 3e^{-t} & e^t \end{bmatrix}$ is a solution matrix. We search for a particular solution in the form $Y(t) = X(t)c(t)$, where $c(t) = \begin{bmatrix} c_1(t) \\ c_2(t) \end{bmatrix}$. By (5.5), one needs to solve the system $X(t)c'(t) = f(t)$, or

$$\begin{bmatrix} e^{-t} & e^t \\ 3e^{-t} & e^t \end{bmatrix} \begin{bmatrix} c_1'(t) \\ c_2'(t) \end{bmatrix} = \begin{bmatrix} te^{-t} \\ -2 \end{bmatrix}.$$

In components,

$$e^{-t} c_1'(t) + e^t c_2'(t) = te^{-t},$$
$$3e^{-t} c_1'(t) + e^t c_2'(t) = -2.$$

Use Cramer's rule to solve this system for $c_1'(t)$ and $c_2'(t)$:

$$c_1'(t) = \frac{\begin{vmatrix} te^{-t} & e^t \\ -2 & e^t \end{vmatrix}}{\begin{vmatrix} e^{-t} & e^t \\ 3e^{-t} & e^t \end{vmatrix}} = \frac{t + 2e^t}{-2} = -\frac{1}{2}t - e^t,$$

$$c_2'(t) = \frac{\begin{vmatrix} e^{-t} & te^{-t} \\ 3e^{-t} & -2 \end{vmatrix}}{\begin{vmatrix} e^{-t} & e^t \\ 3e^{-t} & e^t \end{vmatrix}} = \frac{-2e^{-t} - 3te^{-2t}}{-2} = e^{-t} + \frac{3}{2}te^{-2t}.$$

Integration gives $c_1(t) = -\frac{1}{4}t^2 - e^t$ and $c_2(t) = -e^{-t} - \frac{3}{4}te^{-2t} - \frac{3}{8}e^{-2t}$. In both cases we took the constant of integration to be zero, because one needs only one particular solution. We thus obtain a particular solution

$$Y(t) = X(t)c(t) = \begin{bmatrix} e^{-t} & e^t \\ 3e^{-t} & e^t \end{bmatrix} \begin{bmatrix} -\frac{1}{4}t^2 - e^t \\ -e^{-t} - \frac{3}{4}te^{-2t} - \frac{3}{8}e^{-2t} \end{bmatrix}$$

$$= \begin{bmatrix} -\frac{1}{8}e^{-t}(2t^2 + 6t + 3) - 2 \\ -\frac{3}{8}e^{-t}(2t^2 + 2t + 1) - 4 \end{bmatrix}.$$

Answer: $x(t) = \begin{bmatrix} -\frac{1}{8}e^{-t}(2t^2+6t+3)-2 \\ -\frac{3}{8}e^{-t}(2t^2+2t+1)-4 \end{bmatrix} + c_1 e^{-t} \begin{bmatrix} 1 \\ 3 \end{bmatrix} + c_2 e^t \begin{bmatrix} 1 \\ 1 \end{bmatrix}.$

From (5.5) we express $c'(t) = X^{-1}(t)f(t)$, thus $c(t) = \int_{t_0}^t X^{-1}(s)f(s)\, ds$, with t_0 arbitrary. It follows that $Y(t) = X(t)\int_{t_0}^t X^{-1}(s)f(s)\, ds$ is a particular solution, and the general solution of the nonhomogeneous system (5.1) is then

$$x(t) = X(t)c + X(t) \int_{t_0}^t X^{-1}(s)f(s)\, ds. \tag{5.6}$$

In case matrix $A(t)$ has constant coefficients, $A(t) = A$, the fundamental solution matrix of $x' = Ax$ is given by e^{At}, and this formula becomes

$$x(t) = e^{A(t-t_0)}x(t_0) + \int_{t_0}^t e^{A(t-s)}f(s)\, ds.$$

Indeed, with $X(t) = e^{At}$, one has $X^{-1}(t) = e^{-At}$ and $X(t)X^{-1}(s) = e^{A(t-s)}$. Also, since $x(t_0) = e^{At_0}c$, obtain $c = e^{-At_0}x(t_0)$.

Systems with periodic coefficients
We consider now p-periodic systems

$$x' = A(t)x + f(t), \tag{5.7}$$

and the corresponding homogeneous systems

$$x' = A(t)x. \tag{5.8}$$

Assume that an $n \times n$ matrix $A(t)$ and a vector $f(t) \in R^n$ have continuous entries that are periodic functions of common period $p > 0$, so that $a_{ij}(t+p) = a_{ij}(t)$ and $f_i(t+p) = f_i(t)$ for all i, j, and t. The questions we shall address are whether these systems have p-periodic solutions, so that $x(t + p) = x(t)$ (which means that $x_i(t + p) = x_i(t)$ for all i and t) and whether nonperiodic solutions are bounded as $t \to \infty$.

If $X(t)$ denotes the fundamental solution matrix of (5.8), then the solution of (5.8) satisfying the initial condition $x(0) = x_0$ is

$$x(t) = X(t)x_0.$$

For the nonhomogeneous system (5.7), the solution satisfying the initial condition $x(0) = x_0$, and often denoted by $x(t, x_0)$, is given by

$$x(t) = X(t)x_0 + X(t) \int_0^t X^{-1}(s)f(s)\, ds. \tag{5.9}$$

Indeed, use $t_0 = 0$ in (5.6), then calculate $c = x_0$, by setting $t = 0$.

It is known from the theory of differential equations that a solution $x(t)$ of the system (5.7) is p-periodic if and only if $x(p) = x(0)$, and the same is true for the system (5.8), which can be seen as a particular case of (5.7).

The homogeneous system (5.8) has a p-periodic solution satisfying $x(p) = x(0)$, or $X(p)x_0 = x_0$, provided that the $n \times n$ system of linear equations

$$(I - X(p))x_0 = 0 \tag{5.10}$$

has a nontrivial solution x_0. In such a case $X(p)x_0 = x_0$, so that $X(p)$ has an eigenvalue 1, and the matrix $I - X(p)$ is singular.

Define the vector

$$b = X(p) \int_0^p X^{-1}(s)f(s)\, ds. \tag{5.11}$$

Then $x(p) = X(p)x_0 + b$, by (5.9). The nonhomogeneous system (5.7) has a p-periodic solution $x(t)$ with $x(p) = x(0)$, or $X(p)x_0 + b = x_0$, provided that the system of linear equations

$$(I - X(p))x_0 = b \tag{5.12}$$

has a solution x_0. (If $x_0 \in R^n$ is a solution of (5.12), then $x(t, x_0)$ given by (5.9) is a p-periodic solution of (5.7).)

Case 1 of the following theorem deals with an example of *resonance*, when a periodic forcing term $f(t)$ produces unbounded solutions.

Theorem 6.5.1. *Assume that (5.8) has a p-periodic solution (so that the matrix $I - X(p)$ is singular).*

1. *If vector b does not belong to the range (the column space) of $I - X(p)$, then all solutions of (5.7) are unbounded as $t \to \infty$.*
2. *If vector b belongs to the range of $I - X(p)$, then (5.7) has infinitely many p-periodic solutions. If, moreover, $X(p)$ has an eigenvalue μ with modulus $|\mu| > 1$, then (5.7) has also unbounded solutions (in addition to the p-periodic ones).*

Proof. Let $x(t)$ be any solution of (5.7), represented by (5.9). We shall consider the iterates $x(mp)$. With the vector b as defined by (5.11),

$$x(p) = X(p)x_0 + b.$$

By periodicity of the system (5.7), $x(t + p)$ is also a solution of (5.7), equal to $x(p)$ at $t = 0$. Using (5.9) again yields

$$x(t + p) = X(t)x(p) + X(t) \int_0^t X^{-1}(s)f(s)\, ds.$$

Then

$$x(2p) = X(p)x(p) + b = X(p)(X(p)x_0 + b) + b = X^2(p)x_0 + X(p)b + b.$$

By induction, for any integer $m > 0$,

$$x(mp) = X^m(p)x_0 + \sum_{k=0}^{m-1} X^k(p)b. \tag{5.13}$$

Case 1. Assume that b does not belong to the range of $I - X(p)$. Then the linear system (5.12) has no solutions, and hence its determinant is zero. Since $\det(I - X(p))^T = \det(I - X(p)) = 0$, it follows that the system

$$(I - X(p))^T v = 0 \tag{5.14}$$

has nontrivial solutions. We claim that it is possible to find a nontrivial solution v_0 of (5.14), for which the scalar product with b satisfies

$$b \cdot v_0 \neq 0. \tag{5.15}$$

Indeed, assuming otherwise, b would be orthogonal to the null-space of $(I - X(p))^T$, and then, by the Fredholm alternative, the linear system (5.12) would be solvable, a contradiction. From (5.14), $v_0 = X(p)^T v_0$, then $X(p)^T v_0 = X^2(p)^T v_0$, which gives $v_0 = X^2(p)^T v_0$, and inductively

$$v_0 = X^k(p)^T v_0, \quad \text{for all positive integers } k. \tag{5.16}$$

Using (5.13),

$$x(mp) \cdot v_0 = X^m(p) x_0 \cdot v_0 + \sum_{k=0}^{m-1} X^k(p) b \cdot v_0$$

$$= x_0 \cdot X^m(p)^T v_0 + \sum_{k=0}^{m-1} b \cdot X^k(p)^T v_0 = x_0 \cdot v_0 + mb \cdot v_0 \to \infty,$$

as $m \to \infty$, in view of (5.15). Hence, the solution $x(t)$ is unbounded.

Case 2. Assume that b belongs to the range of $I - X(p)$. Then the linear system (5.12) has a solution $\bar{x}_0$, and $x(t, \bar{x}_0)$ is a p-periodic solution of (5.7). Adding to it nontrivial solutions of the corresponding homogeneous system produces infinitely many p-periodic solutions of (5.7).

Assume now that $X(p)$ has an eigenvalue μ, with modulus $|\mu| > 1$. Since $\bar{x}_0$ is a solution of (5.12),

$$\bar{x}_0 = X(p)\bar{x}_0 + b. \tag{5.17}$$

Then

$$X(p)\bar{x}_0 = X^2(p)\bar{x}_0 + X(p)b.$$

Using here (5.17) yields

$$\bar{x}_0 = X^2(p)\bar{x}_0 + X(p)b + b.$$

Continuing to use the latest expression for $\bar{x}_0$ in (5.17), we inductively obtain

$$\bar{x}_0 = X^m(p)\bar{x}_0 + \sum_{k=0}^{m-1} X^k(p)b,$$

so that $\sum_{k=0}^{m-1} X^k(p)b = \bar{x}_0 - X^m(p)\bar{x}_0$. Using this in (5.13), we get

$$x(mp) = \bar{x}_0 + X^m(p)(x_0 - \bar{x}_0). \tag{5.18}$$

Let now y be an eigenvector of $X(p)$, corresponding to the eigenvalue μ, with $|\mu| > 1$, so that $X(p)y = \mu y$, and then $X^m(p)y = \mu^m y$. Choose x_0 so that $x_0 - \bar{x}_0 = y$, which is $x_0 = \bar{x}_0 + y$. Then

$$x(mp) = \bar{x}_0 + X^m(p)y = \bar{x}_0 + \mu^m y,$$

and $x(mp)$ becomes unbounded as $m \to \infty$. □

The following famous theorem is a consequence.

Theorem 6.5.2 (Massera's theorem). *If the nonhomogeneous system* (5.7) *(with p-periodic $A(t)$ and $f(t)$) has a bounded solution (as $t \to \infty$), then it has a p-periodic solution.*

Proof. We shall prove an equivalent statement obtained by logical contraposition: *If the system* (5.7) *has no p-periodic solution, then all of its solutions are unbounded.* Indeed, if (5.7) has no p-periodic solution, then the vector b does not lie in the range of $I - X(p)$, while the matrix $I - X(p)$ is singular (if this matrix was nonsingular, its range would be all of R^n), and hence the homogeneous system (5.8) has a p-periodic solution. We are in the conditions of Case 1 of the preceding Theorem 6.5.1, and hence all solutions of (5.7) are unbounded. □

The assumption of Theorem 6.5.1 that the homogeneous system (5.8) has a p-periodic solution can be seen as a *case of resonance*. The complementary case, when (5.8) does not have a p-periodic solution, is easy. Then the matrix $I - X(p)$ is nonsingular, so that the system (5.12) has a unique solution for any b, and hence the nonhomogeneous system (5.7) has a unique p-periodic solution for any p-periodic $f(t)$.

The eigenvalues ρ of the matrix $X(p)$ are called *the Floquet multipliers*. ($X(p)$ is known as *the monodromy matrix*.) As we saw above, the homogeneous system (5.8) has a p-periodic solution if and only if one of the Floquet multipliers is $\rho = 1$. By a reasoning similar to the above theorem, one can justify the following statement.

Proposition 6.5.1. *All solutions of* (5.8) *tend to zero, as $t \to \infty$, if and only if all Floquet multipliers satisfy $|\rho| < 1$. If, on the other hand, one of the Floquet multipliers has modulus $|\rho| > 1$, then* (5.8) *has an unbounded solution, as $t \to \infty$.*

Periodic Hamiltonian systems

The motion of n particles on a line with positions at time t given by functions $q_i(t)$ is governed by Newton's second law

$$m_i q_i'' = F_i, \quad i = 1, 2, \ldots, n. \tag{5.19}$$

Here m_i is the mass of particle i, function F_i gives the force acting on the i-th particle. Assume that there is a function $U(t, q_1, q_2, \ldots, q_n)$, so that $F_i = -\frac{\partial U}{\partial q_i}$. The function $U(t, q_1, q_2, \ldots, q_n)$ is called *the potential energy*, it reflects the interactions between the particles (it "couples" the particles). Introduce *the impulses $p_i = m_i q_i'$, the kinetic energy* $\sum_{i=1}^{n} \frac{1}{2m_i} p_i^2$, and *the total energy* (also called *the Hamiltonian function*)

$$H = K + U = \sum_{i=1}^{n} \frac{1}{2m_i} p_i^2 + U(t, q_1, q_2, \ldots, q_n).$$

One can write the equations of motion (5.19) as *a Hamiltonian system*

$$q_i' = \frac{\partial H}{\partial p_i}, \qquad (5.20)$$
$$p_i' = -\frac{\partial H}{\partial q_i}.$$

(Indeed, $\frac{\partial H}{\partial p_i} = \frac{1}{m_i} p_i$, so that the first equation in (5.20) follows by the definition of impulses, while the second equation follows by (5.19).) The system (5.20) has $2n$ equations and $2n$ unknowns $q_1, \ldots, q_n, p_1, \ldots, p_n$, and $H = H(t, q_1, \ldots, q_n, p_1, \ldots, p_n)$. Introduce the vectors $q = \begin{bmatrix} q_1 \\ \vdots \\ q_n \end{bmatrix}$ and $p = \begin{bmatrix} p_1 \\ \vdots \\ p_n \end{bmatrix}$. One can write (5.20) in the vector form as

$$q' = \frac{\partial H}{\partial p}, \qquad (5.21)$$
$$p' = -\frac{\partial H}{\partial q}.$$

If the Hamiltonian function is independent of t, meaning that $H = H(p, q)$, then along the solutions of (5.21)

$$H(p(t), q(t)) = \text{constant}.$$

Indeed, using the chain rule,

$$\frac{d}{dt} H(p(t), q(t)) = \frac{\partial H}{\partial p} p' + \frac{\partial H}{\partial q} q' = \frac{\partial H}{\partial p}\left(-\frac{\partial H}{\partial q}\right) + \frac{\partial H}{\partial q}\frac{\partial H}{\partial p} = 0.$$

Consider the vector $x = \begin{bmatrix} q \\ p \end{bmatrix} \in R^{2n}$ and the block matrix $J = \begin{bmatrix} O & I \\ -I & O \end{bmatrix}$ of size $2n \times 2n$, *the symplectic unit matrix*. Here O is the zero matrix and I the identity matrix, both of size $n \times n$. One may write (5.21) in the form

$$x' = JH_x, \qquad (5.22)$$

where H_x denotes the gradient, $H_x = \begin{bmatrix} H_q \\ H_p \end{bmatrix}$.

Assume that the Hamiltonian function $H(t, x)$ is a quadratic form

$$H(t, x) = \frac{1}{2} \sum_{i,j=1}^{2n} a_{ij}(t) x_i x_j,$$

with a symmetric matrix $A(t) = [a_{ij}(t)]$ of size $2n \times 2n$. Then (5.22) takes the form

$$x' = JA(t)x, \tag{5.23}$$

which is a linear system with p-periodic coefficients. Let $X(t)$ be the fundamental solution matrix of (5.23) of size $2n \times 2n$, so that each column of $X(t)$ is a solution of (5.23), and $X(0) = I$. Observe that

$$X'(t) = JA(t)X(t).$$

Since the trace $\mathrm{tr}(JA(t)) = 0$ (which is justified in Exercises), it follows by Liouville's formula (which is stated and justified in Exercises) that

$$\det X(t) = \det X(0)\, e^{\int_0^t \mathrm{tr}\,(JA(s))\,ds} = 1, \quad \text{for all } t, \tag{5.24}$$

since $\det X(0) = \det I = 1$.

If $x(t)$ and $y(t)$ are two solutions of (5.23), their *symplectic product* $x^T Jy$ remains constant for all t. Indeed,

$$\frac{d}{dt}x^T Jy = (x^T)'Jy + x^T Jy' = x^T A^T J^T Jy + x^T JJAy$$
$$= -x^T AJ^2 y + x^T J^2 Ay = x^T Ay - x^T Ay = 0,$$

since $A^T = A$, $J^T = -J$, and $J^2 = -I$. If $X(t)$ and $Y(t)$ are two solution matrices (meaning that each column of both $X(t)$ and $Y(t)$ is a solution of (5.23)), then similarly

$$X^T(t)JY(t) = \text{constant matrix}. \tag{5.25}$$

Indeed, the (i,j)th entry of $X^T JY$ is $x_i^T Jy_j$, where x_i is the column i of X, y_j is the column j of Y, both solutions of (5.23), and hence $x_i^T Jy_j$ remains constant. In particular, in case of fundamental solution matrix $X(t)$ and $Y(t) = X(t)$, setting $t = 0$ in (5.25) gives

$$X^T(t)JX(t) = J, \quad \text{for all } t. \tag{5.26}$$

A polynomial equation for ρ, with numerical coefficients $a_0, a_1, \ldots, a_m$,

$$P(\rho) \equiv a_0 \rho^m + a_1 \rho^{m-1} + \cdots + a_{m-1}\rho + a_m = 0 \tag{5.27}$$

is called *symmetric* if $a_k = a_{m-k}$ for all integers k, $0 \le k \le m$. For such equations,

$$P\!\left(\frac{1}{\rho}\right) = \frac{1}{\rho^m}(a_0 + a_1\rho + \cdots + a_{m-1}\rho^{m-1} + a_m\rho^m) = \frac{1}{\rho^m}P(\rho) \quad (\rho \ne 0), \tag{5.28}$$

so that if ρ_0 is a root of (5.27), then $\frac{1}{\rho_0}$ is also a root. Conversely, if (5.28) holds, then equation (5.27) is symmetric.

Theorem 6.5.3 (Lyapunov–Poincaré). *Assume that the matrix $A(t)$ is periodic, $A(t + p) = A(t)$ for some $p > 0$ and all t. Then the characteristic equation of the fundamental solution matrix $X(p)$ of (5.23),*

$$f(\rho) \equiv \det[\rho I - X(p)] = 0, \tag{5.29}$$

is symmetric.

Proof. Since $X(p)$ is of size $2n \times 2n$, so is $\rho I - X(p)$, and (5.29) is a polynomial equation of degree $2n$. Observing that $\det J = \det J^{-1} = 1$, and using (5.26), which implies that $X^T(p) = JX^{-1}(p)J^{-1}$, and then using that $\det X^{-1}(p) = 1$ by (5.24), we obtain

$$
\begin{aligned}
f\left(\frac{1}{\rho}\right) &= \det\left[\frac{1}{\rho}I - X(p)\right] = \frac{1}{\rho^{2n}}\det[I - \rho X(p)] \\
&= \frac{1}{\rho^{2n}}\det[I - \rho X^T(p)] = \frac{1}{\rho^{2n}}\det[JIJ^{-1} - \rho JX^{-1}(p)J^{-1}] \\
&= \frac{1}{\rho^{2n}}\det J \det[I - \rho X^{-1}(p)]\det J^{-1} \\
&= \frac{(-1)^{2n}}{\rho^{2n}}\det X^{-1}(p)\det[\rho I - X(p)] \\
&= \frac{1}{\rho^{2n}}\det[\rho I - X(p)] = \frac{1}{\rho^{2n}}f(\rho),
\end{aligned}
$$

so that $f(\rho)$ is a symmetric polynomial. □

This theorem implies that if ρ is a Floquet multiplier of (5.23), then so is $\frac{1}{\rho}$. Hence, it is not possible for a Hamiltonian system (5.23) to have all solutions tending to zero as $t \to \infty$, since the Floquet multipliers cannot all satisfy $|\rho| < 1$. In fact, if there are some solutions of (5.23) tending to zero, then there are other solutions that are unbounded, as $t \to \infty$.

The proof given above is due to I. M. Gel'fand and V. B. Lidskii [10]. We followed the nice presentation in B. P. Demidovič [6].

Exercises

1. Find the general solution of the following systems by guessing the form of a particular solution:
   (a) $x' = \begin{bmatrix} 0 & 1 \\ 1 & 0 \end{bmatrix} x + \begin{bmatrix} 2e^{2t} \\ -e^{2t} \end{bmatrix}$.
       Answer: $x(t) = \begin{bmatrix} e^{2t} \\ 0 \end{bmatrix} + c_1 e^{-t} \begin{bmatrix} -1 \\ 1 \end{bmatrix} + c_2 e^{t} \begin{bmatrix} 1 \\ 1 \end{bmatrix}$.
   (b) $x' = \begin{bmatrix} 0 & 1 \\ 1 & 0 \end{bmatrix} x + \begin{bmatrix} e^{2t} \\ 0 \end{bmatrix}$.
       Answer: $x(t) = \begin{bmatrix} \frac{2}{3}e^{2t} \\ \frac{1}{3}e^{2t} \end{bmatrix} + c_1 e^{-t} \begin{bmatrix} -1 \\ 1 \end{bmatrix} + c_2 e^{t} \begin{bmatrix} 1 \\ 1 \end{bmatrix}$.

(c) $x' = \begin{bmatrix} 0 & -1 \\ 1 & 0 \end{bmatrix} x + \begin{bmatrix} 0 \\ 3\cos 2t \end{bmatrix}$.

Hint. Search for a particular solution in the form

$$x(t) = \begin{bmatrix} A\cos 2t + B\sin 2t \\ C\cos 2t + D\sin 2t \end{bmatrix}.$$

Answer: $x(t) = \begin{bmatrix} \cos 2t \\ 2\sin 2t \end{bmatrix} + c_1 \begin{bmatrix} \cos t \\ \sin t \end{bmatrix} + c_2 \begin{bmatrix} -\sin t \\ \cos t \end{bmatrix}$.

(d) $x' = \begin{bmatrix} 2 & 2 \\ 2 & -1 \end{bmatrix} x + \begin{bmatrix} 2t-1 \\ -t \end{bmatrix}$.

Hint. Search for a particular solution in the form

$$x(t) = \begin{bmatrix} At + B \\ Ct + D \end{bmatrix}.$$

Answer: $x(t) = \begin{bmatrix} -\frac{1}{6} \\ -t+\frac{2}{3} \end{bmatrix} + c_1 e^{-2t} \begin{bmatrix} -1 \\ 2 \end{bmatrix} + c_2 e^{3t} \begin{bmatrix} 2 \\ 1 \end{bmatrix}$.

(e) $x' = \begin{bmatrix} 0 & 1 \\ 1 & 0 \end{bmatrix} x + \begin{bmatrix} e^{2t} \\ -t \end{bmatrix}$.

Hint. Break the search for a particular solution into two pieces $Y(t) = Y_1(t) + Y_2(t)$, where $Y_1(t)$ is a particular solution for the system in part (b), and $Y_2(t)$ is a particular solution for the system $x' = \begin{bmatrix} 0 & 1 \\ 1 & 0 \end{bmatrix} x + \begin{bmatrix} 0 \\ -t \end{bmatrix}$.

Answer: $x(t) = \begin{bmatrix} \frac{2}{3}e^{2t}+t \\ \frac{1}{3}e^{2t}+1 \end{bmatrix} + c_1 e^{-t} \begin{bmatrix} -1 \\ 1 \end{bmatrix} + c_2 e^{t} \begin{bmatrix} 1 \\ 1 \end{bmatrix}$.

(f) $x' = \begin{bmatrix} 0 & -1 \\ 1 & 0 \end{bmatrix} x + \begin{bmatrix} e^{2t} \\ -2e^t \end{bmatrix}$.

2. Solve the following initial value problems:

(a) $x'(t) = \begin{bmatrix} 1 & 2 \\ 4 & 3 \end{bmatrix} x(t) + \begin{bmatrix} 1 \\ -1 \end{bmatrix}$, $x(0) = \begin{bmatrix} 1 \\ -2 \end{bmatrix}$.

Answer: $x_1(t) = \frac{e^{-t}}{3} - \frac{e^{5t}}{3} + 1$, $x_2(t) = -\frac{e^{-t}}{3} - \frac{2e^{5t}}{3} - 1$.

(b)

$$x_1' = 3x_1 + 5x_2 + t, \quad x_1(0) = 0,$$
$$x_2' = -5x_1 - 3x_2, \quad x_2(0) = 0.$$

Answer:

$$x_1(t) = \frac{1}{64}(12t + 4 - 3\sin 4t - 4\cos 4t), \quad x_2(t) = \frac{5}{64}(\sin 4t - 4t).$$

3. Solve the following initial value problems:

(a)

$$x_1' = x_1 + 2x_2 + e^{-t}, \quad x_1(0) = 0,$$
$$x_2' = 4x_1 + 3x_2, \quad x_2(0) = 0.$$

Answer: $x_1(t) = \frac{e^{5t}}{18} + \frac{e^{-t}}{18}(12t - 1)$, $x_2(t) = \frac{e^{5t}}{9} - \frac{e^{-t}}{9}(6t + 1)$.

(b)

$$x_1' = -x_2 + \sin t, \quad x_1(0) = 1,$$
$$x_2' = x_1, \quad\quad\quad x_2(0) = 0.$$

Answer: $x_1(t) = \frac{1}{2}t \sin t + \cos t$, $x_2(t) = \frac{1}{2}(3 \sin t - t \cos t)$.

4. (a) Justify the formula for differentiation of a determinant, namely

$$\frac{d}{dt} \begin{vmatrix} a(t) & b(t) \\ c(t) & d(t) \end{vmatrix} = \begin{vmatrix} a'(t) & b'(t) \\ c(t) & d(t) \end{vmatrix} + \begin{vmatrix} a(t) & b(t) \\ c'(t) & d'(t) \end{vmatrix}.$$

(b) Consider a system

$$x' = A(t)x, \tag{5.30}$$

where a 2×2 matrix $A(t) = [a_{ij}(t)]$ has entries depending on t. Let $X(t) = \begin{bmatrix} x_{11}(t) & x_{12}(t) \\ x_{21}(t) & x_{22}(t) \end{bmatrix}$ be any solution matrix, so that the vectors $\begin{bmatrix} x_{11}(t) \\ x_{21}(t) \end{bmatrix}$ and $\begin{bmatrix} x_{12}(t) \\ x_{22}(t) \end{bmatrix}$ are two linearly independent solutions of (5.30). The determinant $W(t) = |X(t)|$ is called *the Wronskian determinant* of (5.30). Show that *Liouville's formula* holds, namely, for any number t_0,

$$W(t) = W(t_0)e^{\int_{t_0}^{t} \mathrm{tr}A(s)\, ds}, \tag{5.31}$$

where the trace $\mathrm{tr}\, A(t) = a_{11}(t) + a_{22}(t)$.
Hint. Calculate

$$W' = \begin{vmatrix} x_{11}'(t) & x_{12}'(t) \\ x_{21}(t) & x_{22}(t) \end{vmatrix} + \begin{vmatrix} x_{11}(t) & x_{12}(t) \\ x_{21}'(t) & x_{22}'(t) \end{vmatrix}. \tag{5.32}$$

Using (5.30) and properties of determinants, calculate

$$\begin{vmatrix} x_{11}'(t) & x_{12}'(t) \\ x_{21}(t) & x_{22}(t) \end{vmatrix} = \begin{vmatrix} a_{11}x_{11} + a_{12}x_{21} & a_{11}x_{12} + a_{12}x_{22} \\ x_{21} & x_{22} \end{vmatrix}$$

$$= \begin{vmatrix} a_{11}x_{11} & a_{11}x_{12} \\ x_{21} & x_{22} \end{vmatrix} = a_{11}W.$$

Similarly, the second determinant in (5.32) is equal to $a_{22}W$. Then

$$W' = (a_{11} + a_{22})W = (\mathrm{tr}\, A)\, W.$$

Solving this differential equation for W gives (5.31).

(c) Show that Liouville's formula (5.31) holds also for $n \times n$ systems (5.30).

5. (a) Let $Y(t)$ be an $n \times n$ matrix function, and suppose the inverse $Y^{-1}(t)$ exists for t. Show that

$$\frac{d}{dt} Y^{-1}(t) = -Y^{-1} \frac{dY}{dt} Y^{-1}.$$

Hint. Differentiate the identity $Y(t)Y^{-1}(t) = I$.

(b) Let $X(t)$ be the fundamental solution matrix of

$$x' = A(t)x. \qquad (5.33)$$

Show that the fundamental solution matrix of *the adjoint system*

$$z' = -A^T(t)z \qquad (5.34)$$

satisfies $Z(t) = Y^{-1}(t)$, where $Y(t) = X^T(t)$.

Hint. The fundamental solution matrix of (5.33) satisfies $X' = AX$. Then $Y' = YA^T$, or $Y^{-1}Y' = A^T$. Multiply by Y^{-1}, and use part (a).

(c) Assume that the matrix $A(t)$ is p-periodic, so that $A(t+p) = A(t)$ for all t, and suppose that (5.33) has a p-periodic solution. Show that the same is true for (5.34).

Hint. Observe that $X(p)$ has an eigenvalue $\lambda = 1$.

(d) Assume that the homogeneous system (5.33) has a p-periodic solution. Show that the nonhomogeneous system

$$x' = A(t)x + f(t),$$

with a given p-periodic vector $f(t)$, has a p-periodic solution if and only if

$$\int_0^p f(t) \cdot z(t)\, dt = 0,$$

where $z(t)$ is any p-periodic solution of (5.34).

6. Let a matrix A be symmetric, and J be the symplectic unit matrix, both of size $2n \times 2n$.
   (a) Show that the trace $\mathrm{tr}(JA) = 0$.
   Hint. Write A as a block matrix $A = \begin{bmatrix} A_1 & A_2 \\ A_2 & A_3 \end{bmatrix}$, where A_1, A_2, A_3 are $n \times n$ matrices. Then $JA = \begin{bmatrix} A_2 & A_3 \\ -A_1 & -A_2 \end{bmatrix}$.
   (b) Show that $J^{-1} = -J$.
   (c) Show that $|J| = 1$.
   Hint. Expand $|J|$ in the first row, then in the last row, and conclude that $|J|$ is independent of n. When $n = 1$, $|J| = 1$.

6.6 Difference equations

Suppose one day there is a radio communication from somewhere in the Universe. How to test if it was sent by intelligent beings? Perhaps one could send them the digits of π. Or you can try the numbers $1, 1, 2, 3, 5, 8, 13, 21, \ldots$, *the Fibonacci sequence*. These numbers have a long history on our planet, dating back to 1202, with a myriad of applications, particularly to botany. The Fibonacci sequence begins with two ones, $F_1 = 1$, $F_2 = 1$, and then each number is the sum of the preceding two,

$$F_n = F_{n-1} + F_{n-2}, \quad n = 3, 4, \ldots,$$
$$F_1 = 1, \quad F_2 = 1. \tag{6.1}$$

To derive a formula for F_n, let us look for a solution of this *difference equation* in the form $F_n = r^n$, with the number r to be determined. Substitution into the equation, followed by division by r^{n-2}, gives

$$r^n = r^{n-1} + r^{n-2},$$

which simplifies to

$$r^2 = r + 1,$$

a quadratic equation with the roots $r_1 = \frac{1-\sqrt{5}}{2}, r_2 = \frac{\sqrt{5}+1}{2}$ (r_2 is known from antiquity as *the golden section*). We found two solutions of the equation in (6.1), r_1^n and r_2^n. Their linear combination with arbitrary coefficients c_1 and c_2 is also a solution of the difference equation, and we shall obtain the Fibonacci numbers,

$$F_n = c_1 r_1^n + c_2 r_2^n,$$

once c_1 and c_2 are chosen to satisfy the *initial conditions* (the second line in (6.1)). Obtain

$$F_1 = c_1 r_1 + c_2 r_2 = 1,$$
$$F_2 = c_1 r_1^2 + c_2 r_2^2 = 1.$$

Solving this system gives $c_1 = -\frac{1}{\sqrt{5}}, c_2 = \frac{1}{\sqrt{5}}$. We thus obtain *Binet's formula*

$$F_n = \frac{1}{\sqrt{5}}\left[\left(\frac{\sqrt{5}+1}{2}\right)^n - \left(\frac{1-\sqrt{5}}{2}\right)^n\right].$$

Other difference equations can be solved similarly. Their theory is "parallel" to differential equations.

We consider next *matrix recurrence relations* (also known as *matrix difference equations*) of the form

$$x_n = Ax_{n-1}, \quad n = 1, 2, \ldots, \tag{6.2}$$

where $x \in R^m$, and A is an $m \times m$ matrix. *The initial vector* x_0 is prescribed. Solving (6.2) is easy: $x_1 = Ax_0$, $x_2 = Ax_1 = AAx_0 = A^2x_0$, $x_3 = Ax_2 = AA^2x_0 = A^3x_0$, and in general,

$$x_n = A^n x_0. \tag{6.3}$$

We now analyze this solution in case A has a complete set of m linearly independent eigenvectors $x_1, x_2, \ldots, x_m$ with the corresponding eigenvalues $\lambda_1, \lambda_2, \ldots, \lambda_m$, some possibly repeated (recall that such A is diagonalizable). Since $Ax_i = \lambda_i x_i$, it follows that $A^n x_i = \lambda_i^n x_i$, for all i and n. The eigenvectors $x_1, x_2, \ldots, x_m$ form a basis of R^m, which can be used to decompose the initial vector

$$x_0 = c_1 x_1 + c_2 x_2 + \cdots + c_m x_m,$$

with some numbers $c_1, c_2, \ldots, c_m$. Then the solution (6.3) takes the form

$$
\begin{aligned}
x_n &= A^n (c_1 x_1 + c_2 x_2 + \cdots + c_m x_m) \\
&= c_1 A^n x_1 + c_2 A^n x_2 + \cdots + c_m A^n x_m \\
&= c_1 \lambda_1^n x_1 + c_2 \lambda_2^n x_2 + \cdots + c_m \lambda_m^n x_m.
\end{aligned}
\tag{6.4}
$$

In case all eigenvalues of A have modulus $|\lambda_i| < 1$, the solution x_n approaches the zero vector as $n \to \infty$. The difference equation (6.2) is then called *stable*. If there are eigenvalues of A satisfying $|\lambda_i| > 1$, some solutions become unbounded as $n \to \infty$. The difference equation (6.2) is then called *unstable*. For example, if $|\lambda_1| > 1$, and $x_0 = c_1 x_1$ with some $c_1 \neq 0$, then the sequence $A^n x_0 = c_1 \lambda_1^n x_1$ is unbounded. In case all eigenvalues of A satisfy $|\lambda_i| \leq 1$, and at least one has $|\lambda| = 1$, the difference equation (6.2) is called *neutrally stable*.

The following example introduces an important class of difference equations, with applications to probability theory.

Example 1. Suppose that each year 0.8 (or 80 %) of democrat voters remain democrat, while 0.1 switch to republicans and 0.1 to independents (these are *the probabilities*, with $0.8 + 0.1 + 0.1 = 1$). For independent voters, the probabilities are: 0.1 switch to democrats, 0.8 remain independent, and 0.1 switch to republicans. For republican voters, 0 switch to democrats, 0.3 join independents, while 0.7 remain republican. Denoting d_n, i_n, r_n the numbers of democrats, independents and republicans, respectively, after n years, obtain the recurrence relations

$$
\begin{aligned}
d_n &= 0.8d_{n-1} + 0.1i_{n-1}, \\
i_n &= 0.1d_{n-1} + 0.8i_{n-1} + 0.3r_{n-1}, \\
r_n &= 0.1d_{n-1} + 0.1i_{n-1} + 0.7r_{n-1}.
\end{aligned}
\tag{6.5}
$$

The initial numbers d_0, i_0, r_0 are prescribed, and add up to the total number of voters denoted by V,

$$d_0 + i_0 + r_0 = V.$$

Introducing *the transition matrix*

$$A = \begin{bmatrix} 0.8 & 0.1 & 0 \\ 0.1 & 0.8 & 0.3 \\ 0.1 & 0.1 & 0.7 \end{bmatrix},$$

and the vector $x_n = \begin{bmatrix} d_n \\ i_n \\ r_n \end{bmatrix}$, one puts the system (6.5) into the form $x_n = Ax_{n-1}$. Then $x_n = A^n x_0$, where $x_0 = \begin{bmatrix} d_0 \\ i_0 \\ r_0 \end{bmatrix}$, the initial vector. Calculations show that A has an eigenvalue $\lambda_1 = 1$ with a corresponding eigenvector $x_1 = \begin{bmatrix} 1 \\ 2 \\ 1 \end{bmatrix}$, an eigenvalue $\lambda_2 = \frac{7}{10}$ with a corresponding eigenvector $x_2 = \begin{bmatrix} -1 \\ 1 \\ 0 \end{bmatrix}$, and an eigenvalue $\lambda_3 = \frac{3}{5}$ with a corresponding eigenvector $x_3 = \begin{bmatrix} 1 \\ -2 \\ 1 \end{bmatrix}$. By (6.4),

$$x_n = c_1 x_1 + c_2 \left(\frac{7}{10} \right)^n x_2 + c_3 \left(\frac{3}{5} \right)^n x_3 \rightarrow c_1 x_1 = \begin{bmatrix} c_1 \\ 2c_1 \\ c_1 \end{bmatrix},$$

as $n \rightarrow \infty$. Since the sum of the entries of x_n gives the total number of voters V, $c_1 + 2c_1 + c_1 = V$ or $c_1 = \frac{1}{4}V$, we conclude that

$$x_n = \begin{bmatrix} d_n \\ i_n \\ r_n \end{bmatrix} \rightarrow \begin{bmatrix} \frac{1}{4}V \\ \frac{1}{2}V \\ \frac{1}{4}V \end{bmatrix} \quad \text{as } n \rightarrow \infty,$$

so that eventually a quarter of voters will vote democrat, a quarter republican, and the remaining half will be independents. The initial numbers d_0, i_0, r_0 *do not matter in the long run for the percentages of voter distribution*. One says that *the iterates x_n approach a steady state*.

The transition $x_{n-1} \rightarrow x_n$ in the last example is known as a *Markov process*. The matrix A is an example of a *Markov matrix*. These are square $m \times m$ matrices $A = [a_{ij}]$ with nonnegative entries, $a_{ij} \geq 0$, and each column adding up to 1,

$$\sum_{i=1}^{m} a_{ij} = 1, \quad \text{for all } j. \tag{6.6}$$

Theorem 6.6.1. *Any Markov matrix A has an eigenvalue $\lambda = 1$. All other eigenvalues have modulus $|\lambda| \leq 1$.*

Proof. The columns of the matrix $A - I$ add up to zero. Hence $|A - I| = |A - 1I| = 0$, so that $\lambda = 1$ is an eigenvalue of A.

Turning to the second statement, take the modulus of component i of the relation $Ax = \lambda x$ (with $x \neq 0$), then apply the triangle inequality, and use that $a_{ij} \geq 0$,

$$\lambda x_i = \sum_{j=1}^{m} a_{ij} x_j,$$

$$|\lambda||x_i| \leq \sum_{j=1}^{m} |a_{ij}||x_j| = \sum_{j=1}^{m} a_{ij}|x_j|.$$

Sum over i, then switch the order of summation, and use (6.6) to get

$$|\lambda| \sum_{i=1}^{m} |x_i| \leq \sum_{i=1}^{m}\sum_{j=1}^{m} a_{ij}|x_j| = \sum_{j=1}^{m}\left(\sum_{i=1}^{m} a_{ij}\right)|x_j| = \sum_{j=1}^{m} |x_j|.$$

Dividing both sides by $\sum_{i=1}^{m} |x_i| > 0$ gives $|\lambda| \leq 1$. □

We consider next more general matrix recurrence relations of the form

$$x_n = Ax_{n-1} + b, \quad x_0 \text{ is prescribed}, \tag{6.7}$$

where $x \in R^m$; an $m \times m$ matrix A and a vector $b \in R^m$ are given.

Proposition 6.6.1. *Assume that all eigenvalues of a matrix A have modulus $|\lambda_i| < 1$. Then the iterations in (6.7) converge, and $\lim_{n \to \infty} x_n = (I - A)^{-1}b$.*

Proof. The iterates are $x_1 = Ax_0 + b$, $x_2 = Ax_1 + b = A(Ax_0 + b) + b = A^2x_0 + (I + A)b$, $x_3 = A^3x_0 + (I + A + A^2)b$, and in general,

$$x_n = A^n x_0 + (I + A + A^2 + \cdots + A^{n-1})b.$$

An exercise in Section 6.4 used Jordan normal form to conclude that $A^n \to O$, and $\sum_{k=0}^{n-1} A^k b \to (I - A)^{-1}b$. □

In the complex plane (where the eigenvalues λ_i lie), the condition $|\lambda_i| < 1$ implies that all eigenvalues lie inside of the unit circle around the origin.

Gershgorin's circles

Recall that any complex number $x + iy$ can be represented by a point (x, y) in a plane, called *the complex plane*. Real numbers x lie on the x-axis of the complex plane. Eigenvalues λ of a matrix A are represented by points in the complex plane. The modulus $|\lambda|$ gives the distance from the point λ to the origin of the complex plane.

Given an $m \times m$ matrix $A = [a_{ij}]$, define *Gershgorin's circles* $C_1, C_2, \ldots, C_m$ in the complex plane, where C_i has its center at the diagonal entry a_{ii} (the point $(a_{ii}, 0)$), and its radius is $r_i = \sum_{j \ne i} |a_{ij}|$ (the sum of absolute values along the rest of row i), so that $C_i : |\lambda - a_{ii}| = r_i$.

Theorem 6.6.2 (Gershgorin's circle theorem). *Every eigenvalue of A lies in at least one of the circles $C_1, C_2, \ldots, C_m$.*

Proof. Suppose that λ is any eigenvalue of A, and x is a corresponding eigenvector. Let x_k be the largest in modulus component of x, so that $|x_k| \ge |x_j|$ for all j. Considering the component k of $Ax = \lambda x$ leads to

$$(\lambda - a_{kk})x_k = \sum_{j \ne k} a_{kj}x_j.$$

Take the modulus of both sides, then use the triangle inequality to get

$$|\lambda - a_{kk}| \le \sum_{j \ne k} |a_{kj}| \frac{|x_j|}{|x_k|} \le \sum_{j \ne k} |a_{kj}| = r_k,$$

so that λ lies in C_k (possibly on its rim). □

A square matrix A is called *diagonally dominant* if $|a_{ii}| > \sum_{j \ne i} |a_{ij}|$ (or $|a_{ii}| > r_i$) for all i.

Proposition 6.6.2. *Diagonally dominant matrices are nonsingular (invertible).*

Proof. Gershgorin's circle C_i is centered at a_{ii} which lies at the distance $|a_{ii}|$ from the origin of the complex plane. The radius of C_i is smaller than $|a_{ii}|$. Hence, the origin ($\lambda = 0$) is not included in any of Gershgorin's circles, and then A has no zero eigenvalue. □

Proposition 6.6.3. *Assume that a matrix A is symmetric, diagonally dominant, and it has positive diagonal entries, $a_{ii} > 0$ for all i. Then A is positive definite.*

Proof. Since A is symmetric, its eigenvalues are real, and in fact the eigenvalues are positive because all of Gershgorin's circles lie in the right half of the complex plane. □

Jacobi's iterations

For solving an $m \times m$ system of linear equations,

$$Ax = b, \tag{6.8}$$

Gaussian elimination is fast and efficient, provided that m is not too large (say $m \le 100$). Computers have to round off numbers, in order to store them ($\frac{1}{3} \approx 0.33\ldots3$). Therefore, *round-off errors* occur often in numerical operations. These round-off errors may accumulate for large matrices A (that require many numerical operations), making the answers unreliable. Therefore for large systems, one uses *iterative methods* of the form

$$x_n = Cx_{n-1} + d, \tag{6.9}$$

with an appropriate $m \times m$ matrix C, and $d \in R^m$. If the iterates $x_n \in R^m$ converge to the solution of (6.8), beginning with any $x_0 \in R^m$, the method will be *self-correcting with respect to the round-off errors*. Component i of the system (6.8) is $\sum_{i=1}^{m} a_{ij}x_j = b_i$, or

$$a_{ii}x_i + \sum_{j \neq i} a_{ij}x_j = b_i.$$

Solve for x_i (assuming that $a_{ii} \neq 0$ for all i) to get

$$x_i = \frac{1}{a_{ii}}\left(b_i - \sum_{j \neq i} a_{ij}x_j\right), \tag{6.10}$$

and introduce *Jacobi's iterations* $x_n = \begin{bmatrix} x_1^n \\ \vdots \\ x_i^n \\ \vdots \\ x_m^n \end{bmatrix}$, which are

$$x_i^n = \frac{1}{a_{ii}}\left(b_i - \sum_{j \neq i} a_{ij}x_j^{n-1}\right), \quad n = 1, 2, \ldots. \tag{6.11}$$

(Here x_i^n denotes the component i of the vector x_n.)

Proposition 6.6.4. *If the matrix A is diagonally dominant, Jacobi's iterations (6.11) converge to the solution of (6.8) for any initial vector x_0.*

Proof. Observe that $a_{ii} \neq 0$ for diagonally dominant matrices, so that Jacobi's iterations (6.11) are well defined. Put Jacobi's iterations into the matrix form (6.9). Here the matrix C has zero diagonal entries, $c_{ii} = 0$, and the off-diagonal entries $c_{ij} = -\frac{a_{ij}}{a_{ii}}$. The vector d has components $\frac{b_i}{d_{ii}}$. All of Gershgorin's circles for matrix C are centered at the origin of the complex plane, with the radii $r_i = \frac{1}{|a_{ii}|}\sum_{j \neq i} |a_{ij}| < 1$, because A is diagonally dominant. By Gershgorin's circle theorem, all eigenvalues of matrix C lie inside of the unit circle around the origin, so that they satisfy $|\lambda| < 1$. By Proposition 6.6.1, Jacobi's iterations converge.

Denote $x_i = \lim_{n \to \infty} x_i^n$. Passing to the limit in (6.11) gives (6.10), which is equivalent to (6.8), so that Jacobi's iterations converge to the solution x of (6.8). $\qquad\square$

Exercises

1. Show that every third Fibonacci number is even.
2. Show that $\lim_{n \to \infty} \frac{F_{n+1}}{F_n} = \frac{\sqrt{5}+1}{2}$, the golden section.

3. Solve the difference equation $x_n = 3x_{n-1} - 2x_{n-2}$, with the initial conditions $x_0 = 4$, $x_1 = 5$.
   Answer: $x_n = 3 + 2^n$.

4. (a) Show that the difference equation (6.1), defining the Fibonacci numbers, can be put into the matrix form $x_n = Ax_{n-1}$, with $x_n = \begin{bmatrix} F_n \\ F_{n-1} \end{bmatrix}$ and $A = \begin{bmatrix} 1 & 1 \\ 1 & 0 \end{bmatrix}$, for $n = 3, 4, \ldots$, with $x_2 = \begin{bmatrix} 1 \\ 1 \end{bmatrix}$.
   (b) Conclude that $x_n = A^{n-2} \begin{bmatrix} 1 \\ 1 \end{bmatrix}$.
   (c) Diagonalize A, and obtain another derivation of Binet's formula for F_n.

5. Calculate the $n \times n$ *tridiagonal determinant*

$$
D_n = \begin{vmatrix}
1 & -1 & & & & \\
1 & 1 & -1 & & & \\
& 1 & 1 & -1 & & \\
& & & \ddots & \ddots & \vdots \\
& & & & 1 & 1
\end{vmatrix}.
$$

(Ones on the main diagonal, and on the lower subdiagonal, -1's on the upper subdiagonal. All other entries of D_n are zero.)
Hint. Expand D_4 in the first row

$$
D_4 = \begin{vmatrix}
1 & -1 & 0 & 0 \\
1 & 1 & -1 & 0 \\
0 & 1 & 1 & -1 \\
0 & 0 & 1 & 1
\end{vmatrix} = \begin{vmatrix}
1 & -1 & 0 \\
1 & 1 & -1 \\
0 & 1 & 1
\end{vmatrix} + \begin{vmatrix}
1 & -1 & 0 \\
0 & 1 & -1 \\
0 & 1 & 1
\end{vmatrix}.
$$

The first determinant on the right is D_3. Expand in the first column of the second determinant to get D_2. Hence, $D_4 = D_3 + D_2$. By a similar reasoning, $D_n = D_{n-1} + D_{n-2}$ for $n \geq 3$. Also, $D_1 = 1$, $D_2 = 2$. Then $D_n = F_{n+1}$.

6. (a) From the fact that column 1 + column 2 = 2(column 3) (so that the columns are linearly dependent) determine one of the eigenvalues of the following matrix A and the corresponding eigenvector:

$$
A = \begin{bmatrix}
1/6 & 1/3 & 1/4 \\
1/6 & 2/3 & 5/12 \\
2/3 & 0 & 1/3
\end{bmatrix}.
$$

Answer: $\lambda_1 = 0$, corresponding to $x_1 = \begin{bmatrix} 1 \\ 1 \\ -2 \end{bmatrix}$.
   (b) Verify that A is a Markov matrix.
   (c) Without calculating the characteristic polynomial, determine the other eigenvalues of A.
   Answer: $\lambda_2 = \frac{1}{6}$, $\lambda_3 = 1$.

7. Recall *the Hilbert matrix* with entries $a_{ij} = \frac{1}{i+j-1}$.

(a) Set up Jacobi iterations for the 3×3 Hilbert matrix and an arbitrary vector $b \in R^3$.

(b*) Write a computer program for the general $n \times n$ case.

8. Given two vectors $x, y \in R^n$, we write $x > y$ if $x_i > y_i$ for all components. Similarly, $x \geq y$ if $x_i \geq y_i$ for all i. For example, $x \geq \mathbf{0}$ means that $x_i \geq 0$ for all i.

(a) Suppose that $x \in R^n$ satisfies $x \geq \mathbf{0}$ and $x \neq \mathbf{0}$. Assume that an $n \times n$ matrix A has *positive entries*, so that all $a_{ij} > 0$. Show that $Ax > \mathbf{0}$.

(b*) Justify the following *Perron–Frobenius theorem*. Assume that the $n \times n$ matrix A has positive entries. Then the largest in absolute value eigenvalue is positive and simple (it is a simple root of the corresponding characteristic equation). Every component of the corresponding eigenvector can be chosen to be positive (with a proper factor).

(c) Let A be a Markov matrix with positive entries. Show that Theorem 6.6.1 can be sharpened as follows: A has a simple eigenvalue $\lambda = 1$, and all entries of the corresponding eigenvector are positive; moreover, all other eigenvalues satisfy $|\lambda| < 1$.

(d) If the entries of a Markov matrix are only assumed to be nonnegative, show that it is possible to have other eigenvalues on the circle $|\lambda| = 1$ in the complex plane, in addition to $\lambda = 1$.

Hint. The matrix $B = \begin{bmatrix} 0 & 1 \\ 1 & 0 \end{bmatrix}$ has eigenvalues $1, -1$.

9. Let A be an $n \times n$ Markov matrix, $x \in R^n$, and $y = Ax$. Show that the sum of the entries of y is the same as the sum of the entries of x. Conclude that for any Markov process $x_n = Ax_{n-1}$ the sum of the entries of x_n remains the same for all n.

Hint. Observe that $\sum_{i=1}^n \sum_{j=1}^n a_{ij}x_j = \sum_{j=1}^n \sum_{i=1}^n a_{ij}x_j = \sum_{j=1}^n x_j$.

10. Consider a Markov matrix $A = \begin{bmatrix} \frac{1}{2} & 0 & \frac{1}{2} \\ 0 & \frac{1}{2} & \frac{1}{2} \\ \frac{1}{2} & \frac{1}{2} & 0 \end{bmatrix}$.

(a) Show that for any $x_0 \in R^3$, $\lim_{n \to \infty} A^n x_0 = a \begin{bmatrix} 1/3 \\ 1/3 \\ 1/3 \end{bmatrix}$, where a is the sum of the entries of x_0.

Hint. Matrix A has an eigenvalue $\lambda = 1$ with an eigenvector $\begin{bmatrix} 1/3 \\ 1/3 \\ 1/3 \end{bmatrix}$, and the eigenvalues $\pm \frac{1}{2}$.

(b) Consider $A_0 = \begin{bmatrix} 1/3 & 1/3 & 1/3 \\ 1/3 & 1/3 & 1/3 \\ 1/3 & 1/3 & 1/3 \end{bmatrix}$. Show that $\lim_{n \to \infty} A^n = A_0$.

Hint. Show that for any vector $x_0 \in R^3$, $\lim_{n \to \infty} A^n x_0 = A_0 x_0$.

11. Draw Gershgorin's circles for the matrix $A = \begin{bmatrix} -3 & 1 & 1 \\ 1 & 4 & -1 \\ 0 & 2 & 4 \end{bmatrix}$. Is A diagonally dominant?

7 Applications to Calculus and Differential Geometry

Linear Algebra has many uses in diverse areas of science, engineering, economics, image processing, etc. It is perhaps ironic that applications to other branches of mathematics are often neglected. In this chapter we use Hessian matrices to develop Taylor's series for functions of many variables, leading to the second derivative test for extrema. In the process, Sylvester's test is covered, thus adding to the theory of positive definite matrices. Application to Differential Geometry is also "a two-way street," with generalized eigenvalue problem and generalized Rayleigh quotient deepening our understanding of the standard topics.

7.1 Hessian matrix

In this section we use positive definite matrices to determine minima and maxima of functions with two or more variables. But first a useful decomposition of symmetric matrices is discussed.

$A = LDL^T$ decomposition

Assume that Gaussian elimination can be performed for an $n \times n$ matrix A without any row exchanges, and $|A| \neq 0$. Recall that in such a case one can decompose $A = LU$, where L is a lower triangular matrix with the diagonal entries equal to 1, and U is an upper triangular matrix with the diagonal entries equal to the pivots of A, denoted by $d_1, d_2, \ldots, d_n$. Observe that all $d_i \neq 0$, because $|A| = d_1 d_2 \cdots d_n \neq 0$. The decomposition $A = LU$ is unique.

Write $U = DU_1$, where D is a diagonal matrix, with the diagonal entries equal to $d_1, d_2, \ldots, d_n$, and U_1 is another upper triangular matrix with the diagonal entries equal to 1. (Row i of U_1 is obtained by dividing row i of U by d_i.) Then

$$A = LDU_1, \tag{1.1}$$

and this decomposition (known as *LDU decomposition*) is unique.

Now suppose, in addition, that A is symmetric, so that $A^T = A$. Then

$$A = A^T = (LDU_1)^T = U_1^T DL^T,$$

where U_1^T is lower triangular and L^T is upper triangular. Comparison with (1.1) gives $U_1^T = L$ and $L^T = U_1$, since the decomposition (1.1) is unique. We conclude that any symmetric matrix A, with $|A| \neq 0$, can be decomposed as

$$A = LDL^T, \tag{1.2}$$

https://doi.org/10.1515/9783111086507-007

where L is a lower triangular matrix with the diagonal entries equal to 1, and D is a diagonal matrix, provided that no row exchanges are needed in the row reduction of A. The diagonal entries of D are the nonzero pivots of A.

Sylvester's criterion

For an $n \times n$ matrix

$$A = \begin{bmatrix} a_{11} & a_{12} & a_{13} & \cdots & a_{1n} \\ a_{21} & a_{22} & a_{23} & \cdots & a_{2n} \\ a_{31} & a_{32} & a_{33} & \cdots & a_{3n} \\ \vdots & \vdots & \vdots & \ddots & \vdots \\ a_{n1} & a_{n2} & a_{n3} & \cdots & a_{nn} \end{bmatrix},$$

the submatrices

$$A_1 = a_{11}, \; A_2 = \begin{bmatrix} a_{11} & a_{12} \\ a_{21} & a_{22} \end{bmatrix}, \; A_3 = \begin{bmatrix} a_{11} & a_{12} & a_{13} \\ a_{21} & a_{22} & a_{23} \\ a_{31} & a_{32} & a_{33} \end{bmatrix}, \ldots, A_n = A$$

are called *the principal submatrices*. The determinants of the principal submatrices

$$|A_1| = a_{11}, \; |A_2| = \begin{vmatrix} a_{11} & a_{12} \\ a_{21} & a_{22} \end{vmatrix}, \; |A_3| = \begin{vmatrix} a_{11} & a_{12} & a_{13} \\ a_{21} & a_{22} & a_{23} \\ a_{31} & a_{32} & a_{33} \end{vmatrix}, \ldots, |A_n| = |A|$$

are called *the principal minors*.

Theorem 7.1.1 (Sylvester's criterion). *A symmetric matrix A is positive definite if and only if all of its principal minors are positive.*

Proof. Assume that A is positive definite, so that $Ax \cdot x > 0$ for all $x \neq 0$. Let $x = e_1$, the first coordinate vector in R^n. Then $|A_1| = a_{11} = Ae_1 \cdot e_1 > 0$. (Using $x = e_i$, one can similarly conclude that $a_{ii} > 0$ for all i.) Let now $x = \begin{bmatrix} x_1 \\ x_2 \\ 0 \\ \vdots \\ 0 \end{bmatrix}$. Then $0 < Ax \cdot x = A_2 \begin{bmatrix} x_1 \\ x_2 \end{bmatrix} \cdot \begin{bmatrix} x_1 \\ x_2 \end{bmatrix}$, for any vector $\begin{bmatrix} x_1 \\ x_2 \end{bmatrix} \in R^2$. It follows that the 2×2 matrix A_2 is positive definite, and then both eigenvalues of A_2 are positive, so that $|A_2| > 0$ as the product of positive eigenvalues. Using $x = \begin{bmatrix} x_1 \\ x_2 \\ x_3 \\ 0 \\ \vdots \\ 0 \end{bmatrix}$, one concludes similarly that $|A_3| > 0$, and so on.

Conversely, assume that all principal minors are positive. Let us apply Gaussian elimination to A. We claim that all pivots are positive. (We shall show that all diagonal entries obtained in the process of row reduction are positive.) Indeed, the first pivot d_1 is the first principal minor $a_{11} > 0$. If d_2 denotes the second pivot, then

$$0 < |A_2| = d_1 d_2,$$

so that $d_2 > 0$. (Gaussian elimination reduces A_2 to an upper triangular matrix with d_1 and d_2 on the diagonal.) Similarly,

$$0 < |A_3| = d_1 d_2 d_3,$$

implying that the third pivot d_3 is positive, and so on.

Since all pivots of A are positive, no row exchanges are needed in Gaussian elimination, and by (1.2) we can decompose $A = LDL^T$, where L is a lower triangular matrix with the diagonal entries equal to 1, and D is a diagonal matrix. The diagonal entries of D are the positive pivots $d_1, d_2, \ldots, d_n$ of A.

For any $x \neq 0$, let $y = L^T x$. Observe that $y \neq 0$, since otherwise $x = (L^T)^{-1} y = 0$, a contradiction (L^T is invertible, because $|L^T| = 1 \neq 0$). Then

$$Ax \cdot x = LDL^T x \cdot x = DL^T x \cdot L^T x = Dy \cdot y$$
$$= d_1 y_1^2 + d_2 y_2^2 + \cdots + d_n y_n^2 > 0,$$

so that A is positive definite. □

A symmetric matrix A is called *negative definite* if $-A$ is positive definite, which implies that $(-A)x \cdot x > 0$ for all $x \neq 0$, and that all eigenvalues of $-A$ are positive. It follows that for a negative definite matrix $Ax \cdot x < 0$ for all $x \neq 0$, and that all eigenvalues of A are negative.

Theorem 7.1.2. *A symmetric matrix A is negative definite if and only if all of its principal minors satisfy $(-1)^k |A_k| > 0$.*

Proof. Assume that $(-1)^k |A_k| > 0$ for all k. The principal minors of the matrix $-A$ are

$$|(-A)_k| = |-A_k| = (-1)^k |A_k| > 0, \quad \text{for all } k.$$

By Sylvester's criterion, the matrix $-A$ is positive definite. The converse statement follows by reversing this argument. □

The second derivative test

By Taylor's formula, any twice continuously differentiable function can be approximated around an arbitrary point x_0 as

$$f(x) \approx f(x_0) + f'(x_0)(x - x_0) + \frac{1}{2} f''(x_0)(x - x_0)^2, \tag{1.3}$$

for x near x_0. If x_0 is a critical point, where $f'(x_0) = 0$, one has

$$f(x) \approx f(x_0) + \frac{1}{2}f''(x_0)(x - x_0)^2. \tag{1.4}$$

In case $f''(x_0) > 0$, it follows that $f(x) > f(x_0)$ for x near x_0, so that x_0 is a point of local minimum. If $f''(x_0) < 0$, then x_0 is a point of local maximum. Setting $x = x_0 + h$, one can rewrite (1.3) as

$$f(x_0 + h) \approx f(x_0) + f'(x_0)h + \frac{1}{2}f''(x_0)h^2,$$

for |h| small.

Let now $f(x,y)$ be twice continuously differentiable function of two variables. Taylor's approximation near a fixed point (x_0, y_0) uses partial derivatives (here $f_x = \frac{\partial f}{\partial x}$, $f_{xx} = \frac{\partial^2 f}{\partial x^2}$, etc.)

$$f(x_0 + h_1, y_0 + h_2) \approx f(x_0, y_0) + f_x(x_0, y_0)h_1 + f_y(x_0, y_0)h_2$$
$$+ \frac{1}{2}[f_{xx}(x_0, y_0)h_1^2 + 2f_{xy}(x_0, y_0)h_1 h_2 + f_{yy}(x_0, y_0)h_2^2], \tag{1.5}$$

provided that *both $|h_1|$ and $|h_2|$ are small*. If (x_0, y_0) is *a critical point*, where $f_x(x_0, y_0) = f_y(x_0, y_0) = 0$, then

$$f(x_0 + h_1, y_0 + h_2) \approx f(x_0, y_0)$$
$$+ \frac{1}{2}[f_{xx}(x_0, y_0)h_1^2 + 2f_{xy}(x_0, y_0)h_1 h_2 + f_{yy}(x_0, y_0)h_2^2]. \tag{1.6}$$

The second term on the right is $\frac{1}{2}$ times a *quadratic form in h_1, h_2* with the matrix

$$H = \begin{bmatrix} f_{xx}(x_0, y_0) & f_{xy}(x_0, y_0) \\ f_{xy}(x_0, y_0) & fyy(x_0, y_0) \end{bmatrix}$$

called *the Hessian matrix.*

(One can also write $H = \begin{bmatrix} f_{xx}(x_0,y_0) & f_{xy}(x_0,y_0) \\ f_{yx}(x_0,y_0) & fyy(x_0,y_0) \end{bmatrix}$, because $f_{yx}(x_0, y_0) = f_{xy}(x_0, y_0)$. Observe also that $H = H(x_0, y_0)$.)

Introducing the vector $h = \begin{bmatrix} h_1 \\ h_2 \end{bmatrix}$, one can write the quadratic form in (1.6) as $\frac{1}{2}Hh \cdot h$. Then (1.6) takes the form

$$f(x_0 + h_1, y_0 + h_2) \approx f(x_0, y_0) + \frac{1}{2}Hh \cdot h.$$

If the Hessian matrix H is positive definite, so that $Hh \cdot h > 0$ for all $h \neq 0$, then for all h_1 and h_2, with $|h_1|$ and $|h_2|$ small

$$f(x_0 + h_1, y_0 + h_2) > f(x_0, y_0).$$

It follows that $f(x, y) > f(x_0, y_0)$ for all points (x, y) near (x_0, y_0), so that (x_0, y_0) is a point of local minimum. By Sylvester's criterion, Theorem 7.1.1, H is positive definite provided that $f_{xx}(x_0, y_0) > 0$ and

$$f_{xx}(x_0, y_0)f_{yy}(x_0, y_0) - f_{xy}^2(x_0, y_0) > 0.$$

If the Hessian matrix H is negative definite, then for all h_1 and h_2, with $|h_1|$ and $|h_2|$ small

$$f(x_0 + h_1, y_0 + h_2) < f(x_0, y_0),$$

and (x_0, y_0) is a point of local maximum. By Theorem 7.1.2, H is negative definite provided that $f_{xx}(x_0, y_0) < 0$ and

$$f_{xx}(x_0, y_0)f_{yy}(x_0, y_0) - f_{xy}^2(x_0, y_0) > 0.$$

A symmetric matrix A is called indefinite provided that the quadratic form $Ah \cdot h$ takes on both positive and negative values. This happens when A has both positive and negative eigenvalues (as follows by diagonalization of $Ah \cdot h$). If the Hessian matrix $H(x_0, y_0)$ is indefinite, there is no extremum of $f(x, y)$ at (x_0, y_0). One says that (x_0, y_0) is a saddle point. A saddle point occurs provided that

$$f_{xx}(x_0, y_0)f_{yy}(x_0, y_0) - f_{xy}^2(x_0, y_0) < 0.$$

Indeed, this quantity gives the determinant of $H(x_0, y_0)$, which equals to the product of the eigenvalues of $H(x_0, y_0)$, so that the eigenvalues of $H(x_0, y_0)$ have opposite signs.

For functions of more than two variables, it is convenient to use vector notation. If $f = f(x_1, x_2, \ldots, x_n)$, we define a row vector $x = (x_1, x_2, \ldots, x_n)$, and then $f = f(x)$. Taylor's formula around some point $x_0 = (x_1^0, x_2^0, \ldots, x_n^0)$ takes the form

$$f(x) \approx f(x_0) + \nabla f(x_0) \cdot (x - x_0) + \frac{1}{2}H(x_0)(x - x_0) \cdot (x - x_0),$$

for x close to x_0 (i. e., when the distance $\|x - x_0\|$ is small). Here $\nabla f(x_0) = \begin{bmatrix} f_{x_1}(x_0) \\ f_{x_2}(x_0) \\ \vdots \\ f_{x_n}(x_0) \end{bmatrix}$ is the gradient vector, and

$$H(x_0) = \begin{bmatrix} f_{x_1 x_1}(x_0) & f_{x_1 x_2}(x_0) & \cdots & f_{x_1 x_n}(x_0) \\ f_{x_2 x_1}(x_0) & f_{x_2 x_2}(x_0) & \cdots & f_{x_2 x_n}(x_0) \\ \vdots & \vdots & \ddots & \vdots \\ f_{x_n x_1}(x_0) & f_{x_n x_2}(x_0) & \cdots & f_{x_n x_n}(x_0) \end{bmatrix}$$

is the Hessian matrix.

A point x_0 is called critical if $\nabla f(x_0) = 0$, or in components,

$$f_{x_1}(x_1^0, x_2^0, \ldots, x_n^0) = f_{x_2}(x_1^0, x_2^0, \ldots, x_n^0) = \cdots = f_{x_n}(x_1^0, x_2^0, \ldots, x_n^0) = 0.$$

At a critical point,

$$f(x) \approx f(x_0) + \frac{1}{2}H(x_0)(x - x_0) \cdot (x - x_0),$$

for x near x_0. So that x_0 is a point of minimum of $f(x)$ if the Hessian matrix $H(x_0)$ is positive definite, and x_0 is a point of maximum if $H(x_0)$ is negative definite. Sylvester's criterion and Theorem 7.1.2 give a straightforward way to decide. If $H(x_0)$ is indefinite then x_0 is called *a saddle point* (there is no extremum at x_0).

Example. Let $f(x, y, z) = 2x^2 - xy + 2xz + 2y^2 + yz + z^2 + 3y$.
To find the critical points, set the first partials to zero:

$$f_x = 4x - y + 2z = 0,$$
$$f_y = -x + 4y + z + 3 = 0,$$
$$f_z = 2x + y + 2z = 0.$$

This 3×3 linear system has a unique solution $x = -2, y = -2, z = 3$. To apply the second derivative test at the point $(-2, -2, 3)$, calculate the Hessian matrix $H(-2, -2, 3) = \begin{bmatrix} 4 & -1 & 2 \\ -1 & 4 & 1 \\ 2 & 1 & 2 \end{bmatrix}$. Its principal minors, which are 4, 15, 6, are all positive, and hence $H(-2, -2, 3)$ is positive definite by Sylvester's criterion. There is a local minimum at the point $(-2, -2, 3)$, and since there are no other critical points, this is the point of global minimum.

Exercises

1. (a) If a matrix A is positive definite, show that $a_{ii} > 0$ for all i.
   Hint. Observe $a_{ii} = Ae_i \cdot e_i$, where e_i is the coordinate vector.
   (b) If a 5×5 matrix A is positive definite, show that the submatrix $\begin{bmatrix} a_{22} & a_{24} \\ a_{42} & a_{44} \end{bmatrix}$ is also positive definite.
   Hint. Consider $Ax \cdot x$, where $x \in R^5$ has $x_1 = x_3 = x_5 = 0$.
   (c) Show that all other submatrices of the form $\begin{bmatrix} a_{ii} & a_{ij} \\ a_{ji} & a_{jj} \end{bmatrix}$ are positive definite, $1 \le i < j \le 5$.

2. By inspection (just by looking) determine why the following matrices are not positive definite:
   (a) $\begin{bmatrix} 5 & 2 & 1 \\ 2 & 1 & -1 \\ 1 & -1 & -2 \end{bmatrix}$.
   (b) $\begin{bmatrix} 5 & 2 & 0 \\ 2 & 1 & 1 \\ 0 & 1 & 0 \end{bmatrix}$.
   (c) $\begin{bmatrix} 5 & 3 & 1 \\ 3 & 1 & 1 \\ 1 & 2 & 8 \end{bmatrix}$.

(d) $\begin{bmatrix} 4 & 2 & 1 \\ 2 & 1 & -1 \\ 1 & -1 & 2 \end{bmatrix}$.

3. Determine if the following symmetric matrices are positive definite, negative definite, indefinite, or none of the above:

   (a) $\begin{bmatrix} 4 & 2 \\ 2 & 2 \end{bmatrix}$.

   Answer: Positive definite.

   (b) $\begin{bmatrix} -4 & 1 \\ 1 & -3 \end{bmatrix}$.

   Answer: Negative definite.

   (c) $\begin{bmatrix} 4 & 3 \\ 3 & -4 \end{bmatrix}$.

   Answer: Indefinite.

   (d) $\begin{bmatrix} 4 & 2 \\ 2 & 1 \end{bmatrix}$.

   Answer: None of the above. (This matrix is positive semidefinite.)

4. (a) Use Gershgorin's circle theorem to confirm that the following symmetric matrix is positive definite:

$$\begin{bmatrix} 4 & 2 & 0 & -1 \\ 2 & 7 & -1 & 3 \\ 0 & -1 & 6 & 2 \\ -1 & 3 & 2 & 7 \end{bmatrix}.$$

   Hint. Show that all eigenvalues are positive.

   (b) Use Sylvester's criterion on the same matrix.

5. Determine the critical points of the following functions, and examine them by the second derivative test:

   (a) $f(x,y,z) = x^3 + 30xy + 3y^2 + z^2$.

   Answer: Saddle point at $(0,0,0)$, and a point of minimum at $(50, -250, 0)$. (The Hessian matrix $H(0,0,0)$ has eigenvalues $3 - 3\sqrt{101}, 3 + 3\sqrt{101}, 2$.)

   (b) $f(x,y,z) = -x^2 - 2y^2 - z^2 + xy + 2xz$.

   Answer: Saddle point at $(0,0,0)$.

   (c) $f(x,y,z) = -x^2 - 2y^2 - 4z^2 + xy + 2xz$.

   Answer: Point of maximum at $(0,0,0)$.

   (d) $f(x,y) = xy + \frac{20}{x} + \frac{50}{y}$.

   Answer: Point of minimum at $(2,5)$.

   (e) $f(x,y,z) = \frac{y^2}{2x} + 2x + \frac{2z^2}{y} + \frac{4}{z}$.

   Answer: Point of minimum at $(\frac{1}{2}, 1, 1)$, point of maximum at $(-\frac{1}{2}, -1, -1)$.

   (f*) $f(x_1, x_2, \ldots, x_n) = x_1 + \frac{x_2}{x_1} + \frac{x_3}{x_2} + \cdots + \frac{x_n}{x_{n-1}} + \frac{2}{x_n}$, $x_i > 0$ for all i.

   Answer: Global minimum of $(n + 1)2^{\frac{1}{n+1}}$ occurs at the point $x_1 = 2^{\frac{1}{n+1}}$, $x_2 = x_1^2, \ldots, x_n = x_1^n$.

6. Find the maximum value of

$$f(x,y,z) = \sin x + \sin y + \sin z - \sin(x + y + z)$$

over the cube $0 < x < \pi, 0 < y < \pi, 0 < z < \pi$.

Answer: The point of maximum is $(\frac{\pi}{2}, \frac{\pi}{2}, \frac{\pi}{2})$, the maximum value is 4.

7. (The second derivative test) Let (x_0, y_0) be a critical point of $f(x, y)$, so that $f_x(x_0, y_0) = f_y(x_0, y_0) = 0$. Let $D = f_{xx}(x_0, y_0)f_{yy}(x_0, y_0) - f_{xy}^2(x_0, y_0)$. Show that

(a) If $D > 0$ and $f_{xx}(x_0, y_0) > 0$, then (x_0, y_0) is a point of minimum.

(b) If $D > 0$ and $f_{xx}(x_0, y_0) < 0$, then (x_0, y_0) is a point of maximum.

(c) If $D < 0$, then (x_0, y_0) is a saddle point.

8. Find the $A = LDU$ decomposition of the following matrices:

(a) $A = \begin{bmatrix} 1 & 2 \\ 3 & 4 \end{bmatrix}$.

Answer: $L = \begin{bmatrix} 1 & 0 \\ 3 & 1 \end{bmatrix}$, $D = \begin{bmatrix} 1 & 0 \\ 0 & -2 \end{bmatrix}$, $U = \begin{bmatrix} 1 & 2 \\ 0 & 1 \end{bmatrix}$.

(b) $A = \begin{bmatrix} 1 & 2 & 1 \\ -2 & -1 & 1 \\ 1 & -1 & 0 \end{bmatrix}$.

Answer: $L = \begin{bmatrix} 1 & 0 & 0 \\ -2 & 1 & 0 \\ 1 & -1 & 1 \end{bmatrix}$, $D = \begin{bmatrix} 1 & 0 & 0 \\ 0 & 3 & 0 \\ 0 & 0 & 2 \end{bmatrix}$, and $U = \begin{bmatrix} 1 & 2 & 1 \\ 0 & 1 & 1 \\ 0 & 0 & 1 \end{bmatrix}$.

9. Find the $A = LDL^T$ decomposition of the following symmetric matrices:

(a) $A = \begin{bmatrix} 1 & -1 & 1 \\ -1 & 4 & -4 \\ 1 & -4 & 6 \end{bmatrix}$.

Answer: $L = \begin{bmatrix} 1 & 0 & 0 \\ -1 & 1 & 0 \\ 1 & -1 & 1 \end{bmatrix}$, $D = \begin{bmatrix} 1 & 0 & 0 \\ 0 & 3 & 0 \\ 0 & 0 & 2 \end{bmatrix}$.

(b) $A = \begin{bmatrix} 1 & -1 & 2 \\ -1 & 2 & -3 \\ 2 & -3 & 6 \end{bmatrix}$.

Answer: $L = \begin{bmatrix} 1 & 0 & 0 \\ -1 & 1 & 0 \\ 2 & -1 & 1 \end{bmatrix}$, $D = I$.

7.2 Jacobian matrix

For vector functions of multiple arguments, the central role of a derivative is played by the Jacobian matrix studied in this section.

The inverse function of $y = 5x$ is $x = \frac{1}{5}y$. How about $y = x^2$? The inverse function cannot be $x = \pm\sqrt{y}$ because functions have unique values. So that the function $f(x) = x^2$ does not have an inverse function, which is valid for all x. Let us try to invert this function near $x = 1$. Near that point, both x and y are positive, so that $x = \sqrt{y}$ gives the inverse function. (Near $x = 0$ one cannot invert $y = x^2$, since there are both positive and negative x's near $x = 0$, and the formula would have to be $x = \pm\sqrt{y}$, which is not a function.) Observe that $f'(1) = 2 > 0$ so that $f(x)$ is increasing near $x = 1$. It follows that all y's near $f(1) = 1$ come from a unique x, and the inverse function exists. At $x = 0$, $f'(0) = 0$, and there is no inverse function. Recall *the inverse function theorem* from Calculus: if $y = f(x)$ is defined on some interval around x_0, with $y_0 = f(x_0)$, and $f'(x_0) \neq 0$, then the inverse function $x = x(y)$ exists on some interval around y_0.

Now suppose there is a map $(x, y) \rightarrow (u, v)$,

$$u = f(x, y),$$
$$v = g(x, y),$$

(2.1)

given by continuously differentiable functions $f(x, y)$ and $g(x, y)$, and let us try to solve for x and y in terms of u and v. What will replace the notion of the derivative in this case? The matrix of partial derivatives

$$J(x,y) = \begin{bmatrix} f_x(x,y) & f_y(x,y) \\ g_x(x,y) & g_y(x,y) \end{bmatrix}$$

is called *the Jacobian matrix*. Its determinant $|J(x,y)|$ is called *the Jacobian determinant*. Suppose that

$$u_0 = f(x_0, y_0),$$
$$v_0 = g(x_0, y_0),$$

and the Jacobian determinant

$$|J(x_0,y_0)| = \begin{vmatrix} f_x(x_0,y_0) & f_y(x_0,y_0) \\ g_x(x_0,y_0) & g_y(x_0,y_0) \end{vmatrix} \neq 0.$$

The inverse function theorem asserts that for (u,v) lying in a sufficiently small disk around the point (u_0, v_0), one can solve the system (2.1) for $x = \varphi(u,v)$ and $y = \psi(u,v)$, with two continuously differentiable functions $\varphi(u,v)$ and $\psi(u,v)$ satisfying $x_0 = \varphi(u_0, v_0)$ and $y_0 = \psi(u_0, v_0)$. Moreover, the system (2.1) has no other solutions near the point (u_0, v_0). The proof of this theorem can be found, for example, in the book of V. I. Arnold [1].

Example 1. The surface

$$x = u^3 v + 2u + 2,$$
$$y = u + v,$$
$$z = 3u - v^2$$

passes through the point $(2, 1, -1)$ when $u = 0$ and $v = 1$. The Jacobian matrix of the first two of these equations is

$$J(u,v) = \begin{bmatrix} \dfrac{\partial x}{\partial u} & \dfrac{\partial x}{\partial v} \\ \dfrac{\partial y}{\partial u} & \dfrac{\partial y}{\partial v} \end{bmatrix} = \begin{bmatrix} 3u^2 v + 2 & u^3 \\ 1 & 1 \end{bmatrix}.$$

Since the Jacobian determinant

$$|J(0,1)| = \begin{vmatrix} 2 & 0 \\ 1 & 1 \end{vmatrix} = 2 \neq 0,$$

it follows by the inverse function theorem that we can solve the first two equations for u and v as functions of x and y (near the point $(x,y) = (2,1)$), obtaining $u = \varphi(x,y)$ and $v = \psi(x,y)$, and then use these functions in the third equation to express z as a function of x and y. Conclusion: near the point $(2, 1, -1)$ this surface can be represented in the form $z = F(x, y)$ with some function $F(x, y)$.

More generally, for a surface

$$x = x(u, v),$$
$$y = y(u, v),$$
$$z = z(u, v),$$

with given functions $x(u, v), y(u, v), z(u, v)$, assume that the rank of the Jacobian matrix

$$J(u, v) = \begin{bmatrix} \dfrac{\partial x}{\partial u} & \dfrac{\partial x}{\partial v} \\[2mm] \dfrac{\partial y}{\partial u} & \dfrac{\partial y}{\partial v} \\[2mm] \dfrac{\partial z}{\partial u} & \dfrac{\partial z}{\partial v} \end{bmatrix}$$

is two. Then a pair of rows of $J(u, v)$ is linearly independent, and we can express one of the coordinates (x, y, z) through the other two. Indeed, if say rows 1 and 3 are linearly independent, then $\begin{vmatrix} \frac{\partial x}{\partial u} & \frac{\partial x}{\partial v} \\ \frac{\partial z}{\partial u} & \frac{\partial z}{\partial v} \end{vmatrix} \neq 0$, and we can express u, v through x, z by the inverse function theorem. Then from row 2 obtain $y = F(x, z)$ with some function $F(x, z)$ (near some point (x_0, y_0, z_0)).

For a map $R^3 \to R^3$ given by $(f(x, y, z), g(x, y, z), h(x, y, z))$, with some differentiable functions $f(x, y, z), g(x, y, z), h(x, y, z)$, the Jacobian matrix takes the form

$$J(x, y, z) = \begin{bmatrix} f_x(x, y, z) & f_y(x, y, z) & f_z(x, y, z) \\ g_x(x, y, z) & g_y(x, y, z) & g_z(x, y, z) \\ h_x(x, y, z) & h_y(x, y, z) & h_z(x, y, z) \end{bmatrix},$$

and the statement of the inverse function theorem is similar.

Recall that Jacobian determinants also occur in Calculus when one changes coordinates in double and triple integrals. If

$$x = x(u, v),$$
$$y = y(u, v)$$

is a one-to-one map taking a region D in the uv-plane onto a region R of the xy-plane, then

$$\iint_R f(x, y)\, dxdy = \iint_D f(x(u, v), y(u, v)) \, \left| \begin{matrix} \dfrac{\partial x}{\partial u} & \dfrac{\partial x}{\partial v} \\[2mm] \dfrac{\partial y}{\partial u} & \dfrac{\partial y}{\partial v} \end{matrix} \right| \, dudv.$$

Here the absolute value of the Jacobian determinant gives the *magnification factor* of the element of area. This formula is justified in exercises of Section 7.4.

Example 2. If one switches from the Cartesian to polar coordinates, $x = r \cos \theta$, $y = r \sin \theta$, then

$$\begin{vmatrix} \frac{\partial x}{\partial r} & \frac{\partial x}{\partial \theta} \\ \frac{\partial y}{\partial r} & \frac{\partial y}{\partial \theta} \end{vmatrix} = \begin{vmatrix} \cos \theta & -r \sin \theta \\ \sin \theta & r \cos \theta \end{vmatrix} = r,$$

leading to the familiar formula $dxdy = r\, dr d\theta$ for double integrals.

Example 3. Evaluate $\iint_R \sqrt{1 - \frac{x^2}{a^2} - \frac{y^2}{b^2}}\, dxdy$, over an elliptical region R, $\frac{x^2}{a^2} + \frac{y^2}{b^2} \leq 1$, with $a > 0$, $b > 0$.

Use the map $x = au$, $y = bv$, taking R onto the unit disc D, $u^2 + v^2 \leq 1$. The Jacobian determinant is

$$\begin{vmatrix} \frac{\partial x}{\partial u} & \frac{\partial x}{\partial v} \\ \frac{\partial y}{\partial u} & \frac{\partial y}{\partial v} \end{vmatrix} = \begin{vmatrix} a & 0 \\ 0 & b \end{vmatrix} = ab.$$

Then

$$\iint_R \sqrt{1 - \frac{x^2}{a^2} - \frac{y^2}{b^2}}\, dxdy = ab \iint_D \sqrt{1 - u^2 - v^2}\, dudv$$

$$= ab \int_0^{2\pi} \int_0^1 \sqrt{1 - r^2}\, r\, dr d\theta = \frac{2}{3}\pi ab,$$

using polar coordinates in the uv-plane on the second step.

Exercises

1. Consider the map $(x, y) \to (u, v)$, $u = x^3 + y^2$, $v = ye^x + 1$. For a given point (x_0, y_0), calculate the corresponding point (u_0, v_0). Determine if the inverse function theorem (IFT) applies, and if it does, state its conclusion.
   (a) $(x_0, y_0) = (0, 0)$.
      Answer: IFT does not apply.
   (b) $(x_0, y_0) = (0, 1)$.
      Answer: IFT applies. An inverse map $x = \varphi(u, v)$, $y = \psi(u, v)$ exists for (u, v) near $(u_0, v_0) = (1, 2)$. Also, $\varphi(1, 2) = 0$ and $\psi(1, 2) = 1$.
   (c) $(x_0, y_0) = (1, 0)$.
2. (a) Consider a map $(u, v) \to (x, y)$ given by some functions $x = x(u, v)$, $y = y(u, v)$, and a map $(p, q) \to (u, v)$ given by $u = u(p, q)$, $v = v(p, q)$. Together they define a composite map $(p, q) \to (x, y)$, given by $x = x(u(p, q), v(p, q))$, $y = y(u(p, q), v(p, q))$. Justify the chain rule for Jacobian matrices

$$\begin{bmatrix} x_p & x_q \\ y_p & y_q \end{bmatrix} = \begin{bmatrix} x_u & x_v \\ y_u & y_v \end{bmatrix} \begin{bmatrix} u_p & u_q \\ v_p & v_q \end{bmatrix}.$$

(b) Assume that a map $(u, v) \rightarrow (x, y)$, given by $x = f(u, v)$ and $y = g(u, v)$, has *an inverse map* $(x, y) \rightarrow (u, v)$, given by $u = p(x, y)$ and $v = q(x, y)$, so that $x = f(p(x, y), q(x, y)), y = g(p(x, y), q(x, y))$. Show that

$$\begin{bmatrix} p_x & p_y \\ q_x & q_y \end{bmatrix} = \begin{bmatrix} f_u & f_v \\ g_u & g_v \end{bmatrix}^{-1}.$$

3. (a) If one switches from Cartesian coordinates to the spherical ones,

$$x = \rho \sin \varphi \cos \theta,$$
$$y = \rho \sin \varphi \sin \theta,$$
$$z = \rho \cos \varphi,$$

with $\rho > 0, 0 \le \theta \le 2\pi, 0 \le \varphi \le \pi$, show that the absolute value of the Jacobian determinant is

$$\left| \begin{matrix} \frac{\partial x}{\partial \rho} & \frac{\partial x}{\partial \theta} & \frac{\partial x}{\partial \varphi} \\ \frac{\partial y}{\partial \rho} & \frac{\partial y}{\partial \theta} & \frac{\partial y}{\partial \varphi} \\ \frac{\partial z}{\partial \rho} & \frac{\partial z}{\partial \theta} & \frac{\partial z}{\partial \varphi} \end{matrix} \right| = \rho^2 \sin \varphi,$$

and conclude that $dxdydz = \rho^2 \sin \varphi \, d\rho d\theta d\varphi$ for triple integrals.

(b) Evaluate $\iiint_V \sqrt{1 - \frac{x^2}{a^2} - \frac{y^2}{b^2} - \frac{z^2}{c^2}} \, dxdydz$ over an ellipsoidal region V, $\frac{x^2}{a^2} + \frac{y^2}{b^2} + \frac{z^2}{c^2} \le 1$.

(c) Find the volume of the ellipsoid V.

Answer: $\frac{4}{3}\pi abc$.

4. (a) Sketch the parallelogram R bounded by the lines $-x + y = 0, -x + y = 1, 2x + y = 2,$ and $2x + y = 4$.

(b) Show that the map $(x, y) \rightarrow (u, v)$, given by $u = -x + y$ and $v = 2x + y$, takes R onto a rectangle D, $0 \le u \le 1, 2 \le v \le 4$, and the Jacobian determinant $\left| \begin{matrix} u_x & u_y \\ v_x & v_y \end{matrix} \right|$ is -3.

(c) Show that the inverse map $(u, v) \rightarrow (x, y)$ taking D onto R is given by $x = -\frac{1}{3}u + \frac{1}{3}v, y = \frac{2}{3}u + \frac{1}{3}v$, and the Jacobian determinant $\left| \begin{matrix} x_u & x_v \\ y_u & y_v \end{matrix} \right| = -\frac{1}{3}$.

(d) Evaluate $\iint_R (x + 2y) \, dxdy$.
Hint. Reduce this integral to $\frac{1}{3} \iint_D (u + v) \, dudv$.

7.3 Curves and surfaces

We now review the notions of arc length and curvature for curves in two and three dimensions, and of coordinate curves and tangent planes for surfaces.

Curves

Parametric equations of a circle of radius 2 around the origin in the xy-plane can be written as

$$x = 2\cos t,$$
$$y = 2\sin t, \quad 0 \le t \le 2\pi.$$

Indeed, here $x^2 + y^2 = 4\cos^2 t + 4\sin^2 t = 4$. As the parameter t varies over the interval $[0, 2\pi]$, the point (x, y) traces out this circle, moving counterclockwise. The polar angle of the point (x, y) is equal to t (since $\frac{y}{x} = \tan t$). Consider a vector $y(t) = (2\cos t, 2\sin t)$. As t varies from 0 to 2π, the tip of $y(t)$ traces out the circle $x^2 + y^2 = 4$. Thus $y(t)$ *represents this circle*. Similarly, $y_1(t) = (2\cos t, \sin t)$ represents the ellipse $\frac{x^2}{4} + y^2 = 1$.

A vector function $y(t) = (f(t), g(t), h(t))$ with given functions $f(t)$, $g(t)$, $h(t)$, and $t_0 \le t \le t_1$, defines a three-dimensional curve. If a particle is moving on the curve $y(t)$, and t is time, then $y'(t)$ gives *velocity of the particle, and* $\|y'(t)\|$ *its speed*. The distance covered by the particle (or the length of this curve) is the integral of its speed,

$$L = \int_{t_0}^{t_1} \|y'(t)\| \, dt = \int_{t_0}^{t_1} \|y'(z)\| \, dz.$$

(Indeed, this integral is limit of the Riemann sum $\sum_{i=1}^{n} \|y'(t_i)\| \Delta t$, which on each subinterval is the product of speed and time.) If we let the upper limit of integration vary, and call it t, then the resulting function of t,

$$s(t) = \int_{t_0}^{t} \|y'(z)\| \, dz,$$

is called *the arc length function*, and it provides the distance traveled between the time instances t_0 and t. By the fundamental theorem of Calculus,

$$\frac{ds}{dt} = \|y'(t)\|. \tag{3.1}$$

(Both sides of this relation give speed.)

The velocity vector $y'(t)$ is tangent to the curve $y(t)$, $T(t) = \frac{y'(t)}{\|y'(t)\|}$ gives *the unit tangent vector*.

Example 1. Consider a *helix* $y(t) = (\cos t, \sin t, t)$. (The xy-component $(\cos t, \sin t)$ moves on the unit circle around the origin, while $z = t$, so that the curve climbs.) Calculate the velocity vector $y'(t) = (-\sin t, \cos t, 1)$, the speed $\|y'(t)\| = \sqrt{2}$, and the unit tangent vector

$$T(t) = \frac{1}{\sqrt{2}}(-\sin t, \cos t, 1).$$

The arc length function, as measured from $t_0 = 0$, is

$$s = \int_0^t \|\gamma'(z)\| \, dz = \int_0^t \sqrt{2} \, dz = \sqrt{2}t.$$

Let us express $t = \frac{s}{\sqrt{2}}$ and *reparameterize* this helix using *the arc length s as a parameter*

$$\gamma(s) = \left(\cos \frac{s}{\sqrt{2}}, \sin \frac{s}{\sqrt{2}}, \frac{s}{\sqrt{2}} \right).$$

Are there any advantages of the new parameterization? Let us calculate

$$\gamma'(s) = \left(-\frac{1}{\sqrt{2}} \sin \frac{s}{\sqrt{2}}, \frac{1}{\sqrt{2}} \cos \frac{s}{\sqrt{2}}, \frac{1}{\sqrt{2}} \right),$$

$$\|\gamma'(s)\| = \frac{1}{2} + \frac{1}{2} = 1.$$

The speed equals 1 at all points, the curve is now of *unit speed*.

Arc length parameterization always produces unit speed curves, as is shown next.

Theorem 7.3.1. *If s is the arc length function on the curve $\gamma(t)$, and the parameterization $\gamma(s)$ is used, then $\|\gamma'(s)\| = 1$, and therefore $\gamma'(s) = T(s)$, the unit tangent vector.*

Proof. Relate the two parameterizations,

$$\gamma(t) = \gamma(s(t)).$$

By the chain rule, and the formula (3.1),

$$\gamma'(t) = \frac{dy}{ds} \frac{ds}{dt} = \frac{dy}{ds} \|\gamma'(t)\|.$$

Then

$$\gamma'(s) = \frac{\gamma'(t)}{\|\gamma'(t)\|},$$

which is the unit tangent vector $T(s)$. □

Suppose that a particle moves on a sphere. Then its velocity vector $\gamma'(t)$ is perpendicular to the radius vector $\gamma(t)$ at all time.

Proposition 7.3.1. *Assume that $\|\gamma(t)\| = a$ for all t, where a is a number. Then $\gamma'(t) \cdot \gamma(t) = 0$, so that $\gamma'(t) \perp \gamma(t) = 0$ for all t.*

Proof. We are given that

$$y(t) \cdot y(t) = a^2.$$

Differentiate both sides and simplify

$$y'(t) \cdot y(t) + y(t) \cdot y'(t) = 0,$$
$$2y'(t) \cdot y(t) = 0,$$

so that $y'(t) \cdot y(t) = 0.$ □

Let $y(s)$ be a unit speed curve (i. e., s is the arc length). Define *the curvature* of the curve as

$$\kappa(s) = \|y''(s)\| = \|T'(s)\|. \tag{3.2}$$

Since

$$y''(s) = \lim_{\Delta s \to 0} \frac{y'(s + \Delta s) - y'(s)}{\Delta s}, \tag{3.3}$$

and both $y'(s + \Delta s)$ and $y'(s)$ are unit tangent vectors, the curvature measures how quickly the unit tangent vector turns. Since $\|y'(s)\| = 1$ for all s, it follows by Proposition 7.3.1 that $y''(s) \perp y'(s)$. The vector $y''(s)$ is called *the normal vector*. It is perpendicular to the tangent vector, and it points in the direction that the curve $y(s)$ bends to, as can be seen from (3.3).

Example 2. Consider $y(t) = (a \cos t, a \sin t)$, a circle of radius a around the origin. Expect curvature to be the same at all points. Let us switch to the arc length parameter, $s = \int_0^t \|y'(z)\| \, dz = \int_0^t a \, dz = at$, so that $t = \frac{s}{a}$. Then

$$y(s) = \left(a \cos \frac{s}{a}, a \sin \frac{s}{a} \right),$$
$$y'(s) = \left(- \sin \frac{s}{a}, \cos \frac{s}{a} \right),$$
$$y''(s) = \left(-\frac{1}{a} \cos \frac{s}{a}, -\frac{1}{a} \sin \frac{s}{a} \right),$$
$$\kappa(s) = \|y''(s)\| = \frac{1}{a}.$$

The curvature is inversely proportional to the radius of the circle.

Arc length parameterization is rarely available for a general curve $y(t) = (x(t), y(t), z(t))$ because the integral $s = \int_{t_0}^t \sqrt{x'^2 + y'^2 + z'^2} \, dz$ tends to be complicated. Therefore, we wish to express curvature as a function of t, $\kappa = \kappa(t)$. Using the chain rule, the inverse function theorem, and (3.1), express

$$\kappa = \|T'(s)\| = \left\| T'(t)\frac{dt}{ds} \right\| = \frac{\|T'(t)\|}{\frac{ds}{dt}} = \frac{\|T'(t)\|}{\|\gamma'(t)\|}. \tag{3.4}$$

The problem with this formula is that the unit tangent vector

$$T(t) = \left(\frac{x'(t)}{\sqrt{x'^2(t) + y'^2(t) + z'^2(t)}}, \frac{y'}{\sqrt{x'^2 + y'^2 + z'^2}}, \frac{z'}{\sqrt{x'^2 + y'^2 + z'^2}} \right)$$

is cumbersome to differentiate. A convenient formula for the curvature is given next.

Theorem 7.3.2. *One has* $\kappa(t) = \frac{\|\gamma'(t)\times\gamma''(t)\|}{\|\gamma'(t)\|^3}$.

Proof. By the definition of $T(t)$ and (3.1),

$$\gamma'(t) = T(t)\|\gamma'(t)\| = T(t)\frac{ds}{dt}.$$

Using the product rule,

$$\gamma''(t) = T'\frac{ds}{dt} + T\frac{d^2s}{dt^2}.$$

Take the vector product of both sides with T, and use that $T \times T = \mathbf{0}$ to obtain

$$T \times \gamma''(t) = T \times T' \|\gamma'(t)\|.$$

Substitute $T = \frac{\gamma'(t)}{\|\gamma'(t)\|}$ on the left to express

$$\gamma'(t) \times \gamma''(t) = T \times T' \|\gamma'(t)\|^2.$$

Take the length of both sides

$$\|\gamma'(t) \times \gamma''(t)\| = \|T \times T'\| \|\gamma'(t)\|^2 = \|T'\| \|\gamma'(t)\|^2.$$

(Because $\|T \times T'\| = \|T\|\|T'\| \sin\frac{\pi}{2} = \|T'\|$, using that $T' \perp T$ by Proposition 7.3.1.) Then

$$\frac{\|\gamma'(t) \times \gamma''(t)\|}{\|\gamma'(t)\|^3} = \frac{\|T'(t)\|}{\|\gamma'(t)\|} = \kappa(t),$$

in view of (3.4). □

Surfaces

Two parameters, called u and v, are needed to define a surface $\sigma(u, v) = (x(u, v), y(u, v), z(u, v))$, with given functions $x(u, v)$, $y(u, v)$, $z(u, v)$. As the point (u, v) varies in some parameter region D of the uv-plane, the tip of the vector function $\sigma(u, v)$ traces out a surface, which can be alternatively represented in *a parametric form* as

$$x = x(u, v),$$
$$y = y(u, v),$$
$$z = z(u, v).$$

Example 3. Consider $\sigma(u, v) = (u, v + 1, u^2 + v^2)$. Here $x = u, y = v + 1, z = u^2 + v^2$, or $z = x^2 + (y - 1)^2$. The surface is a paraboloid with the vertex at $(0, 1, 0)$ (see a Calculus book if a review is needed).

Example 4. A sphere of radius a around the origin can be described in spherical coordinates as $\rho = a$. Expressing the Cartesian coordinates through the spherical ones gives a parameterization,

$$x = a \sin \varphi \cos \theta,$$
$$y = a \sin \varphi \sin \theta,$$
$$z = a \cos \varphi.$$

Here $0 \le \theta \le 2\pi, 0 \le \varphi \le \pi$. The rectangle $[0, 2\pi] \times [0, \pi]$ is the parameter region D in the $\theta\varphi$-plane.

Example 5. A completely different parameterization of a sphere of radius a around the origin, called *the Mercator projection*, was introduced in the sixteenth century for the needs of naval navigation,

$$\sigma(u, v) = (a \operatorname{sech} u \cos v, a \operatorname{sech} u \sin v, a \tanh u),$$

with $-\infty < u < \infty, 0 \le v \le 2\pi$. It uses *hyperbolic functions* reviewed in Exercises, where it is also shown that the components of $\sigma(u, v)$ satisfy $x^2 + y^2 + z^2 = a^2$.

Example 6. Suppose that a curve

$$y = f(u),$$
$$z = g(u), \quad u_0 \le u \le u_1,$$

in the yz-plane is rotated around the z-axis. Let us parameterize the resulting *surface of revolution*. We need to express (x, y, z) on this surface. The z coordinate is $z = g(u)$. The trace of this surface on each horizontal plane is a circle around the origin of radius $f(u)$. We obtain

$$\sigma(u, v) = (f(u) \cos v, f(u) \sin v, g(u)),$$

with $u_0 \le u \le u_1, 0 \le v \le 2\pi$.

Example 7. Assume that a circle of radius a, centered at the point $(b, 0)$ in the yz-plane, is rotated around the z-axis, $b > a$. The resulting surface is called *torus* (or doughnut, or bagel). Parameterizing this circle as

$$y = b + a \cos \theta,$$
$$z = a \sin \theta, \quad 0 \le \theta \le 2\pi,$$

we obtain a parameterization of this torus, as a surface of revolution,

$$\sigma(\theta, \varphi) = ((b + a \cos \theta) \cos \varphi, (b + a \cos \theta) \sin \varphi, a \sin \theta),$$

with $0 \le \theta \le 2\pi, 0 \le \varphi \le 2\pi$.

At a particular pair of parameters $u = u_0$ and $v = v_0$, we have a point $\sigma(u_0, v_0)$, call it P, on a surface $\sigma(u, v)$ called S. The curve $\sigma(u, v_0)$ depends on a parameter u, it lies on S, and passes through P at $u = u_0$. The curve $\sigma(u, v_0)$ is called *the u-curve* through P. Similarly, *the v-curve* through P is $\sigma(u_0, v)$. The u- and v-curves are known as *the coordinate curves*. The tangent vectors to the u- and v-curves at the point P are respectively $\sigma_u(u_0, v_0)$ and $\sigma_v(u_0, v_0)$.

Example 8. Figure 7.1 shows the graph of the torus

$$\sigma(\theta, \varphi) = ((3 + \cos \theta) \cos \varphi, (3 + \cos \theta) \sin \varphi, \sin \theta)$$

drawn by *Mathematica*, together with the θ-curve $\sigma(\theta, \frac{\pi}{4})$ and the φ-curve $\sigma(\frac{\pi}{4}, \varphi)$ drawn at the point $\sigma(\frac{\pi}{4}, \frac{\pi}{4})$. Observe that *Mathematica* draws other coordinate curves to produce a good-looking graph.

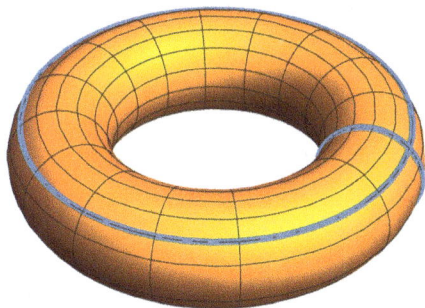

Figure 7.1: Torus, $a = 1$ and $b = 3$.

We shall consider only *regular surfaces, meaning that the vectors $\sigma_u(u_0, v_0)$ and $\sigma_v(u_0, v_0)$ are linearly independent at all points (u_0, v_0).* The plane through P that $\sigma_u(u_0, v_0)$ and $\sigma_v(u_0, v_0)$ span is called *the tangent plane to S at P.* The vector $\sigma_u(u_0, v_0) \times \sigma_v(u_0, v_0)$ is normal to this tangent plane.

Example 9. Let us find the tangent plane to the surface

$$\sigma(u, v) = (u^2 + 1, v^3 + 1, u + v) \quad \text{at the point } P = (5, 2, 3).$$

This surface passes through P at $u_0 = 2$, $v_0 = 1$. Calculate $\sigma_u(2, 1) = (4, 0, 1) = 4\mathbf{i} + \mathbf{k}$, $\sigma_v(2, 1) = (0, 3, 1) = 3\mathbf{j} + \mathbf{k}$, the normal vector $\sigma_u(2, 1) \times \sigma_v(2, 1) = -3\mathbf{i} - 4\mathbf{j} + 12\mathbf{k}$. Obtain

$$-3(x - 5) - 4(y - 2) + 12(z - 3) = 0,$$

which simplifies to $-3x - 4y + 12z = 13$. (Recall that $a(x - x_0) + b(y - y_0) + c(z - z_0) = 0$ gives an equation of the plane passing through the point (x_0, y_0, z_0) with normal vector $a\mathbf{i} + b\mathbf{j} + c\mathbf{k}$.)

Consider a surface $\sigma(u, v)$, with (u, v) belonging to a parameter region D of the uv-plane. Any curve $u = u(t)$, $v = v(t)$ lying in the region D produces a curve $\sigma(u(t), v(t))$ lying on this surface, see Figure 7.2 in the next section for an example.

Exercises

1. (a) Sketch the graph of the ellipse $y(t) = (2\cos t, 3\sin t)$, $0 \le t \le 2\pi$.
    (b) Is t the polar angle here? Hint. Try $t = \frac{\pi}{4}$.
        Answer: No.
2. Find the curvature κ of a planar curve $x = x(t)$, $y = y(t)$.
    Hint. Write this curve as $y(t) = (x(t), y(t), 0)$ and use Theorem 7.3.2.
    Answer: $\kappa(t) = \dfrac{|x'(t)y''(t) - x''(t)y'(t)|}{(x'^2(t) + y'^2(t))^{\frac{3}{2}}}$.
3. Let $(x(s), y(s))$ be a planar curve, s is the arc length. Let $\theta(s)$ be the angle that the unit tangent vector $T(s) = (x'(s), y'(s))$ makes with the x-axis. Justify the following formulas:
    (a) $\kappa(s) = |x'(s)y''(s) - x''(s)y'(s)|$.
    (b) $\theta(s) = \tan^{-1}\dfrac{y'(s)}{x'(s)}$.
    (c) $\kappa(s) = |\theta'(s)|$. (Curvature gives the speed of rotation of $T(s)$.)
4. Find the curvature κ of a planar curve $y = f(x)$.
    Hint. Write this curve as $y(x) = (x, f(x), 0)$ and use Theorem 7.3.2.
    Answer: $\kappa(x) = \dfrac{|f''(x)|}{(1 + f'^2(x))^{\frac{3}{2}}}$.
5. (a) Recall *the hyperbolic cosine*, $\cosh t = \frac{e^t + e^{-t}}{2}$, and *the hyperbolic sine*, $\sinh t = \frac{e^t - e^{-t}}{2}$. Calculate the derivatives $(\sinh t)' = \cosh t$, $(\cosh t)' = \sinh t$.
    (b) Show that $\cosh^2 t - \sinh^2 t = 1$ for all t.
    (c) Using the quotient rule, calculate the derivatives of other hyperbolic functions,
        $\tanh u = \frac{\sinh u}{\cosh u}$ and $\operatorname{sech} u = \frac{1}{\cosh u}$.
        Answer: $(\tanh u)' = \operatorname{sech}^2 u$, $(\operatorname{sech} u)' = -\operatorname{sech} u \tanh u$.
    (d) Show that $\tanh^2 u + \operatorname{sech}^2 u = 1$ for all u.

(e) For the Mercator projection of the sphere of radius a around the origin, $\sigma(u, v) = $
(a $\operatorname{sech} u \cos v$, a $\operatorname{sech} u \sin v$, a $\tanh u$), show that $x^2 + y^2 + z^2 = a^2$.

6. (a) On the unit sphere $\sigma(\theta, \varphi) = (\sin \varphi \cos \theta, \sin \varphi \sin \theta, \cos \varphi)$ sketch and identify
the θ-curve $\sigma(\theta, \frac{\pi}{4}), 0 \le \theta \le 2\pi$.

(b) Find the length of this curve.

Answer: $\sqrt{2}\pi$.

(c) Find an equation of the tangent plane at the point $\sigma(\frac{\pi}{4}, \frac{\pi}{4})$.

7.4 The first fundamental form

The first fundamental form extends the concept of arc length to surfaces, and is used to calculate length of curves, angles between curves, and areas of regions on surfaces.

Consider a surface S given by a vector function $\sigma(u, v)$, with (u, v) belonging to some parameter region D. Any curve $(u(t), v(t))$ in the region D defines a curve on the surface S, $\gamma(t) = \sigma(u(t), v(t))$. The length of this curve between two parameter values of $t = t_0$ and $t = t_1$ is

$$L = \int_{t_0}^{t_1} \|\gamma'(t)\| \, dt = \int_{t_0}^{t_1} ds. \tag{4.1}$$

Here $\|\gamma'(t)\| \, dt = ds$, since $\frac{ds}{dt} = \|\gamma'(t)\|$ by (3.1). Using the chain rule, calculate

$$\gamma'(t) = \sigma_u u'(t) + \sigma_v v'(t),$$

and then

$$\|\gamma'(t)\|^2 = \gamma'(t) \cdot \gamma'(t) = (\sigma_u u'(t) + \sigma_v v'(t)) \cdot (\sigma_u u'(t) + \sigma_v v'(t))$$

$$= \sigma_u \cdot \sigma_u \, u'^2(t) + 2\sigma_u \cdot \sigma_v \, u'(t)v'(t) + \sigma_v \cdot \sigma_v \, v'^2(t).$$

It is customary to denote

$$E = \sigma_u \cdot \sigma_u,$$
$$F = \sigma_u \cdot \sigma_v, \tag{4.2}$$
$$G = \sigma_v \cdot \sigma_v.$$

(Observe that $E = E(u, v), F = F(u, v), G = G(u, v)$.) Then

$$\|\gamma'(t)\| = \sqrt{E \, u'^2(t) + 2F \, u'(t)v'(t) + G \, v'^2(t)}, \tag{4.3}$$

so that the length of $\gamma(t)$ is

$$L = \int_{t_0}^{t_1} \sqrt{E \, u'^2(t) + 2F \, u'(t)v'(t) + G \, v'^2(t)} \, dt. \tag{4.4}$$

(Here $E = E(u(t), v(t))$, $F = F(u(t), v(t))$, $G = G(u(t), v(t))$.) Since $ds = \|\gamma'(t)\| \, dt$, using (4.3) we obtain

$$ds = \sqrt{E \, u'^2(t) + 2F \, u'(t)v'(t) + G \, v'^2(t)} \, dt$$

$$= \sqrt{E \, [u'(t) \, dt]^2 + 2F \, [u'(t) \, dt][v'(t) \, dt] + G \, [v'(t) \, dt]^2}$$

$$= \sqrt{E \, du^2 + 2F \, dudv + G \, dv^2},$$

using the differentials $du = u'(t) \, dt$, $dv = v'(t) \, dt$, so that

$$ds^2 = E \, du^2 + 2F \, dudv + G \, dv^2. \tag{4.5}$$

This quadratic form in the variables du and dv is called *the first fundamental form*.

Example 1. Recall that the unit sphere around the origin, $\rho = 1$, can be represented as $\sigma(\theta, \varphi) = (\sin \varphi \cos \theta, \sin \varphi \sin \theta, \cos \varphi)$. Calculate

$$\sigma_\theta(\theta, \varphi) = (-\sin \varphi \sin \theta, \sin \varphi \cos \theta, 0),$$
$$\sigma_\varphi(\theta, \varphi) = (\cos \varphi \cos \theta, \cos \varphi \sin \theta, -\sin \varphi).$$
$$E = \sigma_\theta \cdot \sigma_\theta = \|\sigma_\theta\|^2 = \sin^2 \varphi(\sin^2 \theta + \cos^2 \theta) = \sin^2 \varphi,$$
$$F = \sigma_\theta \cdot \sigma_\varphi = 0,$$
$$G = \sigma_\varphi \cdot \sigma_\varphi = \|\sigma_\varphi\|^2 = \cos^2 \varphi(\sin^2 \theta + \sin^2 \theta) + \sin^2 \varphi = 1.$$

The first fundamental form is

$$ds^2 = \sin^2 \varphi \, d\theta^2 + d\varphi^2.$$

Example 2. For the *helicoid* $\sigma(u, v) = (u \cos v, u \sin v, v)$, calculate

$$\sigma_u(u, v) = (\cos v, \sin v, 0),$$
$$\sigma_v(u, v) = (-u \sin v, u \cos v, 1),$$
$$E = \sigma_u \cdot \sigma_u = 1,$$
$$F = \sigma_u \cdot \sigma_v = 0,$$
$$G = \sigma_v \cdot \sigma_v = u^2 + 1.$$

The first fundamental form is

$$ds^2 = du^2 + (u^2 + 1) \, dv^2.$$

Assume now that the parameter region D is a rectangle $1 \leq u \leq 6$, $0 \leq v \leq 4\pi$. In Figure 7.2 we used *Mathematica* to draw this helicoid over D. Consider a curve $u = t^2$, $v = 5t$, $1 \leq t \leq 2.3$, which lies in the parameter region D, and hence it produces a curve on the helicoid shown in Figure 7.2. The length of this curve on the helicoid is calculated by the formula (4.4) to be

$$L = \int_1^{2.3} \sqrt{u'^2(t) + (u^2(t) + 1)\, v'^2(t)}\, dt = \int_1^{2.3} \sqrt{4t^2 + 25(t^4 + 1)}\, dt$$

$$\approx 20.39.$$

The integral was approximately calculated using *Mathematica*.

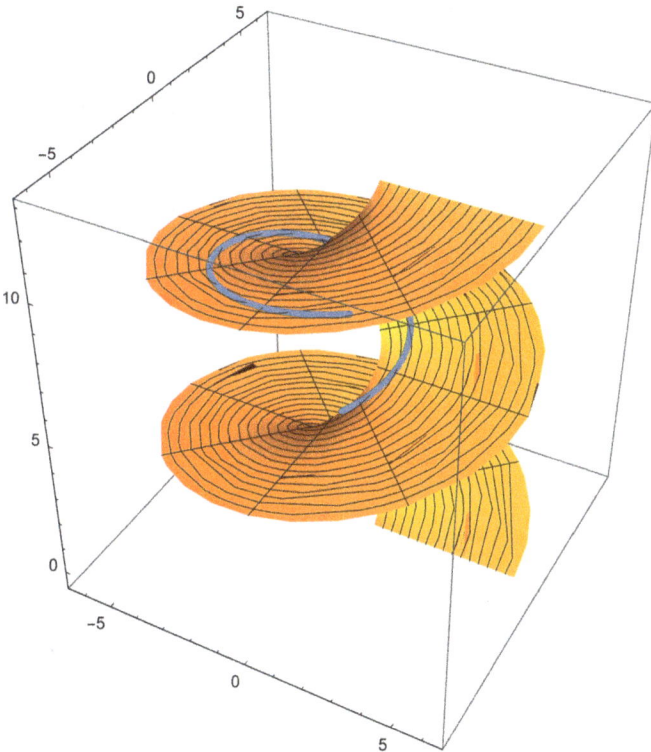

Figure 7.2: A curve on the helicoid from Example 2.

Example 3. Consider the *Mercator projection* of the unit sphere $\sigma(u, v) = (\operatorname{sech} u \cos v, \operatorname{sech} u \sin v, \tanh u)$. Recall that

$$(\operatorname{sech} u)' = \left(\frac{1}{\cosh u}\right)' = -\frac{\sinh u}{\cosh^2 u} = -\operatorname{sech} u \tanh u,$$

and similarly $(\tanh u)' = \operatorname{sech}^2 u$. Calculate

$$\sigma_u(u, v) = (-\operatorname{sech} u \tanh u \cos v, -\operatorname{sech} u \tanh u \sin v, \operatorname{sech}^2 u),$$

$$\sigma_v(u, v) = (-\operatorname{sech} u \sin v, \operatorname{sech} u \cos v, 0),$$

$$E = \|\sigma_u\|^2 = \operatorname{sech}^2 u \tanh^2 u + \operatorname{sech}^4 u$$

$$= \operatorname{sech}^2 u (\tanh^2 u + \operatorname{sech}^2 u) = \operatorname{sech}^2 u,$$

$$F = \sigma_u \cdot \sigma_v = 0,$$

$$G = \|\sigma_v\|^2 = \operatorname{sech}^2 u.$$

The first fundamental form is

$$ds^2 = \operatorname{sech}^2 u \, (du^2 + dv^2).$$

Recall that the angle between two curves at a point of intersection is defined to be the angle between their tangent lines at the point of intersection. On a surface $\sigma = \sigma(u, v)$ consider two curves $\gamma(t) = \sigma(u(t), v(t))$ and $\gamma_1(t) = \sigma(u_1(t), v_1(t))$. Suppose that the curves $(u(t), v(t))$ and $(u_1(t), v_1(t))$ intersect at some point (u_0, v_0) of the parameter region D, with $(u(t_1), v(t_1)) = (u_0, v_0)$ and $(u_1(t_2), v_1(t_2)) = (u_0, v_0)$. If $\mathbf{t}$ and $\mathbf{t_1}$ denote respective tangent vectors to $\gamma(t)$ and $\gamma_1(t)$ at the point of intersection $\sigma(u_0, v_0)$, the angle θ between the curves is given by

$$\cos \theta = \frac{\mathbf{t} \cdot \mathbf{t_1}}{\|\mathbf{t}\| \, \|\mathbf{t_1}\|}.$$

Calculate the tangent vectors

$$\mathbf{t} = \gamma'(t_1) = \sigma_u u'(t_1) + \sigma_v v'(t_1),$$

$$\mathbf{t_1} = \gamma_1'(t_2) = \sigma_u u_1'(t_2) + \sigma_v v_1'(t_2),$$

and then (writing u', v' for $u'(t_1)$, $v'(t_1)$, and u_1', v_1' for $u_1'(t_2)$, $v_1'(t_2)$)

$$\mathbf{t} \cdot \mathbf{t_1} = (\sigma_u u' + \sigma_v v') \cdot (\sigma_u u_1' + \sigma_v v_1')$$

$$= \sigma_u \cdot \sigma_u u' u_1' + \sigma_u \cdot \sigma_v u' v_1' + \sigma_v \cdot \sigma_u u_1' v' + \sigma_v \cdot \sigma_v v' v_1'$$

$$= E u' u_1' + F(u' v_1' + u_1' v') + G v' v_1'.$$

Using the formula (4.3), we conclude

$$\cos \theta = \frac{E u' u_1' + F(u' v_1' + u_1' v') + G v' v_1'}{\sqrt{E u'^2 + 2F u'v' + G v'^2} \, \sqrt{E u_1'^2 + 2F u_1' v_1' + G v_1'^2}}. \tag{4.6}$$

Here E, F, G are evaluated at (u_0, v_0), u' and v' are evaluated at t_1, u_1' and v_1' are evaluated at t_2.

Example 4. For an arbitrary surface $\sigma(u, v)$ let us find the angle between the u-curve $\sigma = \sigma(u, v_0)$, and the v-curve $\sigma = \sigma(u_0, v)$, at any point $\sigma(u_0, v_0)$ on the surface, corresponding to the point $P = (u_0, v_0)$ in the parameter region.

The u-curve can be parameterized as $u(t) = t$, $v(t) = v_0$. At $t = u_0$, it passes through the point $P = (u_0, v_0)$. Calculate $u' = 1$, $v' = 0$. The v-curve can be parameterized as $u_1(t) = u_0$, $v_1(t) = t$. At $t = v_0$, it passes through the same point P. Calculate $u_1' = 0$, $v_1' = 1$. By (4.6),

$$\cos\theta = \frac{F(u_0, v_0)}{\sqrt{E(u_0, v_0)\, G(u_0, v_0)}}.$$

It follows that *the coordinate curves are orthogonal at all points on the surface* $\sigma(u, v)$ *if and only if* $F(u, v) = 0$ *for all* (u, v). That was the case in the Examples 1, 2, and 3 above.

Given two surfaces S and S_1, a map $S \to S_1$ is defined to be a rule assigning to each point of S a unique point on S_1. If $\sigma(u, v)$ gives a parameterization of S, a map from S to S_1 allows one to use the coordinates u and v on S_1 too, representing S_1 with some other vector function $\sigma_1(u, v)$. Indeed, each (u, v) in the parameter region D is mapped by $\sigma(u, v)$ to a unique point on S, and then to a unique point on S_1 by our map. Consequently, a curve $(u(t), v(t))$ on D provides both a curve on S and its image curve on S_1. *A map from S to S_1 is called conformal if given any point P on S, and any two curves on S passing through P, the angle between these curves is the same as the angle between their images on S_1.* (Such maps are rare, but very interesting.)

Theorem 7.4.1. *Let E, F, G and E_1, F_1, G_1 be the coefficients of the first fundamental forms for S and S_1, respectively. The map $\sigma(u, v) \to \sigma_1(u, v)$ is conformal, provided there exists a function $\lambda(u, v)$ such that*

$$E_1(u, v) = \lambda(u, v)E(u, v),$$
$$F_1(u, v) = \lambda(u, v)F(u, v), \tag{4.7}$$
$$G_1(u, v) = \lambda(u, v)G(u, v),$$

for all $(u, v) \in D$. (In other words, the first fundamental forms of S and S_1 are proportional, with a factor $\lambda(u, v)$.)

Proof. Let $(u(t), v(t))$ and $(u_1(t), v_1(t))$ represent two intersecting curves on S and their images on S_1. By (4.6), the cosine of the angle between the curves on S_1 is equal to

$$\frac{E_1 u' u_1' + F_1(u'v_1' + u_1'v') + G_1 v'v_1'}{\sqrt{E_1 u'^2 + 2F_1\, u'v' + G_1\, v'^2}\ \sqrt{E_1\, u_1'^2 + 2F_1\, u_1'v_1' + G_1\, v_1'^2}}.$$

Using the formulas (4.7), then factoring and canceling $\lambda(u, v)$ in both the numerator and denominator, we obtain the formula (4.6), giving $\cos\theta$ for the angle between the curves on S. ☐

Observe that any representation $\sigma(u, v)$ of a surface S can be viewed as a map from a coordinate region D of the uv-plane to S.

Example 5. The Mercator projection

$$\sigma(u, v) = (\text{sech } u \cos v, \text{sech } u \sin v, \tanh u)$$

can be viewed as a map form the strip $-\infty < u < \infty, 0 \le v \le 2\pi$ to the unit sphere around the origin. Its first fundamental form $ds^2 = \text{sech}^2 u \, (du^2 + dv^2)$ is proportional to $du^2 + dv^2$, the first fundamental form of the uv-plane. This map is conformal, by Theorem 7.4.1. This property of Mercator projection made it useful for naval navigation since the sixteenth century. Horizontal lines in the (u, v) coordinate plane are mapped into the meridians on the globe (the unit sphere). Lines making an angle a with the horizontal lines are mapped into curves on the sphere making an angle a with the meridians. These curves are called *loxodromic*, and while they do not give the shortest route, they are easy to maintain using compass.

Exercises

1. Identify the surface, and find the first fundamental form:
   (a) $\sigma(u, v) = (u - v, u + v, u^2 + v^2)$.
      Answer: $z = \frac{1}{2}(x^2 + y^2)$, a paraboloid;
      $ds^2 = (2 + 4u^2) \, du^2 + 4uv \, dudv + (2 + 4v^2) \, dv^2$.
   (b) $\sigma(u, v) = (\sinh u \sinh v, \sinh u \cosh v, \sinh u)$.
      Hint. $y = \pm\sqrt{x^2 + z^2}$, a double cone extending along the y-axis.
   (c) $\sigma(u, v) = (\cosh u, \sinh u, v)$.
      Answer: $x^2 - y^2 = 1$, hyperbolic cylinder;
      $ds^2 = (\sinh^2 u + \cosh^2 u) \, du^2 + dv^2$.
   (d) $\sigma(u, v) = (u + 1, v, u^2 + v^2)$.
      Answer: $z = (x-1)^2 + y^2$, paraboloid; $ds^2 = (1 + 4u^2) \, du^2 + 4uv \, du \, dv + (1 + 4v^2) dv^2$.
   (e) $\sigma(u, v) = (u \cos v, u \sin v, u), u \ge 0$.
      Answer: $z = \sqrt{x^2 + y^2}$, cone; $ds^2 = 2 \, du^2 + u^2 dv^2$.
2. On the cone $\sigma(u, v) = (u \cos v, u \sin v, u)$, $u \ge 0$, sketch the curve $u = e^{2t}$, $v = t$, $0 \le t \le 2\pi$, and find its length.
   Answer: $\frac{3}{2}(e^{4\pi} - 1)$.
3. Find the first fundamental form for the surface $z = f(x, y)$.
   Hint. Here $\sigma(x, y) = (x, y, f(x, y))$.
   Answer: $ds^2 = (1 + f_x^2) dx^2 + f_x f_y \, dxdy + (1 + f_y^2) dy^2$.
4. Identify the surface $\sigma(u, v) = (u, v, u^2 + v^2 + 2u)$, and find the angle between the coordinate curves at any point (u, v).

5. (a) Let a, b, c, d be vectors in R^3. Justify the following vector identity:

$$(a \times b) \cdot (c \times d) = (a \cdot c)(b \cdot d) - (a \cdot d)(b \cdot c).$$

Hint. Write each vector in components. *Mathematica* can help with a rather tedious calculation.

(b) Show that the area of a surface $\sigma(u, v)$ over a parameter region D is given by the double integral $Area = \iint_D \|\sigma_u \times \sigma_v\| \, dA$ (here $dA = du\, dv$).
Hint. Show that the rectangle with vertices $(u, v), (u+\Delta u, v), (u, v+\Delta v), (u+\Delta u, v+\Delta v)$ in D is mapped onto a region on the surface $\sigma(u, v)$ that is approximated by a parallelogram with sides $\sigma_u(u, v)\Delta u$ and $\sigma_v(u, v)\Delta v$.

(c) Conclude that $Area = \iint_D \sqrt{EG - F^2} \, dA$.
Hint. Note that $\|\sigma_u \times \sigma_v\|^2 = (\sigma_u \times \sigma_v) \cdot (\sigma_u \times \sigma_v)$.

(d) Show that the first fundamental form is a positive definite quadratic form for regular surfaces.

(e) On the unit sphere around the origin,

$$\sigma(\theta, \varphi) = (\sin \varphi \cos \theta, \sin \varphi \sin \theta, \cos \varphi),$$

sketch the region $0 \le \varphi \le \frac{\pi}{4}, 0 \le \theta \le 2\pi$, and calculate its area.

6. Let $x = x(u, v), y = y(u, v)$ be a one-to-one map taking a region D in the uv-plane onto a region R of the xy-plane.

(a) Show that the area of the region R is

$$A = \iint_R dx\, dy = \iint_D \left| \begin{matrix} \frac{\partial x}{\partial u} & \frac{\partial x}{\partial v} \\ \frac{\partial y}{\partial u} & \frac{\partial y}{\partial v} \end{matrix} \right| \, du\, dv,$$

so that the absolute value of the Jacobian determinant gives the *magnification factor* of the element of area.
Hint. Consider the surface $\sigma(u, v) = (x(u, v), y(u, v), 0)$.

(b) Justify the change of variables formula for double integrals

$$\iint_R f(x, y) \, dx\, dy = \iint_D f(x(u, v), y(u, v)) \left| \begin{matrix} \frac{\partial x}{\partial u} & \frac{\partial x}{\partial v} \\ \frac{\partial y}{\partial u} & \frac{\partial y}{\partial v} \end{matrix} \right| \, du\, dv.$$

7. (a) Let $\sigma(u, v) = (x(u, v), y(u, v), z(u, v))$ be a vector function, and $u(t), v(t)$ be given functions. Show that

$$\frac{d}{dt} \sigma(u(t), v(t)) = \sigma_u(u(t), v(t))u'(t) + \sigma_v(u(t), v(t))v'(t).$$

(b) Derive the Maclaurin series

$$\sigma(u, v) = \sigma(0, 0) + \sigma_u u + \sigma_v v + \frac{1}{2}(\sigma_{uu}u^2 + 2\sigma_{uv}uv + \sigma_{vv}v^2) + \cdots,$$

with all derivatives evaluated at $(0, 0)$.

Hint. Write Maclaurin series for $g(s) = \sigma(su, sv)$, as a function of s.

7.5 The second fundamental form

The second fundamental form extends the concept of curvature to surfaces. Clearly, the theory is more involved than for curves.

Recall that the unit normal to the tangent plane for a surface $\sigma(u, v)$ is given by $\bar{N} = \frac{\sigma_u \times \sigma_v}{\|\sigma_u \times \sigma_v\|}$. Observe that

$$\sigma_u \cdot \bar{N} = 0, \quad \sigma_v \cdot \bar{N} = 0. \tag{5.1}$$

Given a point $\sigma(u, v)$ and nearby points $\sigma(u + \Delta u, v + \Delta v)$ on a surface, with $|\Delta u|$ and $|\Delta v|$ small, the scalar (inner) product

$$Q = [\sigma(u + \Delta u, v + \Delta v) - \sigma(u, v)] \cdot \bar{N} \tag{5.2}$$

measures how quickly the surface bends away from its tangent plane at the point $\sigma(u, v)$. (If $\sigma(u + \Delta u, v + \Delta v)$ remains on this tangent plane, then $Q = 0$.) By Taylor's formula,

$$\sigma(u + \Delta u, v + \Delta v) - \sigma(u, v)$$
$$\approx \sigma_u(u, v)\Delta u + \sigma_v(u, v)\Delta v$$
$$+ \frac{1}{2}(\sigma_{uu}(u, v)\Delta u^2 + 2\sigma_{uv}(u, v)\Delta u \Delta v + \sigma_{vv}(u, v)\Delta v^2),$$

for $|\Delta u|$ and $|\Delta v|$ small. In view of (5.1),

$$Q \approx \frac{1}{2}(\sigma_{uu}(u, v) \cdot \bar{N} \Delta u^2 + 2\sigma_{uv}(u, v) \cdot \bar{N} \Delta u \Delta v + \sigma_{vv}(u, v) \cdot \bar{N} \Delta v^2)$$
$$= \frac{1}{2}(L \Delta u^2 + 2M \Delta u \Delta v + N \Delta v^2),$$

using the standard notation,

$$L = \sigma_{uu}(u, v) \cdot \bar{N},$$
$$M = \sigma_{uv}(u, v) \cdot \bar{N}, \tag{5.3}$$
$$N = \sigma_{vv}(u, v) \cdot \bar{N}.$$

The quadratic form in the variables du and dv,

$$L(u, v)\, du^2 + 2M(u, v)\, du\, dv + N(u, v)\, dv^2,$$

is called *the second fundamental form*.

Example 1. Consider a plane $\sigma(u, v) = \mathbf{a}+u\mathbf{p}+v\mathbf{q}$, passing through the tip of vector $\mathbf{a}$, and spanned by vectors $\mathbf{p}$ and $\mathbf{q}$. Calculate $\sigma_u(u, v) = \mathbf{p}$, $\sigma_v(u, v) = \mathbf{q}$, $\sigma_{uu}(u, v) = \sigma_{uv}(u, v) = \sigma_{vv}(u, v) = \mathbf{0}$. Hence, $L = M = N = 0$. The second fundamental form of a plane is zero.

Example 2. Consider a paraboloid $\sigma(u, v) = (u, v, u^2 + v^2)$ (the same as $z = x^2 + y^2$). Calculate

$$\sigma_u(u, v) = (1, 0, 2u),$$
$$\sigma_v(u, v) = (0, 1, 2v),$$
$$\sigma_{uu}(u, v) = \sigma_{vv}(u, v) = (0, 0, 2),$$
$$\sigma_{uv}(u, v) = (0, 0, 0),$$
$$\sigma_u(u, v) \times \sigma_v(u, v) = (-2u, -2v, 1),$$
$$\bar{N} = \frac{\sigma_u(u, v) \times \sigma_v(u, v)}{\|\sigma_u(u, v) \times \sigma_v(u, v)\|} = \frac{1}{\sqrt{4u^2 + 4v^2 + 1}}(-2u, -2v, 1),$$
$$L = \sigma_{uu}(u, v) \cdot \bar{N} = \frac{2}{\sqrt{4u^2 + 4v^2 + 1}},$$
$$M = \sigma_{uv}(u, v) \cdot \bar{N} = 0,$$
$$N = \sigma_{vv}(u, v) \cdot \bar{N} = \frac{2}{\sqrt{4u^2 + 4v^2 + 1}}.$$

The second fundamental form is $\frac{2}{\sqrt{4u^2+4v^2+1}}(du^2 + dv^2)$.

If $\gamma(t)$ is a unit speed curve, recall that the vector $\gamma''(t)$ is normal to the curve, and $\|\gamma''(t)\| = \kappa$ gives the curvature. Consider now a unit speed curve $\gamma(t) = \sigma(u(t), v(t))$ on a surface $\sigma(u, v)$, where $(u(t), v(t))$ is a curve in the uv-plane of parameters. Define *the normal curvature* of $\gamma(t)$ as

$$\kappa_n = \gamma''(t) \cdot \bar{N}. \tag{5.4}$$

To motivate this notion, think of an object of unit mass moving on the curve $\gamma(t) = \sigma(u(t), v(t))$ lying on a surface S given by $\sigma(u, v)$. Then $\gamma''(t)$ gives force, and $\gamma''(t) \cdot \bar{N}$ is its normal component, or the force with which the object and surface S act on each other.

Proposition 7.5.1. *If $L(u, v)$, $M(u, v)$, and $N(u, v)$ are the coefficients of the second fundamental form, and $\sigma(u(t), v(t))$ is a unit speed curve, then*

$$\kappa_n = L(u, v)u'^2(t) + 2M(u, v)u'(t)v'(t) + N(u, v)v'^2(t). \tag{5.5}$$

Proof. Using the chain rule, calculate

$$\gamma'(t) = \sigma_u u' + \sigma_v v',$$
$$\gamma''(t) = (\sigma_u)' u' + \sigma_u u'' + (\sigma_v)' v' + \sigma_v v''$$
$$= (\sigma_{uu} u' + \sigma_{uv} v') u' + \sigma_u u'' + (\sigma_{vu} u' + \sigma_{vv} v') v' + \sigma_v v''$$
$$= \sigma_{uu} u'^2 + 2\sigma_{uv} u' v' + \sigma_{vv} v'^2 + \sigma_u u'' + \sigma_v v''.$$

Then we obtain the formula (5.5) for $\kappa_n = \gamma''(t) \cdot \bar{N}$ by using the definitions of L, M, N and (5.1). □

Let $\bar{\gamma}(t) = \sigma(\bar{u}(t), \bar{v}(t))$ be another unit speed curve passing through the same point $P = \sigma(u_0, v_0)$ on the surface $\sigma(u, v)$ as does the curve $\gamma(t) = \sigma(u(t), v(t))$, so that $P = \gamma(t_1) = \bar{\gamma}(t_2)$ for some t_1 and t_2. Assume that $\bar{\gamma}'(t_2) = \gamma'(t_1)$. We claim that then $\bar{u}'(t_2) = u'(t_1)$ and $\bar{v}'(t_2) = v'(t_1)$. Indeed, $\gamma'(t_1) = \sigma_u(u_0, v_0) u' + \sigma_v(u_0, v_0) v'$ and $\bar{\gamma}'(t_2) = \sigma_u(u_0, v_0) \bar{u}' + \sigma_v(u_0, v_0) \bar{v}'$. We are given that at P,

$$\sigma_u(u_0, v_0) \bar{u}' + \sigma_v(u_0, v_0) \bar{v}' = \sigma_u(u_0, v_0) u' + \sigma_v(u_0, v_0) v',$$

which implies that

$$\sigma_u(\bar{u}' - u') + \sigma_v(\bar{v}' - v') = 0.$$

(Here u', v' are evaluated at t_1, while $\bar{u}'$, $\bar{v}'$ at t_2.) Since the vectors σ_u and σ_v are linearly independent (because we consider only regular surfaces), it follows that $\bar{u}' - u' = 0$ and $\bar{v}' - v' = 0$, implying the claim. Then by (5.5) it follows that *the normal curvature is the same for all unit speed curves on a surface, passing through the same point with the same tangent vector.*

Write the formula (5.5) in the form

$$\kappa_n = L(u, v) u'^2(s) + 2M(u, v) u'(s) v'(s) + N(u, v) v'^2(s) \qquad (5.6)$$

to stress the fact that it uses the arc length parameter s (the same as saying unit speed curve). What if t is an arbitrary parameter?

Proposition 7.5.2. *If $E(u, v)$, $F(u, v)$, $G(u, v)$, $L(u, v)$, $M(u, v)$, and $N(u, v)$ are the coefficients of the first and second fundamental forms, and $\sigma(u(t), v(t))$ is any curve on a surface $\sigma(u, v)$, then its normal curvature is*

$$\kappa_n = \frac{L(u, v) u'^2(t) + 2M(u, v) u'(t) v'(t) + N(u, v) v'^2(t)}{E(u, v) u'^2(t) + 2F(u, v) u'(t) v'(t) + G(u, v) v'^2(t)}. \qquad (5.7)$$

Proof. Use the chain rule,

$$\frac{du}{ds} = \frac{du}{dt}\frac{dt}{ds} = \frac{\frac{du}{dt}}{\frac{ds}{dt}},$$

and a similar formula $\frac{dv}{ds} = \frac{\frac{dv}{dt}}{\frac{ds}{dt}}$ in (5.6) to obtain

$$\kappa_n = \frac{L(u,v)u'^2(t) + 2M(u,v)u'(t)v'(t) + N(u,v)v'^2(t)}{(\frac{ds}{dt})^2}. \tag{5.8}$$

Dividing the first fundamental form by dt^2 gives

$$\left(\frac{ds}{dt}\right)^2 = E(u,v)\left(\frac{du}{dt}\right)^2 + 2F(u,v)\frac{du}{dt}\frac{dv}{dt} + G(u,v)\left(\frac{dv}{dt}\right)^2.$$

Use this formula in (5.8) to complete the proof. □

Observe that κ_n in (5.7) is the ratio of two quadratic forms with the matrices $A = \begin{bmatrix} L & M \\ M & N \end{bmatrix}$ in the numerator, and $B = \begin{bmatrix} E & F \\ F & G \end{bmatrix}$ in the denominator. At any fixed point (u_0, v_0), with a curve $(u(t_0), v(t_0)) = (u_0, v_0)$, the matrices A and B have numerical entries, while the direction vector $(u'(t_0), v'(t_0))$ is just a pair of numbers, call them (ξ, η) and denote $x = \begin{bmatrix} \xi \\ \eta \end{bmatrix}$. Then (5.7) takes the form

$$\kappa_n = \frac{Ax \cdot x}{Bx \cdot x}, \tag{5.9}$$

a ratio of two quadratic forms.

In case $B = I$, this ratio is the Rayleigh quotient of A, studied earlier, and its extreme values are determined by the eigenvalues of A. For the general case, one needs to study *generalized eigenvalue problems*, and *generalized Rayleigh quotients*.

Observe that the quadratic form $Bx \cdot x$ is positive definite. Indeed, $E = \|\sigma_u\|^2 > 0$ for regular surfaces, and $EG - F^2 > 0$ by an exercise in Section 7.4. The matrix B is positive definite by Sylvester's criterion.

Generalized eigenvalue problem

If A and B are two $n \times n$ matrices, $x \in R^n$, and

$$Ax = \lambda Bx, \quad x \neq 0, \tag{5.10}$$

we say that x is *a generalized eigenvector*, and λ is *the corresponding generalized eigen-value*. (Observe that generalized eigenvectors here are not related at all to those of Chapter 6.) Calculations are similar to those for the usual eigenvalues and eigenvectors (where $B = I$). Write (5.10) as

$$(A - \lambda B)x = 0.$$

This homogeneous system will have nontrivial solutions provided that

$$|A - \lambda B| = 0.$$

Let $\lambda_1, \lambda_2, \ldots, \lambda_n$ be the solutions of this *characteristic equation*. Then solve the system

$$(A - \lambda_1 B)x = 0$$

for generalized eigenvectors corresponding to λ_1, and so on.

Assume that a matrix B is positive definite. *We say that two vectors $x, y \in R^n$ are B-orthogonal provided that*

$$Bx \cdot y = 0.$$

A vector $x \in R^n$ is called a B-unit if

$$Bx \cdot x = 1.$$

Proposition 7.5.3. *Assume that a matrix A is symmetric, and a matrix B is positive definite. Then the generalized eigenvalues of $Ax = \lambda Bx$ are real, and generalized eigenvectors corresponding to different generalized eigenvalues are B-orthogonal.*

Proof. Generalized eigenvalues satisfy

$$B^{-1}Ax = \lambda x, \quad x \neq 0.$$

The matrix $B^{-1}A$ is symmetric, therefore its eigenvalues λ are real.

Turning to the second part, assume that y is another generalized eigenvector

$$Ay = \mu By, \quad y \neq 0, \tag{5.11}$$

and $\mu \neq \lambda$. Take the scalar products of (5.10) with y, and of (5.11) with x, and then subtract the results

$$Ax \cdot y - Ay \cdot x = \lambda Bx \cdot y - \mu By \cdot x.$$

Since $Ax \cdot y = x \cdot A^T y = x \cdot Ay = Ay \cdot x$, the expression on the left is zero. Similarly, on the right we have $By \cdot x = Bx \cdot y$, and therefore

$$0 = (\lambda - \mu)Bx \cdot y.$$

Since $\lambda - \mu \neq 0$, it follows that $Bx \cdot y = 0$. □

We shall consider *generalized Rayleigh quotients* $\frac{Ax \cdot x}{Bx \cdot x}$ only for 2×2 matrices that occur in the formula $\kappa_n = \frac{Ax \cdot x}{Bx \cdot x}$ for the normal curvature.

Proposition 7.5.4. *Assume that a 2×2 matrix A is symmetric, and a 2×2 matrix B is positive definite. Let $k_1 < k_2$ be the generalized eigenvalues of*

$$Ax = \lambda Bx,$$

and x_1, x_2 corresponding generalized eigenvectors. Then

$$\min_{x \in R^2} \frac{Ax \cdot x}{Bx \cdot x} = k_1, \quad \text{achieved at } x = x_1,$$

$$\max_{x \in R^2} \frac{Ax \cdot x}{Bx \cdot x} = k_2, \quad \text{achieved at } x = x_2.$$

Proof. By scaling of x_1 and x_2, obtain $Bx_1 \cdot x_1 = Bx_2 \cdot x_2 = 1$. Since x_1 and x_2 are linearly independent (a multiple of x_1 is a generalized eigenvector corresponding to k_1, and not to k_2), they span R^2. Given any $x \in R^2$, decompose

$$x = c_1 x_1 + c_2 x_2, \quad \text{with some numbers } c_1, c_2.$$

Using that $Bx_1 \cdot x_2 = 0$ by Proposition 7.5.3, obtain

$$Bx \cdot x = (c_1 Bx_1 + c_2 Bx_2) \cdot (c_1 x_1 + c_2 x_2) = c_1^2 Bx_1 \cdot x_1 + c_2^2 Bx_2 \cdot x_2 = c_1^2 + c_2^2.$$

Similarly, (recall that $Ax_1 = k_1 Bx_1, Ax_2 = k_2 Bx_2$)

$$Ax \cdot x = (c_1 Ax_1 + c_2 Ax_2) \cdot (c_1 x_1 + c_2 x_2)$$
$$= (c_1 k_1 Bx_1 + c_2 k_2 Bx_2) \cdot (c_1 x_1 + c_2 x_2) = k_1 c_1^2 + k_2 c_2^2.$$

Hence,

$$\min_{x \in R^2} \frac{Ax \cdot x}{Bx \cdot x} = \min_{(c_1, c_2)} \frac{k_1 c_1^2 + k_2 c_2^2}{c_1^2 + c_2^2} = k_1,$$

the minimum occurring when $c_1 = 1$ and $c_2 = 0$, or when $x = x_1$, by the properties of Rayleigh quotient (or by a direct argument, see Exercises). Similar argument shows that $\max_{x \in R^2} \frac{Ax \cdot x}{Bx \cdot x} = k_2$, and the maximum is achieved at $x = x_2$. $\square$

Exercises

1. (a) Find the second fundamental form for a surface of revolution $\sigma(u, v) = (f(u) \cos v, f(u) \sin v, g(u))$, assuming that $f(u) > 0$ and $f'^2(u) + g'^2(u) = 1$.

(This surface is obtained by rotating the unit speed curve $x = f(u)$, $z = g(u)$ in the xz-plane around the z-axis.)

Answer: $(f'g'' - f''g')\, du^2 + fg'\, dv^2$.

(b) By setting $f(u) = \cos u$, $g(u) = \sin u$, find the second fundamental form for the unit sphere.

Answer: $du^2 + \cos^2 u\, dv^2$.

(c) By setting $f(u) = 1$, $g(u) = u$, find the second fundamental form for the cylinder $x^2 + y^2 = 1$.

Answer: dv^2.

2. Find the generalized eigenvalues and the corresponding generalized eigenvectors of $Ax = \lambda Bx$:

(a) $A = \begin{bmatrix} -1 & 0 \\ 0 & 2 \end{bmatrix}$, $B = \begin{bmatrix} 3 & 0 \\ 0 & 4 \end{bmatrix}$.

Answer: $\lambda_1 = -\frac{1}{3}$, $x_1 = \begin{bmatrix} 1 \\ 0 \end{bmatrix}$; $\lambda_2 = \frac{1}{2}$, $x_2 = \begin{bmatrix} 0 \\ 1 \end{bmatrix}$.

(b) $A = \begin{bmatrix} 1 & 2 \\ 2 & 1 \end{bmatrix}$, $B = \begin{bmatrix} 2 & 1 \\ 1 & 2 \end{bmatrix}$.

Answer: $\lambda_1 = -1$, $x_1 = \begin{bmatrix} -1 \\ 1 \end{bmatrix}$; $\lambda_2 = 1$, $x_2 = \begin{bmatrix} 1 \\ 1 \end{bmatrix}$.

(c) $A = \begin{bmatrix} 0 & 1 \\ 1 & 0 \end{bmatrix}$, $B = \begin{bmatrix} 2 & -1 \\ -1 & 2 \end{bmatrix}$.

Answer: $\lambda_1 = -\frac{1}{3}$, $x_1 = \begin{bmatrix} -1 \\ 1 \end{bmatrix}$; $\lambda_2 = 1$, $x_2 = \begin{bmatrix} 1 \\ 1 \end{bmatrix}$.

(d) $A = \begin{bmatrix} 0 & 1 & 1 \\ 1 & 0 & 1 \\ 1 & 1 & 0 \end{bmatrix}$, $B = \begin{bmatrix} 2 & 1 & 0 \\ 1 & 2 & 1 \\ 0 & 1 & 2 \end{bmatrix}$.

Answer: $\lambda_1 = -\frac{\sqrt{5}+1}{2}$, $x_1 = \begin{bmatrix} -\frac{\sqrt{5}+1}{2} \\ 1 \end{bmatrix}$;

$\lambda_2 = -\frac{1}{2}$, $x_2 = \begin{bmatrix} -1 \\ 0 \\ 1 \end{bmatrix}$; $\lambda_3 = \frac{\sqrt{5}-1}{2}$, $x_3 = \begin{bmatrix} \frac{\sqrt{5}-1}{2} \\ 1 \end{bmatrix}$.

3. Let B be a positive definite matrix.

(a) Show that the vector $\frac{x}{\sqrt{Bx \cdot x}}$ is B-unit, for any $x \neq 0$.

(b) Assume that vectors $x_1, x_2, \ldots, x_p$ in R^n are mutually B-orthogonal ($Bx_i \cdot x_j = 0$, for $i \neq j$). Show that they are linearly independent.

4. Show that the generalized eigenvalues of $B^{-1}x = \lambda A^{-1}x$ are the reciprocals of the eigenvalues of BA^{-1}.

5. Let $k_1 < k_2$, and c_1, c_2 any numbers with $c_1^2 + c_2^2 \neq 0$.

(a) Show that

$$k_1 \leq \frac{k_1 c_1^2 + k_2 c_2^2}{c_1^2 + c_2^2} \leq k_2.$$

(b) Conclude that

$$\min_{(c_1, c_2)} \frac{k_1 c_1^2 + k_2 c_2^2}{c_1^2 + c_2^2} = k_1,$$

and the minimum occurs at $c_1 = 1$, $c_2 = 0$.

7.6 Principal curvatures

At any point on a regular surface $\sigma(u, v)$, the tangent plane is spanned by the vectors $\sigma_u(u, v)$ and $\sigma_v(u, v)$, so that any vector $\mathbf{t}$ of the tangent plane can be written as

$$\mathbf{t} = \xi\sigma_u + \eta\sigma_v, \quad \text{with some numbers } \xi, \eta. \tag{6.1}$$

Vectors σ_u and σ_v form a basis of the tangent plane, while (ξ, η) give the coordinates of $\mathbf{t}$ with respect to this basis. Let $x = \begin{bmatrix} \xi \\ \eta \end{bmatrix}$. Then the normal curvature in the direction of $\mathbf{t}$ was shown in (5.9) to be $\kappa_n = \frac{Ax \cdot x}{Bx \cdot x}$, where the matrices $A = \begin{bmatrix} L & M \\ M & N \end{bmatrix}$ and $B = \begin{bmatrix} E & F \\ F & G \end{bmatrix}$ involve the coefficients of the second and the first fundamental forms, respectively. *The minimum and maximum values of κ_n are called the principal curvatures.* Let $k_1 < k_2$ be the generalized eigenvalues of

$$Ax = \lambda Bx, \tag{6.2}$$

and $x_1 = \begin{bmatrix} \xi_1 \\ \eta_1 \end{bmatrix}$, $x_2 = \begin{bmatrix} \xi_2 \\ \eta_2 \end{bmatrix}$ corresponding generalized eigenvectors. According to Proposition 7.5.4, the principal curvatures are k_1 and k_2. The following vectors in the tangent plane

$$\mathbf{t}_1 = \xi_1\sigma_u + \eta_1\sigma_v,$$
$$\mathbf{t}_2 = \xi_2\sigma_u + \eta_2\sigma_v \tag{6.3}$$

are called *the principal directions. The product $K = k_1k_2$ is called the Gaussian curvature.*

Theorem 7.6.1. *Assume that $k_1 \neq k_2$. Then*
(i) *The principal directions $\mathbf{t}_1$ and $\mathbf{t}_2$ are perpendicular.*
(ii) *If the generalized eigenvectors of (6.2), x_1 and x_2, are B-unit, then the principal directions $\mathbf{t}_1$ and $\mathbf{t}_2$ are unit vectors.*

Proof. (i) Recall that the matrix B is positive definite. Using (6.3), and the coefficients of the first fundamental form,

$$\mathbf{t}_1 \cdot \mathbf{t}_2 = E\xi_1\xi_2 + F\xi_1\eta_2 + F\eta_1\xi_2 + G\eta_1\eta_2 = Bx_1 \cdot x_2 = 0, \tag{6.4}$$

since x_1 and x_2 are B-orthogonal by Proposition 7.5.3.
(ii) Following the derivation of (6.4), obtain

$$\mathbf{t}_1 \cdot \mathbf{t}_1 = E\xi_1^2 + 2F\xi_1\eta_1 + G\eta_1^2 = Bx_1 \cdot x_1 = 1,$$

and similarly $\mathbf{t}_2 \cdot \mathbf{t}_2 = 1$. □

Example. Let us find the principal curvatures and principal directions for the cylinder $\sigma(u, v) = (\cos v, \sin v, u)$. Recall that $E = 1, F = 0, G = 1, L = 0, M = 0, N = 1$, so that $A = \begin{bmatrix} 0 & 0 \\ 0 & 1 \end{bmatrix}$ and $B = \begin{bmatrix} 1 & 0 \\ 0 & 1 \end{bmatrix}$. Since here $B = I$, the generalized eigenvalue problem becomes $Ax = \lambda x$, with the eigenvalues (the principal curvatures) $k_1 = 0$ and $k_2 = 1$, and the corresponding eigenvectors $x_1 = \begin{bmatrix} \xi_1 \\ \eta_1 \end{bmatrix} = \begin{bmatrix} 1 \\ 0 \end{bmatrix}, x_2 = \begin{bmatrix} \xi_2 \\ \eta_2 \end{bmatrix} = \begin{bmatrix} 0 \\ 1 \end{bmatrix}$. The principal directions are

$$\mathbf{t}_1 = \sigma_u \xi_1 + \sigma_v \eta_1 = \sigma_u = (0, 0, 1),$$
$$\mathbf{t}_2 = \sigma_u \xi_2 + \sigma_v \eta_2 = \sigma_v = (-\sin v, \cos v, 0).$$

The vector $\mathbf{t}_1$ is vertical, while $\mathbf{t}_2$ is horizontal, tangent to the unit circle.

We show next that knowledge of the principal curvatures κ_1 and κ_2, and of the principal directions $\mathbf{t}_1$ and $\mathbf{t}_2$, makes it possible to calculate the normal curvature of any curve on a surface.

Theorem 7.6.2 (Euler's theorem). *Let y be a unit speed curve on a surface $\sigma(u, v)$, with its unit tangent vector $\mathbf{t}$ making an angle θ with $\mathbf{t}_1$. Then the normal curvature of y is*

$$\kappa_n = k_1 \cos^2 \theta + k_2 \sin^2 \theta.$$

Proof. Let $x = \begin{bmatrix} \xi \\ \eta \end{bmatrix}, x_1 = \begin{bmatrix} \xi_1 \\ \eta_1 \end{bmatrix}$, and $x_2 = \begin{bmatrix} \xi_2 \\ \eta_2 \end{bmatrix}$ be the coordinates of $\mathbf{t}, \mathbf{t}_1, \mathbf{t}_2$, respectively, so that (6.1) and (6.3) hold. Recall that x_1 and x_2 are B-orthogonal by Proposition 7.5.3. By scaling we may assume that x_1 and x_2 are B-unit ($Bx_1 \cdot x_1 = Bx_2 \cdot x_2 = 1$), and then, by Theorem 7.6.1, the vectors $\mathbf{t}_1$ and $\mathbf{t}_2$ are orthogonal and unit. Decompose

$$\mathbf{t} = (\mathbf{t} \cdot \mathbf{t}_1)\mathbf{t}_1 + (\mathbf{t} \cdot \mathbf{t}_2)\mathbf{t}_2 = \cos \theta \mathbf{t}_1 + \sin \theta \mathbf{t}_2$$
$$= \cos \theta(\sigma_u \xi_1 + \sigma_v \eta_1) + \sin \theta(\sigma_u \xi_2 + \sigma_v \eta_2)$$
$$= (\xi_1 \cos \theta + \xi_2 \sin \theta)\sigma_u + (\eta_1 \cos \theta + \eta_2 \sin \theta)\sigma_v.$$

In the coordinates of $\mathbf{t}, \mathbf{t}_1, \mathbf{t}_2$, this implies

$$x = \begin{bmatrix} \xi_1 \cos \theta + \xi_2 \sin \theta \\ \eta_1 \cos \theta + \eta_2 \sin \theta \end{bmatrix} = \begin{bmatrix} \xi_1 \\ \eta_1 \end{bmatrix} \cos \theta + \begin{bmatrix} \xi_2 \\ \eta_2 \end{bmatrix} \sin \theta = x_1 \cos \theta + x_2 \sin \theta.$$

Using that x_1 and x_2 are B-orthogonal and B-unit,

$$Bx \cdot x = (Bx_1 \cos \theta + Bx_2 \sin \theta) \cdot (x_1 \cos \theta + x_2 \sin \theta)$$
$$= Bx_1 \cdot x_1 \cos^2 \theta + Bx_2 \cdot x_2 \sin^2 \theta = \cos^2 \theta + \sin^2 \theta = 1,$$

and similarly,

$$Ax \cdot x = (Ax_1 \cos \theta + Ax_2 \sin \theta) \cdot (x_1 \cos \theta + x_2 \sin \theta)$$
$$= (k_1 Bx_1 \cos \theta + k_2 Bx_2 \sin \theta) \cdot (x_1 \cos \theta + x_2 \sin \theta)$$
$$= k_1 Bx_1 \cdot x_1 \cos^2 \theta + k_2 Bx_2 \cdot x_2 \sin^2 \theta = k_1 \cos^2 \theta + k_2 \sin^2 \theta.$$

Since $\kappa_n = \frac{Ax \cdot x}{Bx \cdot x}$, the proof follows. $\qquad\square$

We discuss next the geometrical significance of the principal curvatures. Let P be a point $\sigma(u_0, v_0)$ on a surface $\sigma(u, v)$. We can declare the point (u_0, v_0) to be the new origin in the uv-plane. Then $P = \sigma(0, 0)$. We now declare P to be the origin in the (x, y, z) space where the surface $\sigma(u, v)$ lies. Then $P = \sigma(0, 0) = (0, 0, 0)$. Let $\mathbf{t}_1$ and $\mathbf{t}_2$ be the principal directions at P, $x_1 = \begin{bmatrix} \xi_1 \\ \eta_1 \end{bmatrix}$ and $x_2 = \begin{bmatrix} \xi_2 \\ \eta_2 \end{bmatrix}$ their coordinates with respect to σ_u and σ_v basis, and k_1, k_2 the corresponding principal curvatures. We assume that x_1 and x_2 are B-unit, and therefore $\mathbf{t}_1$ and $\mathbf{t}_2$ are unit vectors. We now direct the x-axis along $\mathbf{t}_1$, the y-axis along $\mathbf{t}_2$, and the z-axis accordingly (along $\mathbf{t}_1 \times \mathbf{t}_2$). The tangent plane at P is then the xy-plane. Denote $u = x\xi_1 + y\xi_2$, $v = x\eta_1 + y\eta_2$. Since

$$u\sigma_u + v\sigma_v = (x\xi_1 + y\xi_2)\sigma_u + (x\eta_1 + y\eta_2)\sigma_v$$
$$= x(\xi_1\sigma_u + \eta_1\sigma_v) + y(\xi_2\sigma_u + \eta_2\sigma_v) = x\mathbf{t}_1 + y\mathbf{t}_2, \qquad (6.5)$$

it follows that the point $(x, y, 0)$ in the tangent plane at P is equal to $u\sigma_u + v\sigma_v$.

For $|x|$ and $|y|$ small, the point (u, v) is close to $(0, 0)$, so that the point $\sigma(u, v)$ lies near $(0, 0, 0)$. Then, neglecting higher order terms and using (6.5) gives

$$\sigma(u, v) = \sigma(0, 0) + u\sigma_u + v\sigma_v + \frac{1}{2}(u^2\sigma_{uu} + 2uv\sigma_{uv} + v^2\sigma_{vv})$$
$$= (x, y, 0) + \frac{1}{2}(u^2\sigma_{uu} + 2uv\sigma_{uv} + v^2\sigma_{vv}),$$

with all derivatives evaluated at $(0, 0)$.

Consider the vector

$$w = \begin{bmatrix} u \\ v \end{bmatrix} = \begin{bmatrix} x\xi_1 + y\xi_2 \\ x\eta_1 + y\eta_2 \end{bmatrix} = xx_1 + yx_2.$$

We now calculate the z coordinate of the vector $\sigma(u, v)$. Neglecting higher order terms, we obtain (here $\bar{N} = (0, 0, 1)$)

$$z = \sigma(u, v) \cdot \bar{N} = \frac{1}{2}(Lu^2 + 2Muv + Nv^2) = \frac{1}{2}Aw \cdot w$$
$$= \frac{1}{2}A(xx_1 + yx_2) \cdot (xx_1 + yx_2) = \frac{1}{2}(k_1 xBx_1 + k_2 yBx_2) \cdot (xx_1 + yx_2)$$
$$= \frac{1}{2}k_1 x^2 + \frac{1}{2}k_2 y^2,$$

since the vectors x_1 and x_2 are B-orthogonal and B-unit. We conclude that near point P the surface $\sigma(u,v)$ coincides with the quadric surface $z = \frac{1}{2}k_1x^2 + \frac{1}{2}k_2y^2$, neglecting the terms of order greater than two.

If k_1 and k_2 are both positive or both negative, the point P is called an *elliptic point* on $\sigma(u,v)$. The surface looks like a paraboloid near P. If k_1 and k_2 are of opposite signs, the point P is called a *hyperbolic point* on $\sigma(u,v)$. The surface looks like a saddle near P.

Consider now a surface $z = f(x,y)$. Assume that $f(0,0) = 0$, so that the origin $O = (0,0,0)$ lies on the surface, and that $f_x(0,0) = f_y(0,0) = 0$, so that the xy-plane gives the tangent plane to the surface at O. Writing $\sigma(x,y) = (x,y,f(x,y))$, calculate $\sigma_x(0,0) = (1,0,0)$, $\sigma_y(0,0) = (0,1,0)$, so that $E = 1$, $F = 0$, $G = 1$, and the matrix of the first fundamental form at O is $B = I$. Here $\bar{N}$ is $(0,0,1)$, and then

$$L = \sigma_{xx}(0,0) \cdot \bar{N} = f_{xx}(0,0),$$
$$M = \sigma_{xy}(0,0) \cdot \bar{N} = f_{xy}(0,0),$$
$$N = \sigma_{yy}(0,0) \cdot \bar{N} = f_{yy}(0,0),$$

and the matrix of the second fundamental form at O is

$$A = H(0,0) = \begin{bmatrix} f_{xx}(0,0) & f_{xy}(0,0) \\ f_{xy}(0,0) & f_{yy}(0,0) \end{bmatrix},$$

which is the Hessian matrix at $(0,0)$. The normal curvature

$$\kappa_n = \frac{H(0,0)w \cdot w}{w \cdot w},$$

in the direction of vector $w = \begin{bmatrix} x \\ y \end{bmatrix}$, is the Rayleigh quotient of the Hessian matrix. We conclude that the eigenvalues of the Hessian matrix $H(0,0)$ give the principal curvatures, and the corresponding eigenvectors are the principal directions at O. (Observe that here $x_1 = t_1$ and $x_2 = t_2$, so that the principal directions coincide with their coordinate vectors.)

Exercises

1. Show that the Gaussian curvature satisfies $K = k_1k_2 = \frac{LN-M^2}{EG-F^2}$.

   Hint. The principal curvatures k_1 and k_2 are roots of the quadratic equation

   $$\begin{vmatrix} L - kE & M - kF \\ M - kF & N - kG \end{vmatrix} = 0.$$

   Write this equation in the form $ak^2 + bk + c = 0$, and use that $k_1k_2 = \frac{c}{a}$.

2. (a) For the torus,

$$\sigma(\theta, \varphi) = ((a + b \cos \theta) \cos \varphi, (a + b \cos \theta) \sin \varphi, b \sin \theta),$$

with $a > b > 0$, show that the first and the second fundamental forms respectively are

$$b^2 \, d\theta^2 + (a + b \cos \theta)^2 d\varphi^2 \quad \text{and} \quad b \, d\theta^2 + (a + b \cos \theta) \cos \theta \, d\varphi^2.$$

(b) Show that the principal curvatures are $k_1 = \frac{1}{b}$, $k_2 = \frac{\cos \theta}{a + b \cos \theta}$.

(c) Which points on a doughnut are elliptic, and which are hyperbolic?

Bibliography

[1]   V. I. Arnol'd, Ordinary Differential Equations. The M.I.T. Press, Cambridge Mass.-London, 1973. Translated from the Russian and edited by Richard A. Silverman.

[2]   R. Bellman, Stability Theory of Differential Equations. McGraw-Hill Book Company, Inc., New York–Toronto–London, 1953.

[3]   R. Bellman, Introduction to Matrix Analysis, 2nd ed. SIAM, Philadelphia, 1997.

[4]   R. Bronson, Matrix Methods: An Introduction, 2nd ed. Academic Press, 1970.

[5]   M. P. Do Carmo, Differential Geometry of Curves and Surfaces, 2nd ed. Dover, 2016.

[6]   B. P. Demidovič, Lectures on the Mathematical Theory of Stability, Izdat. Nauka, Moscow 1967 (in Russian).

[7]   H. Dym, Linear Algebra in Action, Graduate Studies in Mathematics, 78. American Mathematical Society, Providence, RI, 2007.

[8]   S. H. Friedberg, A. J. Insel and L. E. Spence, Linear Algebra, 3rd ed. Prentice Hall, 1997.

[9]   I. M. Gelfand, Lectures on Linear Algebra. Dover, Timeless (originated in 1948).

[10] I. M. Gel'fand and V. B. Lidskii, On the structure of the regions of stability of linear canonical systems of differential equations with periodic coefficients. Usp. Mat. Nauk (N. S.) 10, no. 1(63), 3–40 (1955) (Russian).

[11] P. Korman, Lectures on Differential Equations. AMS/MAA Textbooks, 54. MAA Press, Providence, RI, 2019.

[12] E. Kosygina, C. Robinson and M. Stein, ODE Notes. Online: sites.math.northwestern.edu/ clark/285/ 2006-07/handouts/odenotes.

[13] P. D. Lax, Linear Algebra and Its Applications, 2nd ed. Wiley-Interscience, 2007.

[14] J. M. Ortega, Numerical Analysis. A Second Course. Computer Science and Applied Mathematics. Academic Press, New York–London, 1972.

[15] A. Pressley, Elementary Differential Geometry. Springer Undergraduate Mathematics Series. Springer-Verlag London, Ltd., London, 2001.

[16] G. Strang, Linear Algebra and Its Applications, 4th ed. Brooks/Cole, Cengage Learning, 2008.

[17] D. J. Wright, Introduction to Linear Algebra. McGraw-Hill Education – Europe, 1999.

[18] Yutsumura, Problems in Linear Algebra. Online: yutsumura.com.

[19] 3Blue1Brown, Linear Algebra. An introduction to visualizing what matrices are really doing. Online: 3blue1brown.com/topics/linear-algebra.

https://doi.org/10.1515/9783111086507-008

Index

https://doi.org/10.1515/9783111086507-009

www.ingramcontent.com/pod-product-compliance
Lightning Source LLC
Chambersburg PA
CBHW061400210326
41598CB00035B/6052